How the Snake Lost its Legs

Curious Tales from the Frontier of Evo-Devo

How did the zebra really get its stripes, and the giraffe its long neck? What is the science behind camel humps, leopard spots, and other animal oddities? Such questions have fascinated us for centuries, but the expanding field of evo-devo (evolutionary developmental biology) is now providing, for the first time, a wealth of insights and answers.

Taking inspiration from Kipling's *Just So Stories*, this book weaves emerging insights from evo-devo into a narrative that provides startling explanations for the origin and evolution of traits across the animal kingdom. The author's unique and engaging style makes this narrative both enlightening and entertaining, guiding students and researchers through even complex concepts and encouraging a fuller understanding of the latest developments in the field. The first five chapters cover the first bilaterally symmetric animals, flies, butterflies, snakes, and cheetahs. A final chapter surveys recent results about a menagerie of other animals.

Lewis I. Held, Jr. is Associate Professor of Biology at Texas Tech University, Lubbock, Texas, USA. He has taught developmental biology and human embryology to pre-medical students for twenty-seven years, and received the 2010 Professing Excellence Award and the 1995 President's Excellence in Teaching Medal (Texas Tech University). He is also the author of *Quirks of Human Anatomy* (Cambridge, 2009), *Imaginal Discs* (Cambridge, 2002), and *Models for Embryonic Periodicity* (Karger, 1992).

Frontispiece Legless or partially legless lizards, some of which resemble snakes to a remarkable extent. Numbers next to legs in each panel (and in parentheses in this legend) are the number of digits per foreleg or hindleg respectively if present. (a) *Lerista dorsalis* (4+4). (b) *Hemiergis decresiensis* (3+3). (c) *Lerista desertorum* (2+3). (d) *Lerista timida* (3+3). (e) *Lerista edwardsae* (0+2). (f) *Anomalopus verreauxii* (3+1). (g) *Ophioscincus truncatus* (0+0). (h) *Pygopus lepidopodus* (0+0). Leg loss is common among lizards and may have occurred for the same reason as leg loss in snakes. All photographs were taken by Mark Hutchinson, South Australia Museum, Adelaide, and are used by permission. *For color plate see color section.*

How the Snake Lost Its Legs

Curious Tales from the Frontier of Evo-Devo

LEWIS I. HELD, JR.
Texas Tech University, USA

CAMBRIDGE
UNIVERSITY PRESS

University Printing House, Cambridge CB2 8BS, United Kingdom

Cambridge University Press is part of the University of Cambridge.

It furthers the University's mission by disseminating knowledge in the pursuit of education, learning and research at the highest international levels of excellence.

www.cambridge.org
Information on this title: www.cambridge.org/9781107621398

First published 2014

A catalogue record for this publication is available from the British Library

Library of Congress Cataloguing in Publication data
Held, Lewis I., 1951–
How the snake lost its legs : curious tales from the frontier of evo-devo /
Lewis I. Held, Jr.
pages cm
Includes bibliographical references and index.
ISBN 978-1-107-03044-2 (hardback) – ISBN 978-1-107-62139-8 (paperback)
1. Evolution (Biology) 2. Developmental biology. I. Title.
QH366.2.H435 2014
576.8 – dc23 2013039531

ISBN 978-1-107-03044-2 Hardback
ISBN 978-1-107-62139-8 Paperback

Contents

Color plate section appears between pages 116 and 117.

Preface

The most famous accounts of animal origins in Western literature are those of Rudyard Kipling (1865–1936). His *Just So Stories* (1902) offered fabulous tales about how the leopard got its spots, how the elephant got its trunk, and so forth. It remains one of the most popular children's books of all time.

Fables certainly make good bedtime stories, but they are poor substitutes for real understanding. The foundations for modern biology were forged by Charles Darwin (1809–1882) in his *Origin of Species* (1859). Ever since, scientists have been amassing facts about how animal traits arose. Initially, the data came from fossils, embryos, and anatomy. Then, in the first half of the twentieth century, genetics merged with evolution to reveal the causes behind the changes [1420].

Development flirted with evolution for more than a century [46] before their union was consummated around the time of Stephen Jay Gould's *Ontogeny and Phylogeny* (1977) [828,2311]. Their scion was named "evo-devo" [1559]. Evo-devo initially relied on only a few model organisms [2088], including the fly, the worm, and the mouse. The genetic circuitry of these animals turned out to be surprisingly universal, despite their superficial differences [955].

In 1998 the first animal genome was sequenced – the nematode *Caenorhabditis elegans* [1] – followed in 2000 by flies and humans [81], and more genomes have been added at an increasing pace with each passing year [317]. As pieces of the evo-devo puzzle have come together, various authors have attempted to convey the "big picture" to the general public [1488]. Most notable are Sean Carroll's *Endless Forms Most Beautiful* [339] and his more advanced *From DNA to Diversity* [344]. Sean's books would make good primers for anyone who is not already well versed in genetics.

The genome of each species can be thought of as a kind of cookbook for concocting an adult anatomy, starting in most cases with a single fertilized egg. It is also a history book, because it retains remnants of recipes that were previously used by the organism's ancestors. If we could read all those recipes, then we could decipher development and retrace evolution. Decoding DNA beyond the protein level is hard, however, Because genes interact in complex ways [116].

How the Snake Lost its Legs was written at the behest of Martin Griffiths, Life Sciences Commissioning Editor at Cambridge University Press. Originally, he had in mind a book along the lines of Wallace Arthur's majestic *Evolution: A Developmental Approach* (2011) [75]. When I balked at the idea of writing a textbook per se, Martin relented, and we agreed on a more casual format where readers could sample whatever topics caught their fancy. My aim was to blend Darwin's rigor with Kipling's whimsy into a book that could amuse the darling child in all of us. The final product is hence an homage to both *Just So Stories* and *Origin of Species*. It celebrates animal oddities through facts rather than fables.

Truth can be stranger than fiction, and evo-devo is a case in point. Evo-devo defies our intuition in the same way that magicians amaze children by appearing to turn one thing into another.

Once upon a time, for example, snakes really did walk like lizards [2395], whales really did waddle like hippos [1790], and dolphins may have pranced like antelopes [2195], given how they swim as if they were galloping [505]. Relics of these marvels can be seen in their respective embryos, and traces of these transitions can be gleaned from their respective genomes [503]. Researchers around the world have been excavating such insights at an increasing pace, and the impetus to share them with the wider world outside our ivory towers has become almost irresistible. Hence this book.

Chapter 1 introduces concepts that readers will need in order to navigate the rest of the book. It traces key genes and cellular signals back to the common ancestor of all bilaterally symmetric animals. Chapter 2 provides a primer on gene circuitry using the animal whose genetics we know the best – the fruit fly [965]. Chapter 3 examines how some of these circuits arose during evolution and were co-opted for other duties in a related insect – the butterfly. Chapter 4 focuses on one of the strangest animals that has ever lived – the snake, whose quirks reveal its history and thereby allow us to see how natural selection shapes anatomy. Chapter 5 considers animal coat patterns in general, using the cheetah as a handy case study. Finally, Chapter 6 is a potpourri of recent discoveries about a zoo-full of animals that stretches from A to Z. Every chapter is packed with unsolved mysteries that could furnish offbeat topics for term papers, thesis projects, or journal clubs.

Scott Gilbert [783] and others [994,1473,1602,2377] have advocated using evo-devo to teach about evolution in general. With this goal in mind, I have cited all sources (numbers in brackets) as links to the literature. Only by consulting original articles can students appreciate the richness of data, the nuances of arguments, or the thrill of discoveries. The References section offers instructors a reservoir that they can tap for tidbits to spice up their lectures. Two outstanding "must reads" concern snakes, elephants, and manatees in one case [886] and tigers, cream horses, and silver chickens in the other [2470].

To enhance its didactic utility, I have punctuated the narrative with take-home lessons. To ensure accessibility, I have culled as much jargon as possible, compiled a glossary (albeit a spartan one), used a case-study approach, and perforated the book with "rabbit-holes." Readers can enter this Wonderland by studying the schematics, browsing the bestiary (Chapter 6), sampling the glossary, or just looking at the pictures. Like Alice, readers can then explore whatever piques their curiosity.

Because so many key genes were discovered and characterized in flies [761], the conventions of fly nomenclature will be followed here. Thus, gene names are italicized and can begin with either a capital (due to the first mutant allele being dominant; e.g., *Serrate*) or a lower-case letter (recessive; e.g., *engrailed*). Protein names are in roman type and always capitalized (e.g., Serrate or Engrailed). Key concepts are set in **boldface** and are defined in the Glossary. Quotes are set apart from narrative text as indented, blocked paragraphs in smaller, non-serif typeface. The abbreviation "MY" denotes millions of years, and the slang term "app" means an application, as in computer software.

Reconstructing past events can be as tricky as solving a Sherlock Holmes mystery [823] because there is some margin of error in every inference [171]. Authors must assemble the evidence for their arguments as methodically as a lawyer [1330], and readers must weigh the soundness of all theories as carefully as a juror [1529]. Some lessons can be learned from cases where lovely ideas were subsequently overturned in various investigations of fish [247], stick insects [2231], insect wings [1106], turtles [1343], treehoppers [1474], leopards [1687], and acoel flatworms [1408]. The benchmarks listed below may also offer some helpful guidance in this regard.

Plausibility is the first criterion that any evo-devo scenario must meet [1278], but it cannot be the last [1097], lest we run the risk of believing stories that seem valid only because alternatives have not been adequately assessed [778]. Too often we assume that a structure must have served the same function in the past as it serves today [1340]. In fact, structures need not have served any function at all when they first arose [853].

Testability is the second hurdle that any proposal must meet [186], and every attempt will be made to frame hypotheses so that they can be easily tested. Indeed, this venue is where the evo-devo approach has proven to be so valuable [1106]. The field has matured to the point where we can now manipulate the genes that we think caused specific changes in evolution to see whether we can re-create those events in the laboratory [1073,2291].

Ultimately, fidelity to phylogeny is the best way to be sure that the trends we deduce are real [824]. Evolutionary trees furnish a framework within which developmental deviations can be plotted [1907], and the more robust the trees the better [95]. Only by knowing the historical sequence of species (A begat B begat C), for instance, can we show that a certain structure was lost (by B) and regained (by C) [1193].

Kipling's iconic rubric of "how the [blank] got its [blank]" is retained here despite there being no such thing as *the* leopard or *the* zebra. Indeed, there are *many* kinds of leopards and *several* species of zebras. The old notion of Platonic types was disproven in 1859 and supplanted by the Darwinian concept of variations as the raw material for evolution [1418]. There is no need to re-fight that war.

Another shorthand that needs to be clarified concerns genes. When I refer to "cuticle-stiffening genes" for beetle elytra in Chapter 6, for example, I do not mean that the genes themselves have stiffening properties, but rather that the expression of such genes leads to hardening via agents such as enzymes. Genes alone do not cause form! Likewise, when I say in Chapter 4 that "snakes were inventive" in solving an anatomical problem, I am not implying some sort of serpentine Sanhedrin that met to plot an adaptive strategy, though I am a big fan of Gary Larson's *Far Side* cartoons [1252].

The final delusion that must be dispelled is that evo-devo is only about animals. Plants offer enchanting stories as well [506,2505], but including them would have diluted the Kipling conceit.

Drafts of the entire book were kindly critiqued by Richard Blanton, John (Trey) Fondon, Joseph and Anne Frankel, Ellen Larsen, Jack Levy, and Jeffrey Thomas. Individual chapters were vetted by scholars familiar with topics in Chapter 1 (Richard Campbell, John Gerhart, Thurston Lacalli, and Chris Lowe), Chapter 2 (Joel Atallah, Richard Campbell, Jean-Baptiste Coutelis, Charles Géminard, Artyom Kopp, Stéphane Noselli, and Astrid Petzoldt), Chapter 3 (Vernon French, Ullasa Kodandaramaiah, Fred Nijhout, Jeff Oliver, and Antónia Monteiro), Chapter 4 (Martin Cohn, Jonathan Cooke, Mark Hutchinson, André Pires da Silva, Kurt Schwenk, Oscar Tarazona, John Wiens, and Joost Woltering), Chapter 5 (Jonathan Bard, Jonathan Cooke, and James Murray), and Chapter 6 (Peter Bryant, Richard Campbell, Bonnie Dalzell, and Adam Wilkins).

Other colleagues tutored me in the finer points of bats (Robert Bradley, Tigga Kingston, and Caleb Phillips), birds (Arhat Abzhanov, Nancy McIntyre, and Ken Schmidt), centipedes (Michael Akam), jerboas (Cliff Tabin), leopards (Kristofer Helgen), lizards (Gad Perry), seahorses (Sam Van Wassenbergh), skunks (Adam Ferguson), squirrels (Cody Thompson), and, of course, snakes (David Cundall and Lou Densmore). I accept responsibility for any flaws that remain.

Photos were generously supplied by Colleen Aldous, Jonathan Bard, Greg Barsh, Mark Hutchinson, Shigeru Kondo, Stephen Nash, Stephanos Roussos, Sharlene Santana, and Frederick Stangl. Encouragement was provided by my mother, Minnie Held, my sister, Linda Wren, and my friends George and Ann Asquith, Cal and Melanie Barnes, Sam Braudt, Hugh Brazier, Jason Cooper, Rob Posteraro, Frank Thames, Bill Tydeman, Dean Victory, and Roger Wolcott.

1 The first two-sided animal

Evo-devo's greatest revelation thus far has been that all bilaterally symmetric animals use largely the same genes to construct their bodies, including ours, despite stark differences in gross anatomy [969]. These shared genes have been traced to a multicellular ancestor that lived ~600 million years (MY) ago [405]. That ancestor has been dubbed the "Urbilaterian" [518], where "Ur-" denotes a progenitor, and "bilaterian" indicates that its two sides were mirror images. How did we come to this realization?

The first hint of a deep developmental unity among evolutionarily diverse animals came in 1976 when cockroaches and salamanders were shown to regenerate their legs according to the same cellular rules [707]. However, it seemed absurd to think that those rules might imply a leg-bearing common ancestor, because arthropods and chordates (their respective phyla) were thought to have evolved legs independently of one another [632].

Then, in the 1980s geneticists kept finding the same 180 base-pair motif in virtually all animal genomes [1432]. It was dubbed the "homeobox" (homeosis box) because mutations in the genes that contain it cause **homeosis** (transformation of one body part into another) [1756]. Even more surprising was that insects and vertebrates use homologous clusters of homeobox genes in the same colinear sequence on their chromosomes [1650] to organize the anterior–posterior (A–P) axis of their body (Fig. 1.1). These linked genes were termed "*Hox*" (*Homeobox*) genes [294]. Thus, by definition, all *Hox* genes are homeobox genes, but relatively few homeobox genes reside in a ***Hox* Complex**. Several *Hox* genes (*Scr*, *Ubx*, *abd-A*, and *Abd-B*) were instrumental in insect evolution (see Chapters 2 and 3).

In the 1990s insects and vertebrates were found to also use a common genetic circuit to specify the dorsal–ventral (D–V) axis [1622]. Strangely, however, the signaling molecules that specify the back (D) side of an insect match those that specify the belly (V) side of a vertebrate, and vice versa [754]. The notion that arthropods and chordates have the same gross architecture but are upside-down versions of each other had been proposed in 1822 by Geoffroy Saint-Hilaire based on anatomical similarities alone [1711], but his idea was ridiculed at the time. Now, ~170 years later, he has been gloriously vindicated by molecular genetic analyses.

Bilaterian animals are classified based upon whether the mouth (stoma) end of the gut arises first (protostomes) or second (deuterostomes) during development [1624], so it seemed most plausible to infer that the D–V flip occurred when urbilaterians split into these two groups – either in the protostome lineage leading to arthropods or in the deuterostome lineage leading to chordates [1016].

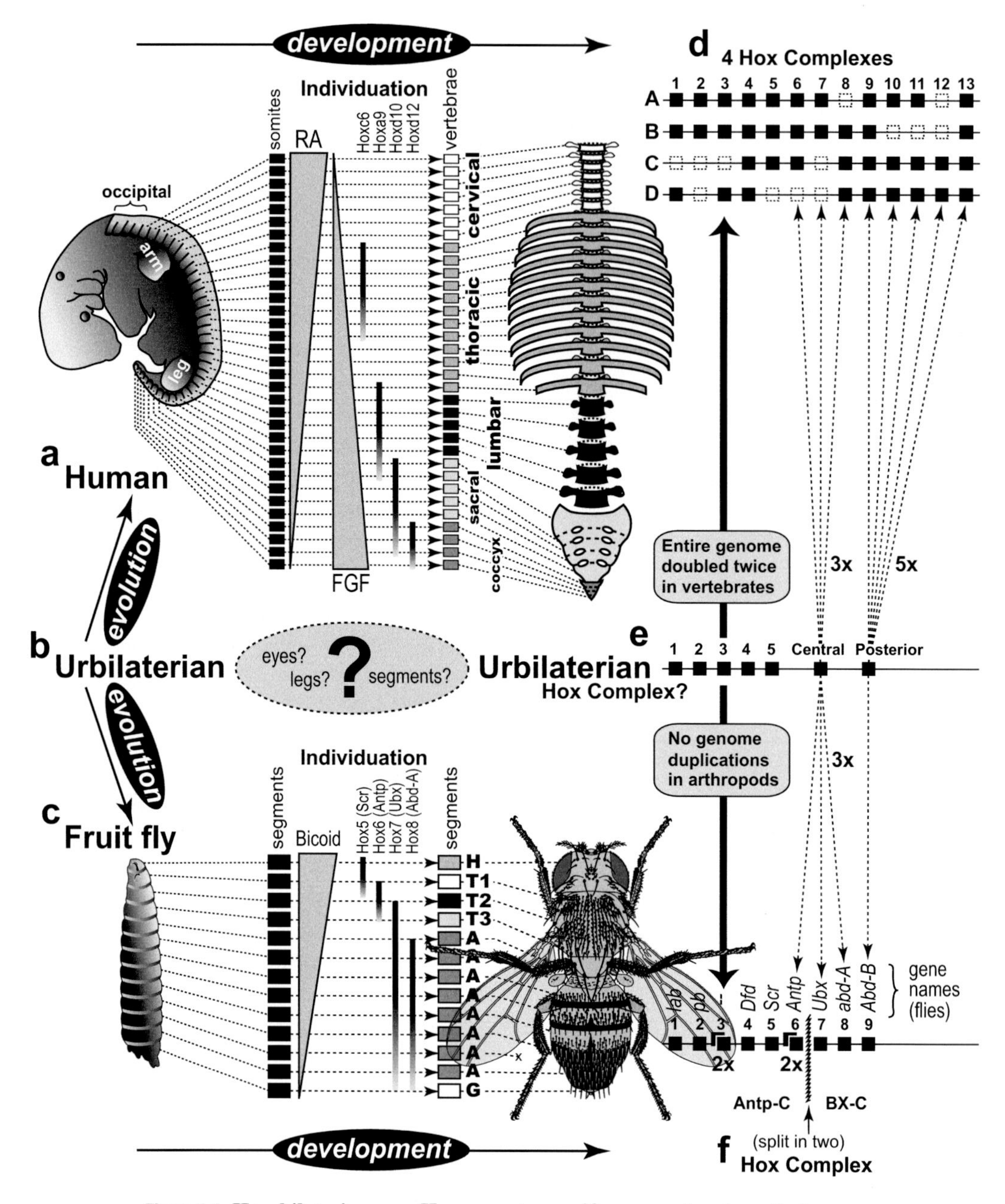

Figure 1.1 How bilaterians use *Hox* genes to specify area codes along their anterior–posterior (A–P) axis [354].

a–c. **Deep homology** of chordate (**a**) versus arthropod (**c**) metamerism based on their common usage of a *Hox*-based code along the body column – a usage that has been traced to a common urbilaterian ancestor (**b**) [294].

a. Differentiation of vertebrae in humans (31-day-old embryo at left). Our 33 vertebrae arise as mesodermal somites [1971]. The first four (occipital) somites form the skull base. All

Why the chordate flipped upside down

In 2005 this lovely theory was, to paraphrase Thomas Huxley [841], attacked by an ugly little fact. We already knew that the gene *nodal* is expressed on the left (but not right) side of the human body at a critical stage to specify our normal asymmetry (heart on the

Figure 1.1 (*cont.*)

somites start alike (black blocks) but diverge caudally to form 7 cervical, 12 thoracic, 5 lumbar, 5 sacral, and 4 coccygeal vertebrae (shaded blocks) [1042]. They acquire identities (**individuation**) via diffusible signals (**morphogens**) secreted from opposite ends of the A–P axis. RA (retinoic acid) [2031] and FGF (fibroblast growth factor) [573] form reciprocal gradients (triangles; cf. Fig. 1.3) [111,799]. *Hox* genes are activated at specific thresholds [473,1151,1240] to furnish area codes [453,673,2366] in the presomitic mesoderm [325]. Expression levels (fading bars) of *Hoxc6*, *a9*, *d10*, and *d12* [286,326,1051] are shown. These zones plus others [875,1971] provide area codes for vertebral types. Modified from ref. [967].

b. Urbilaterian progenitor of all bilaterally symmetric phyla. Its ***Hox* Complex** has been deduced (**e**) [294], but we still don't know what the animal looked like ("?") nor how it might have used morphogen gradients to designate its area codes. The installing of these codes is thought to have made bilaterian anatomy so **evolvable** that it sparked the Cambrian Explosion (Fig. 1.2). Usage of RA along the A–P axis [1955] goes back at least to protochordates [1189,1980], and RA pathway components exist in other deuterostomes [316] and protostomes [30], so RA might have functioned in urbilaterians [179,1976,2055], but, if so, then protostomes must have replaced it [534,1360,1387] as some chordates did also [315]. Other possible morphogens that have been proposed for the urbilaterian A–P axis are Wnt [1622,2402] and TGFβ [52,720,1447].

c. Differentiation of segments in flies. A larva is depicted, though **individuation** actually happens in the embryo [1720,2407]. Instead of RA or FGF, flies regulate the ***Hox* Complex** using Bicoid [876,1433,1773] – a morphogen that evolved relatively recently from a renegade *Hox3* gene [955,2104]! Bicoid acts indirectly via gap genes (not shown) [927,1068,1659] to activate *Hox5, 6, 7, 8*, and *9* (omitted) at different thresholds, and these expression zones, which fade posteriorly, dictate segment identities (shades): head (H, actually six segments [937]), first–third thoracic (T1–T3), abdominal (A), and genital (G). Despite overlaps of *Hox* expression, *Hox* genes obey a dominance hierarchy [572,2063,2482], so their area codes are not combinatorial [354], and the same is true in vertebrates (**a**). Expression domains are from refs. [342,1743] (*Scr* and *Antp*) and [354,1543] (*Ubx* and *abdA*). Adult males (e.g., the fly at right) lack the seventh abdominal segment ("x") [678].

d–f. Evolution of the ***Hox* Complex**, starting with urbilaterians (**e**). Based on refs. [35,745,1276]. For arthropod versus chordate nomenclature see refs. [337,761,1205,1650]. The homeobox motif itself pre-dates metazoans [531].

d. The four clusters of *Hox* genes in humans (and mice) that arose via two genome duplications at the dawn of vertebrates [314], followed by losses of some of the redundant paralogs (dotted squares) [2090,2509].

e. Deduced complement of *Hox* genes in urbilaterians [294].

f. *Hox* genes in the fruit fly *Drosophila melanogaster* [1311]. Overlaps denote local gene duplications (2×) [1756,2104], but neither of the *Hox3* genes (*zen* or *bcd*) nor the *Antp* paralog (*ftz*) participates in the *Hox* code *sensu stricto* [779], so their names are omitted. Abbreviations: *lab* (*labial*), *pb* (*proboscipedia*), *Dfd* (*Deformed*), *Scr* (*Sex combs reduced*), *Antp* (*Antennapedia*), *Ubx* (*Ultrabithorax*), *abd* (*abdominal*).

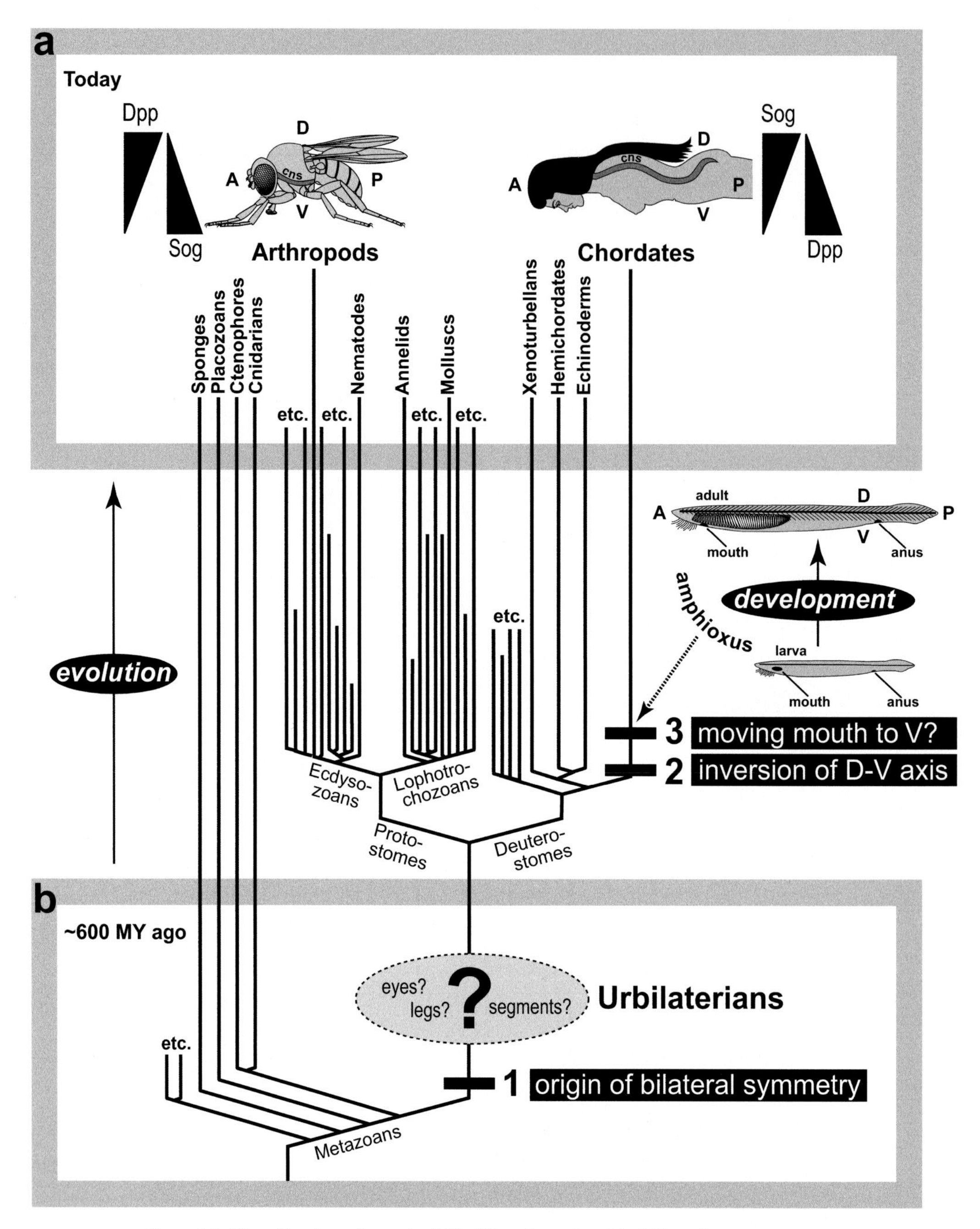

Figure 1.2 How the dorsal–ventral (D–V) axis evolved in bilaterians.

a. Body axes in modern arthropods (left) versus chordates (right). The D–V axis of chordates (e.g., humans) is upside-down relative to that of arthropods (e.g., flies) and other phyla. Thus, for example, the central nervous system (cns), which is depicted as a thin gray stripe, is ventral in arthropods but dorsal in chordates. Triangles (cf. Fig. 1.3) are reciprocal

left, spleen on the right, etc.) [371]. Now we learned that sea urchins express *nodal* on the opposite (right) side [568]. The problem was that echinoderms (the urchin phylum) are deuterostomes like us, yet they were siding with the protostomes [1249]!

This realignment was confirmed by inspecting a third deuterostome phylum [770]. In 2006 a study of gene expression in hemichordates showed that their D–V axis has a protostome-like orientation. Thus, chordates were unmasked as the topsy-turvy heretics of the bilaterian world [1331]. Somewhere within our phylum one of our ancestors must have turned over [1900]... but which one?

In 2007 the most basal chordate alive today – amphioxus [759] – was found to resemble more advanced chordates not only in its left-side expression of *nodal* but also its vertebrate-like orientation of D–V polarity indicators [1197]. Thus, the D–V inversion must have taken place before amphioxus evolved – near the very dawn of the chordate phylum (Fig. 1.2).

Intriguingly, the amphioxus mouth starts out on the left side of its face [746] and then moves ventrally during metamorphosis [1225]. This Picasso-esque asymmetry seems as strange as the migration of one eye during flatfish development [2520], and it suggests that amphioxus is retracing its evolution during its development [1225].

Despite the appeal of this **recapitulation** scenario, we know too little about the ecological habits of early bilaterians [631] or their protochordate descendants [1226] to guess the orientation of their mouths. Tube-shaped trace fossils suggest that they were burrowers [1740], but they might have been detritus scavengers, suspension feeders, or pelagic swimmers [1225]. D–V polarity would be irrelevant for a tunnel dweller but might be useful for a migrant forager. Any attempt to add further details to the D–V inversion event is unfortunately confounded by many complicating factors:

Figure 1.2 (*cont.*)

gradients of the **morphogen** Dpp (Decapentaplegic = BMP4 in vertebrates) [941,1298,1815] and its antagonist, Sog (Short gastrulation = Chordin in vertebrates) [29,754,1521]. Arthropods and chordates are two of ~30 extant bilaterian phyla [517,664,1160,2256]. Phyla are denoted by lines that reach today (extant) or stop short (extinct). Abbreviations: A (anterior), P (posterior), D (dorsal), V (ventral).

b. Urbilaterians evolved a plane of bilateral symmetry ~600 MY ago (black bar 1) [106,1625]. This event was presumably rewarded by natural selection because it afforded faster locomotion. Their D–V axis is thought to have arisen with the same orientation as in flies, with Dpp secreted dorsally and Sog ventrally. This polarity was retained in arthropods and most other phyla [194,1215] but not in chordates [1331]. The inversion (black bar 2) evidently occurred shortly before amphioxus, whose D–V polarity matches that of vertebrates [1980]. Indeed, amphioxus [2131] appears to re-enact the process (**recapitulation**) [1224,1225,1228]: its mouth starts out on the left side of the larval face [1722] and migrates ventrally (black bar 3) during metamorphosis [527]. The stylized phylogenetic tree shows (1) the **adaptive radiation** of protostomes (mouth before anus) [21,900] and deuterostomes (anus before mouth; only four extant phyla) [221,2155] in the Cambrian Explosion ~543 MY ago [433,631] and (2) four phyla of pre-bilaterian metazoans [1375,1753,2184]. Modified from ref. [967] based on the references cited above.

1. Amphioxus swims like a fish but lives in burrows with its body vertical and its mouth pointing up, and it filters seawater to get food instead of hunting [180], so its mouth angle may not matter. Indeed, a D–V axis inversion could have easily happened in such a context of relaxed selection [768].
2. Members of another protochordate group (ascidians) swim with their mouth pointing up [1273], so there appears to have been no overwhelming pressure to reorient the mouth to the ventral side.
3. Amphioxus may have specialized too much at its head end after it branched from the first chordates to be of much use in deducing ancestral features [2280], but ascidians are no more reliable [1227] since their genomes are highly modified [1047], and some of their traits derive from miniaturization [1274]. Nor do hemichordates help [330] because they have two nerve cords (one dorsal and one ventral) [308,1909].
4. Echinoderm and hemichordate larvae rotate as they swim – making their mouth angle irrelevant from a functional perspective [1223], and the same may have been true for basal chordates [1434].
5. As in amphioxus, the mouth moves ventrally in vertebrate development, but it does so without ever deviating from the midline: the mouth is simply pushed down by the growing brain [390]. If mouth repositioning is so easy [1397], then this whole debate may be a tempest in a teapot.
6. Chordates may not have moved their mouths at all. Molecular evidence suggests that they may have just closed the old (dorsal) mouth after creating a new (ventral) one [390]. In other words, we can't be sure that the mouths of protostomes and deuterostomes are homologous [1225].

These caveats, plus those in the next section, alert us to the perils of trying to reconstruct past events when the same data can be explained in many different ways.

Take-home lessons

1. Structures and functions are often so interdependent that anatomy cannot be assessed fully without knowing an organism's ecology.
2. Simple anatomy does not necessarily mean a species is primitive, because secondary losses can occur. Molecular pedigrees are more reliable.
3. **Homology** can be tricky to assess even when the structures being compared look and function alike. The standard of proof is high.

What the urbilaterian looked like

Our inability to envision the inversion has been frustrating, but what is even more discouraging is the futility of our attempts to delineate the urbilaterian itself. All we know for sure is that it had a two-ended gut (useful for high-throughput feeding) and a nascent mesoderm (wedged between the endoderm and ectoderm) for a total of three germ layers [784]. Putative fossils of the correct age turned out to be geologic artifacts [168], and others are simply not old enough [280]. Nor have we been able to rely on "living fossils" – animals that retain ancestral traits [2333]. Acoel flatworms were thought to be primitive bilaterians [957], but they turned out to be degenerate deuterostomes [1752].

In the absence of direct evidence, researchers have resorted to an indirect "triangulation" strategy to reconstruct the progenitor [1160,1673]: wherever homologous genes play the same role in the same part of the body in two distant bilaterian phyla (descended separately from the ancestor), the common antecedent gene is presumed to have acted similarly in urbilaterians [1813]. A low-resolution image of the ancestor has thereby emerged as a three-dimensional quilt of gene expression domains [27,1160] – thus allowing renderings to be sketched [517,2282].

Collectively, these shared genes comprise a "toolkit" [629,1391]. Most of the genes in that kit can be placed into one of two categories [630]. One group is involved in signaling pathways (Table 1.1) that establish axes of the body or its appendages via diffusible proteins called **morphogens** [1146,1242]. The other group consists primarily of **selector genes** [1373,1601] that dictate regional identity by encoding transcription factors (Table 1.2) [130,2094]. These components work together: morphogens create polarized patterns of different cell types by using transcription factors as their downstream agents (Fig. 1.3) [339,412].

The most famous selector gene is called *eyeless* in flies and *Pax6* in mice [761]. The *eyeless/Pax6* gene is expressed in developing eyes in both cases, and an artificially supplied mouse *Pax6* gene can induce extra eyes in flies [903] – thus proving its **deep homology** as an eye-identity gene in both phyla (arthropods and chordates) [2045]. We can therefore infer that their urbilaterian common ancestor probably had a light-sensing organ of some kind [659].

The next question was whether this organ was a compound eye like that in flies (or trilobites) [290] or a simple eye like that in vertebrates (or spiders) [1237] or, perhaps, just an eyespot that was incapable of forming an image [761]. Unfortunately, the triangulation approach cannot settle this question [1954]. Nevertheless, it appears likely that the first "eye" contained both rhabdomeric and ciliary photoreceptors [1621], despite their mutual exclusivity in most extant phyla [646,1234].

Such quandaries in interpreting the *eyeless/Pax6* homology have led some authors to conjecture that while the urbilaterian expressed *eyeless/Pax6* on its face, it lacked eyes of any kind [434,761]. According to this view, eyes evolved later in various clades by using *eyeless/Pax6* to specify different types of eyes, just as *Hox* genes merely offered an abstract scaffold [2072] upon which many diverse structures were later built (Fig. 1.1) [237,1488].

Of one thing we can be certain: it didn't take long for complex eyes to evolve [1344,1970]: the most fearsome sea monster of the Cambrian oceans ~515 MY ago was a meter-long arthropod with 16,000 facets per eye [1266,1728].

Table 1.1 Some intercellular signaling pathways conserved among bilaterian phyla[a]

Pathway[b]	Signal[c]	Receptor[d]	Transcription factor[e]	Quirks[f]	References[g]
Hedgehog	Hh/Shh	Patched, etc.	Ci/Gli3. TF-R becomes TF-A in presence of signal by blocking of TF proteolysis.	Gradient is shaped by lipid adducts on the Hh/Shh protein and by glypicans on responding cells. Unlike flies, vertebrates transduce the signal via cilia.	[47,98,99, 374,1103, 1439,2276]
Notch	Delta, etc.	Notch	Su(H)/CSL. Inner part of Notch goes to nucleus as cofactor, forming a composite TF-A.	Contact-mediated signaling only, though distant cells can interact via cell protrusions. Lateral inhibition forces cells into alternative states and sharpens boundaries.	[48,414,507, 511,890, 2098,2466]
RTK family	EGF, FGF, etc.	EGFR, FGFR, etc.	Pnt/ETS1, etc. Transduction entails Ras and MAP kinase.	Signaling is normally diffusion-mediated but can be contact-mediated (≈ Notch) via cell protrusions.	[476,1275, 1744,1902]
TGFβ family	Dpp/BMP4, etc.	Punt & Tkv, etc.	Mad/Smad1, etc.	Dpp regulates organ growth as well as cuticular patterning in flies – e.g., wing cells die if they do not get a proper dose of Dpp during development.	[90,532,581, 1553,2178, 2458]
Wnt	Wg/Wnt	Frizzled, etc.	Pan/TCF. TF-R becomes TF-A in presence of signal by blocking of cofactor (Arm/β-Cat) proteolysis.	A variant version of this pathway controls planar cell polarity in animals. Transduction **co-opts** a cell-adhesion agent (Arm/β-Cat).	[299,406,812, 1348,1606, 2006]

[a] Only the major pathways that are used in spatial patterning are listed [516,767,1065]. The retinoic acid pathway [1846,1956,2031] is omitted because of its paucity in protostomes [30,313,316,2055]. Protein synonyms (e.g., Dpp/BMP4) are "fly/vertebrate" homologs [1221]. Abbreviations: Arm (Armadillo), β-Cat (β-Catenin), BMP4 (Bone Morphogenetic Protein 4), Ci (Cubitus interruptus), CSL (CBF1-Su(H)-Lag1), Dpp (Decapentaplegic), EGF (Epidermal Growth Factor), EGFR (Epidermal Growth Factor Receptor), ETS1 (E-Twenty Six 1), FGF (Fibroblast Growth Factor), FGFR (Fibroblast Growth Factor Receptor), Gli3 (Glioma-associated oncogene family zinc finger 3), Hh (Hedgehog), Mad (Mothers against Dpp), MAP (Mitogen-Activated Protein), Pan (Pangolin), Pnt (Pointed), RTK (Receptor Tyrosine Kinase), Shh (Sonic hedgehog), Smad1 (Small-MAD 1), Su(H) (Suppressor of Hairless), TCF (T-Cell Factor), TF (Transcription Factor: TF-A activator; TF-R repressor), TGFβ (Transforming Growth Factor β), Tkv (Thick veins), Wg (Wingless), Wnt (Wingless-Int1).

[b] Three pathways (Hedgehog, TGFβ, and Wnt) are named for their ligands, while the other two (Notch and RTK) are named for their receptors.

[c] These five signals comprise the most commonly used "words" employed by animal cells to communicate so as to collaborate in the construction of anatomical patterns [767]. Four of the five signals can act as diffusible **morphogens** [1880]. Delta cannot do so because it is bound to the cell surface, though cell protrusions can extend its reach well beyond the confines of adjacent cells [414,511].

[d] As for which came first – the signals or the receptors – the answer may not be simple [13,239,780]. For example, the Notch receptor paradoxically contains modified repeats of the EGF ligand that are needed for Notch to bind its own ligands [2474].

[e] For primers on how transcription factors work see refs. [130,284,1272,2094]. For the nature of TF binding sites see refs. [1284,2121,2435]. For binding specificity see ref. [1465].

[f] For more idiosyncrasies of the Hedgehog pathway see refs. [1065,1689].

[g] For additional pathways see refs. [1166,1200] or www.cellsignal.com. For crosstalk see refs. [48,890,1569,2148]. For pathway origins see refs. [188,1021,1760,1881,2182]; for Hedgehog in particular see refs. [17,1065,1658,1689]; for Notch see ref. [755]; for RTK see refs. [1153,1154,1302]; for TGFβ see ref. [1712]; and for Wnt see refs. [16,406,1658]. For how pathways repress target genes instead of activating them, see refs. [19,1735].

Table 1.2 Pattern features caused by conserved genes in flies and vertebrates[a]

Patterning role	Gene[b]	Protein function[c]	References
Segment identity	*Hox* genes	Transcription factors: homeobox class	[761]
Segment boundary	*Notch*	Receptor for contact-mediated signal	[482,517,1864]
Segment partitioning	*engrailed*	Transcription factor: homeobox class	[496,1015]
Dorsal–ventral axis: signal	*dpp/BMP4*	Morphogen	[1018]
Dorsal–ventral axis: inhibitor	*sog/chordin*	Morphogen antagonist	[193,1960,2495]
Appendage: distal outgrowth	*Dll*	Transcription factor: homeobox class	[636,1715,1800]
Appendage: proximal identity	*hth/Meis1*	Transcription factor: homeobox class	[1466,1800]
Appendage: dorsal identity	*ap/Lhx*	Transcription factor: LIM homeobox class	[1335,1860,2426]
Appendage: joints	*odd-skipped*	Transcription factor: zinc finger class	[872]
CNS: brain induction	*hh/Shh*	Morphogen	[1705]
CNS: dorsal fates	*dpp/TGFβ*	Morphogen	[1914]
CNS: forebrain	*orthodenticle/Otx*	Transcription factor: paired homeobox class	[11,67,999]
CNS: lateral zonation	*msh/Msx*, etc.	Transcription factor: homeobox class	[67]
CNS: neurotrophin	*spätzle2/BDNF*	Growth factor: cystine-knot class	[2515]
CNS: proneural identity	*ac-sc/Mash1*	Transcription factor: bHLH class	[1804,1830]
CNS: proneural antagonist	*hairy/Hes1*	Transcription factor: bHLH class	[1830]
CNS: lateral inhibition	*Notch*	Receptor for contact-mediated signal	[1804]
CNS: asymmetric mitosis	*numb*	Receptor-mediated endocytosis	[462]
CNS: neuron identity	*knot/COE*	Transcription factor: HLH class	[1074]
PNS: mechanosensor arrays	*numb*	Receptor-mediated endocytosis	[590,1804]
Eye: organ identity	*ey/Pax6*	Transcription factor: paired homeobox class	[763,1388,2226]
Eye: tissue growth	*hth/Meis1*	Transcription factor: homeobox class	[956]
Eye: photoreceptor identity	*atonal/Math5*	Transcription factor: bHLH class	[264,1090,2147]
Eye: post-photoreceptor neurons	*Vsx*	Transcription factor: paired homeobox class	[624,1641]
Eye: non-neuronal tissue	*Mitf*	Transcription factor: bHLH class	[914]
Ear: mechanosensors	*atonal/Math1*	Transcription factor: bHLH class	[1384]
Ear: mechanosensors	*Notch*	Receptor for contact-mediated signal	[590]
Heart: organ identity	*tinman/Nkx2*	Transcription factor: homeobox class	[207,643]
Heart: organ identity	*H15*	Transcription factor: T-box class	[878]

(*cont.*)

Table 1.2 (*cont.*)

Patterning role	Gene[b]	Protein function[c]	References
Gut: stem cell renewal	*wg/Wnt; hh/Shh*	Morphogens	[1761,2165]
Gut/liver: organ identity	*serpent/GATA1*	Transcription factor: zinc finger class	[1838]
Sex determination	*dsx/Dmrt1*	Transcription factor: unusual zinc finger	[736,870,1404]

[a] This table looks deceptively dull, but in fact it conveys the most shocking insight ever to emerge from the field of evo-devo: the bodies of flies and vertebrates – indeed of all bilaterally symmetric animals – are built by the same genetic gadgetry, despite our seemingly different anatomies [969]. This shared circuitry is due to our descent from a common urbilaterian ancestor. Other **deep homologies** [5,785,2045] among bilaterian phyla include skin [1084], muscles [4], tendons [1988], taste sensors [2480], smell sensors [989,1860,2352], sound sensors [144,1583], pain sensors [1120], motor neurons [1509,2204], foregut identity [68], liver function [522,691], hematopoiesis [639,1371], wound healing [564,1552,2323], organ growth [255,558,619], circadian clocks [647], apoptosis [718], apico-basal polarity [2201], planar cell polarity [373,812,871], body asymmetry [1592,2272], excretory physiology [2353], and even a propensity to drown our sorrows in alcohol [1307,1871,2478]! Abbreviations for patterning roles: CNS (central nervous system) and PNS (peripheral nervous system). For details of CNS, PNS, and retinal homologies see refs. [66,777,998,1299,2136], ref. [1933], and refs. [436,1196,1640,2481,2523] respectively. For general reviews see refs. [339,630,2045,2282,2316] as well as Tom Brody's *Interactive Fly* website.

[b] Gene synonyms (e.g., *dpp/BMP4*) are "fly/vertebrate" orthologs [1221]. Most of these genes – especially those in the *Hox* family [1509,1738] – operate in an abstract (structure-independent) manner [493,965,1830,2072]. They therefore suggest that anatomy is "computed" by the genome [237,961]. Abbreviations: *ac* (*achaete*), *ap* (*apterous*), BDNF (Brain-derived Neurotrophic Factor), bHLH (basic helix-loop-helix), *BMP4* (*Bone Morphogenetic Protein 4*), *COE* (*collier/olfactory-1/early B cell factor*), *Dll* (*Distal-less*), *Dmrt1* (*dsx and mab-3-related transcription factor 1*), *dpp* (*decapentaplegic*), *dsx* (*doublesex*), *ey* (*eyeless*), *Hes1* (*Hairy and enhancer of split 1*), *hh* (*hedgehog*), HLH (helix-loop-helix), *Hox* (*Homeobox*), *hth* (*homothorax*), *Lhx* (*LIM homeobox*), *Mash1* (*Mammalian achaete-scute homolog 1*), *Math5* (*Mammalian atonal homolog 5*), *Meis1* (*Myeloid ecotropic viral integration site 1*), *Mitf* (*Microphthalmia-associated transcription factor*), *msh* (*muscle segment homeobox*), *Msx* (*msh homeobox*), *Nkx2* (*Nk2 homeobox*), *Otx* (*orthodenticle homeobox*), *Pax6* (*Paired-homeobox 6*), *sc* (*scute*), *shh* (*sonic hedgehog*), *sog* (*short gastrulation*), *TGFβ* (*Transforming Growth Factor β*), *Vsx* (*Visual system homeobox*), *wg* (*wingless*), *Wnt* (*Wingless-Int1*).

[c] "Contact-mediated" means non-diffusing. For primers on **morphogens** see refs. [1590,1880]. Other articles explain transcription factors in terms of their overall function [130,1037,2094], their types [344,750,1702], their enhancer binding sites [1284,2121,2435], their combinatorial grammar [284,1822,2312,2410,2476], and how they specify organ identity via **selector genes** [344,1373,1601]. For remaining mysteries see ref. [1745].

> Two interpretations are possible. One is that the common ancestor of fruitflies and mice had eyes . . . Alternatively, the homology may be more abstract: *ey/Pax6*, or the ancestral gene from which they evolved, might have specified some activity only in a particular location in the body (the top front of the head). Then the use of the same gene in mice and fruitflies would reflect only the fact that the two animals grow eyes in a similar body region . . . At some level, homology must exist between mice and fruitfly eyes; the question is whether the homology is at the level of eyes, or head regions. [1856]

> All eyes of all kinds require visual pigment genes, and this is the [ancestral] role of *Pax6*; the gene was later coopted for use in the different morphogenetic programs that produce the different structures on which the pigment cells are mounted in different creatures. [630]

The eyes-early versus eyes-late dialectic has grown into two schools of thought about urbilaterian anatomy in general [106,1653]. The Complex-body School posits that urbilaterians had eyes [761], legs [2044,2160], and segments [200,457], plus a central nervous system [526,2183] that was subdivided into distinct regions along two perpendicular axes [107,998]. The Simple-body School contends that urbilaterians may have had primitive eyespots [763] but no segments [385], legs [2160,2427], or advanced structures of any kind [2426], and that the nervous "system" was just a diffuse neural net [1478].

> The evolution of given body parts probably began with the installation of cell differentiation programs to deploy certain specific cell types in a certain position in an organism, initially in a very simple morphological context. Later in evolution the transcriptional regulators of these differentiation gene batteries would have been coopted for use in increasingly complex, clade-specific programs of gene regulation that control pattern formation processes. [630]

One virtue of the Simple-body School is that phyla which lack segments (e.g., nematodes) or a centralized nervous system (e.g., echinoderms?) would never have possessed them to begin with [287], so we would be spared the torment of trying to explain secondary losses that are hard to rationalize for useful traits [72,1096]. For example, if urbilaterians had legs, then why did (subsequent) basal chordates lose them? Indeed, amphioxus lacks any traces of the fins that later became tetrapod legs [2045].

This virtue notwithstanding, it is still hard to see how snakes and centipedes can be using what seems to be the same *Notch*-based oscillator to slice new segments from a rear growth zone [386] if they did not inherit this clockwork from a *segmented* common ancestor [1855]. Likewise, it is hard to imagine how vertebrates and arthropods could have *independently* hit upon the same obtuse process of "resegmentation," whereby the phase of their metameres shifts by half a unit of wavelength [1485] – i.e., somites splitting to make vertebrae [1368] or parasegments splitting to form segments [1250].

How the urbilaterian got its symmetry

How did two-sidedness evolve in the first place? Again we find ourselves on shaky ground, squinting to discern a distant ghost through a dense fog [664,1395]. Bilaterians are thought to have evolved from radially symmetric cnidarians [113,2064], but some cnidarians exhibit bilateral tendencies [210,665,1104], and ctenophores (a more basal taxon) are bi*radially* symmetric [1396,1398], so two-sidedness could have arisen gradually [209,1397]. This process might be re-enacted in annelid development (a putative case of **recapitulation**), where the two-sided trunk grows out of a radially symmetric head [265,2011].

The origin of two-sidedness in bilaterians is only one example – albeit a profound one – of a new axis being shoehorned into an old geometry [1949,2385]. Other instances include (1) the evolutionary conversion of radial flowers (e.g., daisies) to bilaterally symmetric ones (e.g., orchids) in angiosperms [402,1031,1782], (2) the conversion of radial to bilateral feathers in birds [2491], and (3) the insertion of a newly hinged jaw into a jawless fish head [1217,1451].

This last event led to the quasi-mirror symmetry of our upper-versus-lower teeth [185,528]. Any parent who has ever paid an orthodontist to straighten a child's teeth

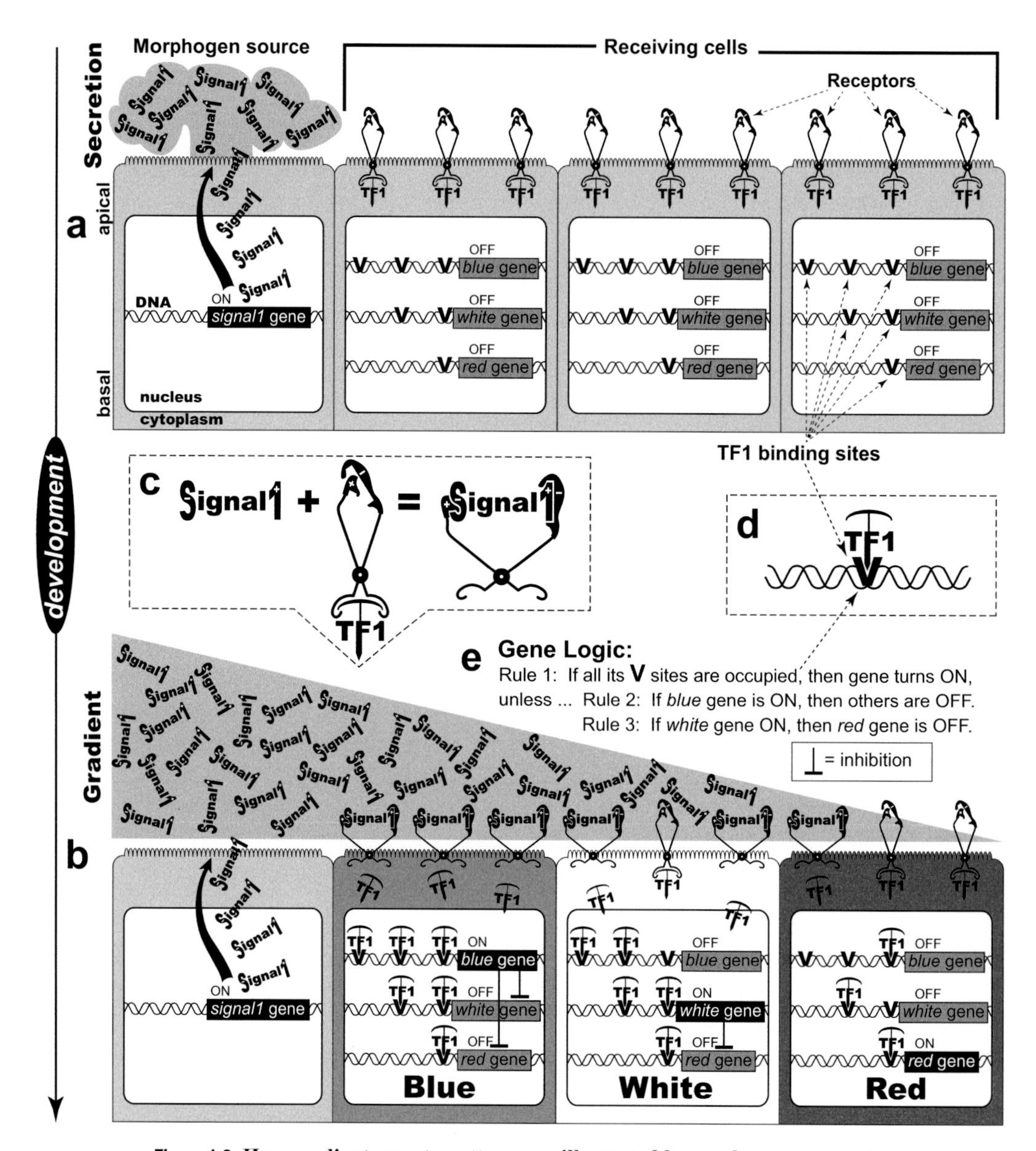

Figure 1.3 How gradients create patterns, as illustrated by an abstract example.

a. Four cells are cartooned. The first one secretes Signal1 (a protein). The other three receive the signal via cell-surface receptors that are schematized as hinged clamps in a closed (unoccupied) state. The receiving cells gauge their distance from the signaling cell (i.e., their position in the array) by measuring the intensity (concentration) of the signal – a diffusible chemical **morphogen**. This intensity (**b**) then dictates cellular fates via gene circuitry (**e**) [1590,1880].

b. Diffusion of Signal1 forms a concentration gradient (triangle) [244]. The frequency with which Signal1 is bound by receptors decreases with distance, as does the likelihood that TF1 (Transcription Factor 1) will be bound by cognate *cis*-regulatory DNA sites. Sites near the *red*, *white*, and *blue* genes get saturated at progressively higher TF1 thresholds

knows only too well that evolution has not yet perfected the occlusion (**morphological integration**) of our upper and lower teeth [1842].

The common theme among these diverse examples is the inception of a linear signaling source along a new mirror plane [1459,1460], which then emits morphogens in symmetric gradients [1622,1898]. How such gradients create patterns will be examined in the next chapter with regard to the upper and lower layers of the fly's wing (morphogen = Wg) and its quasi-symmetric front and back halves (morphogen = Dpp) – a remarkable case of a Cartesian coordinate system.

Figure 1.3 *(cont.)*

(cf. [997,1910]). Some morphogens diffuse in the extracellular space as shown here [212,882,1567,2511], but others do not [98,1953,2421,2514]. In either case, diffusion is so slow that diffusion-based devices are only practical at a small scale (i.e., 10s or 100s but not 1000s of cells) [121,1242,1566,2029].

c. Magnified view of Signal1 binding a receptor, which forces the receptor to release TF1 as in the Notch pathway [890]. Binding is depicted as (1) a jigsaw-puzzle fit of protrusions into cavities [1083] and (2) an attraction between opposite charges (S^- to arm^+, and 1^+ to arm^-), but actual binding interactions can involve more components [2261] (e.g., heterodimeric ligands or receptors [1509]).

d. Magnified view of TF1 bound by a *cis*-regulatory site (**V**) in the DNA. TF binding sites tend to be ~10 base pairs long [2121,2476]. TF1's "anchor" here becomes an "umbrella" that will later bind RNA polymerase and its associated complex of basal transcription agents (not shown).

e. Imaginary circuitry that controls gene expression in responding cells, based on the actual posterior prevalence hierarchy of *Hox* genes [572,2063,2482]. The net result of the gradient is to elicit a "French flag" of states along the axis (blue, white, red), as proposed in Lewis Wolpert's classic model [2438,2439,2442]. This process has been called "painting by genes" [27]. Dissecting the Boolean grammar of *cis*-enhancers is an active area of investigation on the evo-devo frontier [2476]. This schematic was inspired by similar diagrams in refs. [515,1249,2362]. N.B.: Developmental patterns can be created in many ways other than morphogen gradients (cf. Table 5.1) [961].

2 The fly

How the fly got its gyroscopes

Flies are the hummingbirds of the insect world [2342]. They dart and hover and seem to delight in evading pursuers [543]. In contrast, their butterfly cousins are typically more like pelicans that flap and glide and languidly survey the scenery below [483], though butterflies do have a few aeronautic tricks of their own [2099,2497].

What gives flies their midair agility? It is a pair of gyroscopes called halteres [1784]. Halteres are club-shaped outgrowths of the thorax that oscillate up and down when the fly is airborne [542]. Deprived of these powerful stabilizers, flies spin out of control and crash clumsily to earth, often landing helplessly on their backs as if drunk [689].

Halteres evolved from hindwings, which are still present in butterflies and most other insects [336]. Indeed, flies belong to an order called Diptera because they only have two (di-) wings (-ptera) [1839].

Could a four-winged insect have evolved into a two-winged one without making itself so unsteady that it could not fly? Perhaps [254], because removing the hindwings of a butterfly does not prevent it from flying, and such amputees can even steer along prescribed flight paths, albeit more slowly [1946].

Given this redundancy in the airfoils of four-winged insects, could halteres have evolved from forewings instead of hindwings? Yes. A comparable conversion of the forewings occurred in strepsipterans – an obscure order of twisted-wing parasites [1623]. From the standpoint of dipterans, strepsipterans look as silly as horses with their saddles strapped on backward [2382]. Like strepsipterans, coleopterans (beetles) fly with only their hindwings, but they turned their forewings into protective covers (elytra) instead of halteres [56,337].

How did wings become halteres? They shrank at their base and swelled at their tip. The inertia of the swollen tip henceforth bent the flexible stalk whenever the body turned in a new direction, and sensors at the base detected this torque. The full story is worth telling, but it will require some background.

The wings and halteres of an adult fly develop from pockets called imaginal discs that begin as clusters of undifferentiated cells on the flanks of the embryo (Fig. 2.1). "Imaginal" means adult, and "disc" signifies the round, flat shape of these sacs. There are 19 discs in all [965]. They grow inside the larva until metamorphosis, whereupon they turn inside out, replace the larval skin, and secrete a new cuticle. The wing disc grows

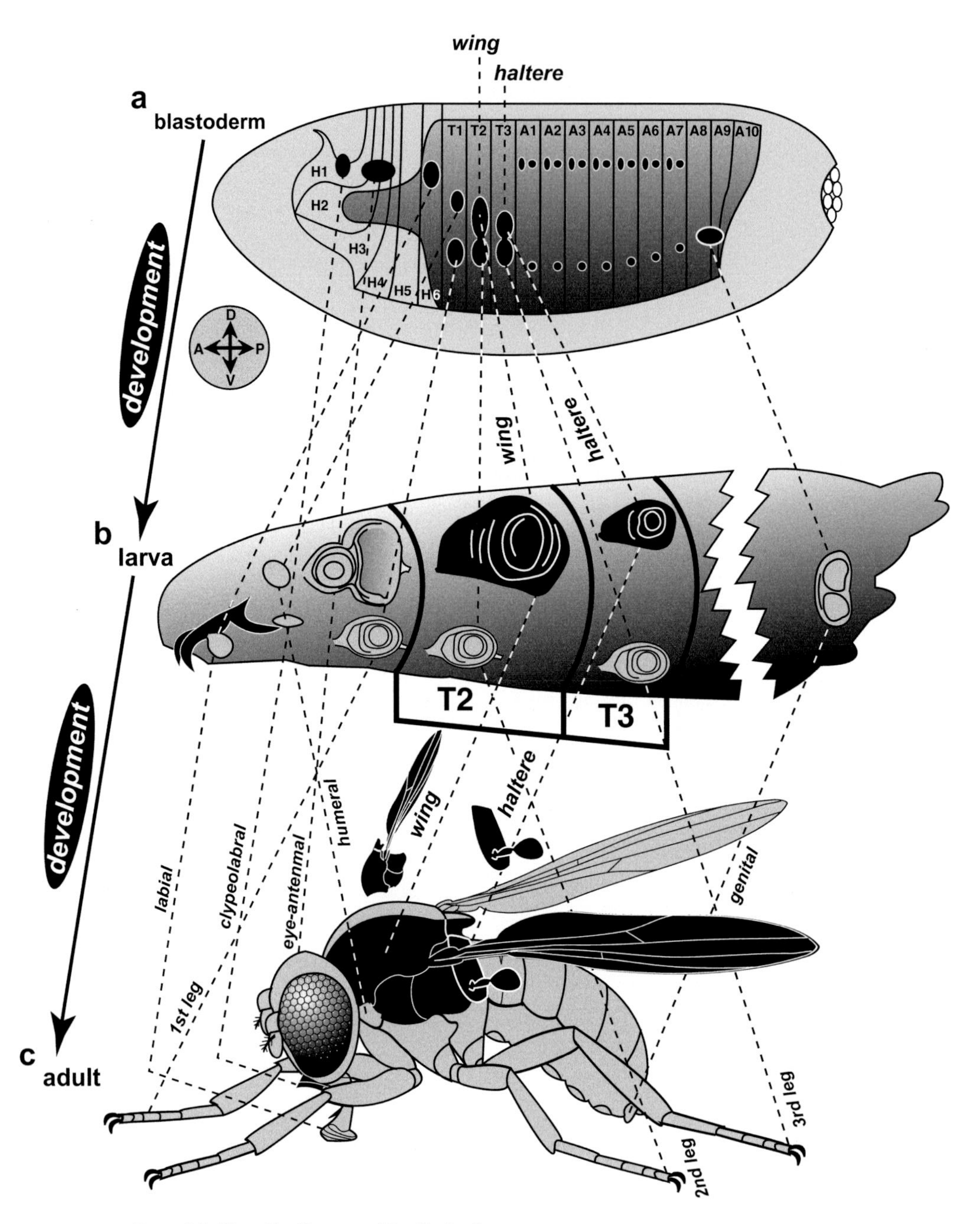

Figure 2.1 How the fly assembles its body.

a. Embryo (left half) at blastoderm stage, when it is a hollow ellipsoid. Destinies of different areas are mapped. The darkly shaded flank will form the larval skin (**b**). The adult skin (**c**) is quilted from 19 "imaginal discs" that begin as spots (black ovals). These spots pucker inward to form pockets that grow inside the larva. Wing and haltere discs are initially joined to the leg disc below them. Vertical lines delineate body segments: six head (H1–H6), three thoracic (T1–T3), and ten abdominal (A1–A10) segments. Peripheral tissue (light gray) moves inside more than the haltere disc [947] and produces a distinctive set of structures, including veins and bristles.

How does the fly make a wing, starting with only a few dozen imaginal disc cells? It begins by exploiting the zonation of the embryo. Fly embryos express the gene *engrailed* (*en*) in zebra-like stripes – one stripe per segment [550]. The wing disc arises at a spot straddling the edge of the *en*-ON stripe in the second thoracic segment (T2). Thus, at its inception the wing disc contains two types of cells: *en*-ON cells in its posterior (P) half and *en*-OFF cells in its anterior (A) half. Next, a line is drawn along the A/P boundary via the gene *decapentaplegic* (*dpp*). The full recipe, whose early steps are listed below, is depicted in Fig. 2.2(a–i).

1. En turns ON *hedgehog* (*hh*) in the P compartment of the disc.
2. All P cells secrete Hh protein, but they cannot respond to it.
3. Hh crosses the A/P line a few cell diameters into A territory.
4. The A cells that receive Hh turn ON *dpp*.

The A/P boundary is now distinguished by a thin stripe of *dpp*-ON cells. Dpp diffuses much further than Hh and spreads in both A and P directions from the A/P line. The concentration of Dpp diminishes with distance from the stripe of producing cells, so that any cell can determine how far it is from the A/P line by the local Dpp concentration. In this way, wing cells effectively acquire an x coordinate.

All that is needed now to create a two-dimensional canvas is a y coordinate. That coordinate is supplied by the **morphogen** Wingless (Wg), which is secreted along the boundary between the dorsal (D) and ventral (V) compartments via a comparable process (Fig. 2.2j–o). The wing margin arises at this D/V boundary.

Thus, flies discovered Cartesian coordinates long before René Descartes. The fly wing is only one example among many where orthogonal axes furnish a framework

Figure 2.1 (*cont.*)

to form internal organs. Black dots in A1–A7 are superficial nests of cells (not discs) that make the adult abdomen, though A7 is aborted in males [678]. White ovals at the rear of the embryo become germ cells. Compass: A, anterior; P, posterior; D, dorsal; V, ventral. Adapted from refs. [937,965].

b. Larva. Most of the abdomen is omitted (broken line), remaining segments are distorted (e.g., T2 and T3 are actually the same width), and relative sizes and sites of discs are only approximate. Wing and haltere discs are shaded black for emphasis. Unlike other insects (cf. Figs. 3.8 and 3.9), advanced flies tuck most of their head into their thorax [880] to form a headless, worm-like larva, and only restore the head to the surface during metamorphosis to give the adult a cockpit full of fancy sensory gadgetry.

c. Adult. The wing disc grows larger than the haltere disc and makes more of the adult surface (black). Insets (between **b** and **c**) clarify the boundary between them. The wing pouch (white lines = concentric folds) inflates and flattens to make the wing; outer folds make half the thorax. Likewise, the center of the haltere disc makes the haltere proper (club), while outer tissue makes a flank plate. Overall, the haltere develops like a small version of the wing. Bristles are omitted for clarity. (Aside from their role as sense organs [1134], bristles help save flies from drowning in rainstorms by serving as anti-wetting agents [541].) From ref. [965].

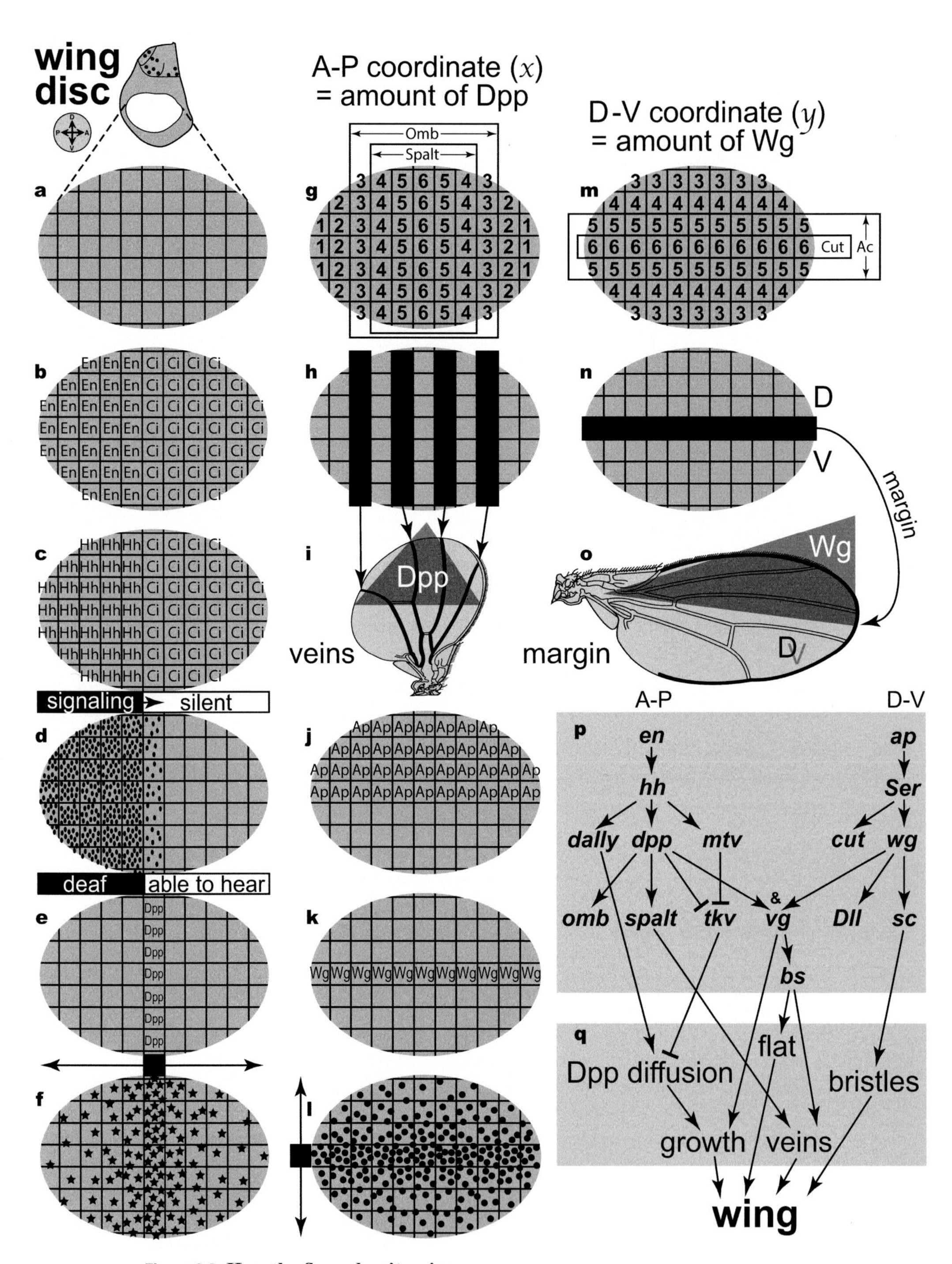

Figure 2.2 How the fly makes its wing.

a–i. Stages of patterning along the anterior–posterior (A–P or x) axis. Adapted from refs. [104,965]. (**a**) A right wing disc is depicted with its pouch removed, enlarged, and rendered as an oval. Dots above the pouch denote bristle precursors on the thorax. Squares represent

Figure 2.2 (*cont.*)

cells, though real cells are less regular and more hexagonal. A mature pouch actually has 1000s of cells, but fewer are shown because most events (**b–o**) occur when it has 10s or 100s of cells. The points of the compass (A, P, D, V) indicate cardinal directions (anterior, posterior, dorsal, ventral). (**b**) P cells express Engrailed (En), while A ones express Cubitus interruptus (Ci), each of which is a transcription factor made by a **selector gene** (*en* or *ci*). (**c**) En turns ON *hedgehog*, which encodes a **morphogen** (Hh). (**d**) Hh proteins (small ovals) are made throughout the P region and cross into A territory, where Ci transduces the Hh signal. Because P cells lack Ci, they are insensitive ("deaf") to their own signal, so they don't respond (by making Dpp). Anterior cells can hear the signal, but the signal only diffuses a short way into their territory, so that the responding cells occupy a narrow stripe along the A/P boundary. (**e**) Hh turns ON *decapentaplegic*, which makes a longer-range **morphogen** (Dpp). (**f**) Dpp proteins (stars) are secreted by a stripe of cells at the A/P interface. Dpp diffuses in both directions (arrows) to the edge of the wing pouch, eventually forming a stable gradient. All cells can hear the Dpp signal. (**g**) Cells assess their distance (x coordinate) from the A/P boundary by the amount of Dpp, denoted by numbers inside squares. Dpp turns ON *optomotor-blind* (*omb*) and *spalt* at successively higher thresholds. Thin arrows indicate breadth of expression (not diffusion). (**h**) Wing veins (black bars) arise at particular sites along the x axis. The inner two veins are elicited by Hh (not shown) [965], while the outer ones are induced just outside the Spalt region by signals that are evoked by either Spalt (A vein) or Omb (P vein) [435]. (**i**) Right wing seen from below. The Dpp gradient is symmetric, and so are the veins (black lines) to a first approximation. The two crossveins are specified separately [1379].

j–o. Stages of patterning along the dorsal–ventral (D–V or y) axis. Adapted from ref. [965]. (**j**) Dorsal cells express the transcription factor Apterous (Ap), which, like En, is made by a homeobox-containing **selector gene** (*ap*). (**k**) Ap turns ON *wingless* (*wg*), which encodes a **morphogen**, at the edge. (**l**) Wg proteins (solid circles) are secreted by a stripe of cells at the D/V interface. They diffuse widely in both directions, eventually creating a stable gradient. All cells can hear the Wg signal. (**m**) Cells assess their distance from the D/V line (y coordinate) by the amount of Wg, denoted by numbers inside squares. Wg turns ON *achaete* (*ac*) and *cut* at successively higher thresholds. Thin arrows indicate breadth of expression (not diffusion). (**n**) Cells at the D/V interface form a wing margin (black bar). During metamorphosis, the wing pouch folds along this line to bring the D and V surfaces together [149]. (**o**) Right wing seen from above (D surface up). If the margin (black line) were to continue secreting Wg (as it did in the disc), then Wg would diffuse proximally toward the wing base. The margin sprouts bristles (due to *ac* and *sc*) along its anterior edge.

p. Circuitry for wing patterning initiated along the A–P or D–V axis [104]. Arrows indicate activation of one gene by another. T-bars denote inhibition. Ampersand (&) shows *dpp* and *wg* acting in combination. Dpp diffusion relies upon *dally* and indirectly upon *master of thick veins* (*mtv*), which acts by inhibiting an inhibitor, namely *thick veins* (*tkv*). Tkv is a Dpp receptor. The *vestigial* (*vg*) gene is jointly regulated by Dpp and Wg [104,1149]. It enables veins to form via *blistered* (*bs*) [508,1657,1873]. *Serrate* (*Ser*) mediates *ap*'s activation of *wg* and sharpens the margin together with *cut*, while *scute* (*sc*) and *achaete* (*ac*, not shown) promote bristle formation. Unlike its role in the leg outgrowth, *Distal-less* (*Dll*) does not mediate wing outgrowth [821]. Butterflies make their wings in basically the same way [343]. Based mainly on refs. [470,965,2350] with data added from refs. [515,723,1365,1520, 2018].

q. Features of wing development that depend on the circuitry in **p**. "Flat" denotes a flattening of the wing blade resulting from the apposition and adhesion of the dorsal and ventral halves of the inflated wing pouch [149].

for patterning [2443]. The theoretician Lewis Wolpert predicted the universality of this geometry decades before it was documented in developing embryos [2440].

Once every wing cell has its (x, y) coordinate, any pattern can be "painted" onto this surface by using subsidiary genes, just as an artist would dab colors from a palette [412]. For example, the x coordinate evokes a *spalt*-ON zone that straddles the *dpp*-ON stripe, and two of the four wing veins then develop at the edges of this zone. Dozens of genes erect the wing's scaffolding, and they comprise an intricate genetic circuit (Fig. 2.2p) [965]. Overall, the wing is built by a bootstrapping process that increases complexity gradually, starting with binary states – for example, *en*-ON versus *en*-OFF.

The sort of elaboration that we see in the wing disc is called **emergence**, and it is ubiquitous in embryos [187,1241]. It endows their organs with a seemingly magical aura of spontaneity [114,412,814]. Order and organization seem to come from nowhere, one shape changes seamlessly into another, and organs take on a life of their own [339,1357].

This principle is nicely illustrated by John Conway's famous video game *Life* [108,747], which popularized the field of cellular automata [748,2437]. Given different initial conditions ($\approx$ heterogeneities), the same set of state-change rules ($\approx$ circuitry) can lead to wildly dissimilar patterns [2437].

Now that wing development has been presented in some detail, we can return to the question of how haltere development differs and how these differences are implemented genetically. Amazingly, the distinctions between the two discs all implicate a single culpable gene [2110]. The haltere expresses the *Hox* gene *Ultrabithorax* (*Ubx*) [143,2379], whereas the wing does not [258]. *Ubx* turns out to be pivotal in steering disc development away from a wing and toward a haltere:

1. *Ubx* is necessary for haltere development. If the haltere disc is deprived of *Ubx* function, then it makes a wing instead of a haltere [1311,1314]. If *Ubx* function is removed from only portions of the haltere, then those parts adopt a wing-like character [297,2380].
2. *Ubx* is sufficient for haltere development. If *Ubx* is mutationally or artificially turned ON in the wing disc, then that disc forms a haltere instead of a wing [1311,2378].

Considering its power to transform a wing into a haltere and to do so on a regionally autonomous basis, *Ubx* was initially thought to endow each cell with haltere versus wing identity directly [355]. Subsequent analyses showed that *Ubx* instead acts indirectly via target genes [980,1733] at key points in the A–P and D–V patterning pathways [1520].

1. Along the A–P (x) axis, Ubx directly reduces *dpp* expression [1365] and indirectly impedes Dpp mobility [515] by inhibiting *dally* [470] and *mtv* [469]. Ubx also directly blocks *spalt* [723,2350] and *bs* [2350], thereby aborting vein formation.
2. Along the D–V (y) axis, Ubx inhibits *wg* both directly [2018,2350] and indirectly by blocking Wg from being transduced to turn ON *vg*, thus reducing growth [471,1519]. Ubx directly suppresses *sc* [2350], leading to bristle loss. Ubx also turns OFF *cut*, which removes the margin [2018]. Because *bs* is required to stick the D and V surfaces of the wing together [1525], its arrest by Ubx explains why the haltere inflates to make a round balloon instead of collapsing to form a flat blade [2350].

Thus, *Ubx* behaves more like an intrusive meddler than an aloof executive [26]. It intervenes at many levels, rather than using a chain of command [353,354]. The overall circuitry may be hierarchical, but *Ubx*'s influence upon it is not [2350]. *Ubx* reaches deeply into the bowels of the wing's gene network and short-circuits it so as to convert a flat airfoil into a knobby gyroscope. It does something similar to turn a T2 leg into a T3 leg [2113,2234], though the reshaping in that case is much less dramatic.

In the realm of the T3 body segment, therefore, *Ubx* wields absolute power as a **master gene** [1255]. The effect of "unplugging" *Ubx* is to produce the iconic "bithorax" (two-thorax, four-wing) phenotype, where halteres become wings. Because halteres evolved from wings, this transformation constitutes an **atavism** – i.e., the restoration of an ancestral anatomy.

If a binary genetic switch can be flipped so easily to turn back the clock of evolution, might evolution have invented halteres all at once as well? One famous geneticist certainly thought so [548,803]. In his 1940 book *The Material Basis of Evolution* [801] Richard Goldschmidt argued that saltatory mutations are a major driving force of evolution. His notion of **hopeful monsters** has been largely discounted [25], but such mutants may have played a role in certain lineages [829,1858,2191]. With regard to the haltere, the evidence supports a transformation that may have begun abruptly but then proceeded gradually to reach its present state [2350].

1. The fossil record shows that proto-dipteran hindwings diminished in size before they adopted a knobby shape [1839].
2. Butterflies express *Ubx* as strongly in their hindwings as flies do in their halteres [2351], so we may comfortably conclude that *Ubx* was expressed in the T3 segment of pre-dipteran insects long before the haltere evolved [2018]. This lack of temporal congruence refutes the notion that *Ubx* per se causes halteres.
3. Ubx alters the expression of its target genes by binding to nearby *cis*-regulatory sites, and multiple Ubx-binding sites are required to control certain genes (e.g., *spalt*) [723]. Its binding sites could not have arisen all at once by any known mutational mechanism.

The mutations that sculpted the haltere must have been primarily in DNA sequences next to *Ubx*'s target genes, not in the *Ubx* gene itself. (Ubx had earlier become a repressor via an alanine-rich motif that it acquired near its homeodomain [721,1889].) In theory, any gene could have come under *Ubx*'s influence if nearby mutations had hit upon the "magic word" to retain Ubx proteins [501]. Such captures must have occurred by other genes besides the targets in Fig. 2.3b, but those links did not endure because they failed to confer any selective advantage.

If the haltere was derived from the hindwing, then evidence of its history might be discernible from its anatomy. Indeed, researchers have found nests of tiny sense organs at the base of the haltere that match those of the wing and appear to have remained unaltered over the eons (Fig. 2.3f). The identity of these sensilla has been verified by various approaches, including genetic mosaics [1542], partial homeosis [238,423], bithorax phenocopy [1346], and manipulation of *Ubx* expression as a function of time [1872,1904]. Finding such ancient relics (~225 MY old [2110]) embedded in

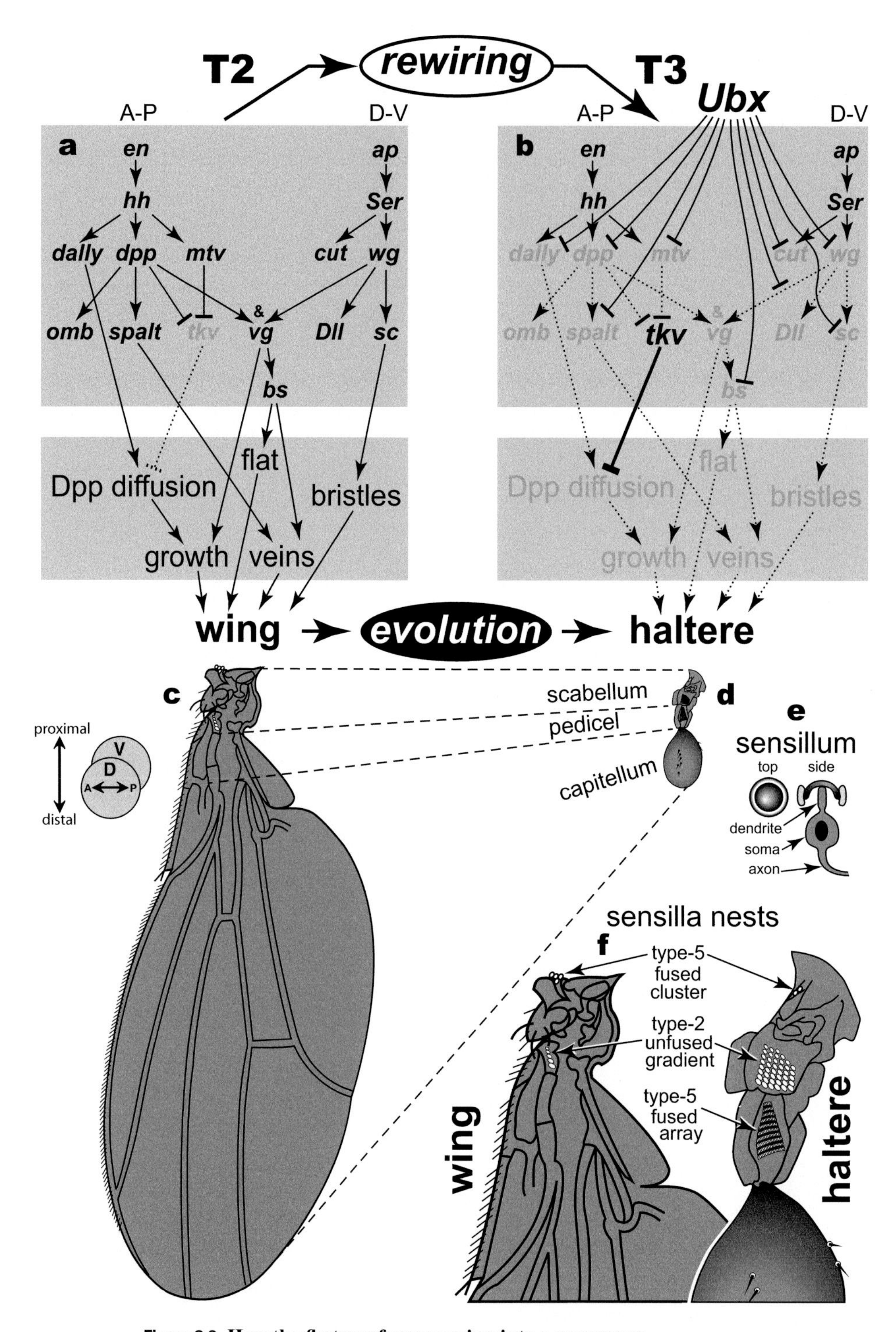

Figure 2.3 How the fly transforms a wing into a gyroscope.

the cuticle has been as gratifying as finding a pharaoh's tomb with its treasures intact (boldface added):

> It has long been recognized that the haltere is homologous to the wing. **It is pleasing, therefore, to find this homology apparent** even at the level of resolution afforded by the [scanning electron microscope]. For example, the sensilla of the [anterior notal wing process] field of the wing closely resemble the two metathoracic papillae of the haltere, both being composed of fused, type-5 sensilla. The d.Rad.A sensilla of the wing and the dorsal scabellar sensilla of the haltere are likewise similar to one another. Both fields possess type-2 sensilla arranged in longitudinal rows, and increasing in diameter distally.

Figure 2.3 *(cont.)*

a. Circuitry for wing development (from Fig. 2.2). In the panel above are genes (italicized), and in the panel below (roman type) are the relevant features at the cell or tissue level. Anatomical diagrams are based on refs. [270,423].

b. Effects of the *Hox* gene *Ubx*, as verified by (1) disabling *Ubx* in the haltere or (2) expressing *Ubx* ectopically in the wing. Flies acquired halteres ~225 MY ago by reshaping their hindwings [2110]. This remarkable transformation was accomplished with a single tool: the repressor protein Ubx [721,1283]. Ubx rewired the circuitry of the hindwing disc as shown here. Arrows denote activation; T-barred lines denote inhibition. Faded letters signify genes that are inhibited by *Ubx*. Dotted lines are weakened links. The only gene that is stimulated by *Ubx* is *tkv*, but that activation is indirect. Overall, Ubx transforms the wing into a haltere by (1) stunting its growth, (2) aborting its veins, (3) erasing its edge bristles, and (4) causing what's left of its wing blade to inflate like a water balloon. This inflated bulb tends to stay put during pitch and yaw rolling (Newton's first law) – thus bending the stalk and allowing the resulting strains to be sensed by tiny sensilla. T2 and T3 denote the second and third thoracic body segments.

c. Dorsal surface of a left wing. The ventral surface (not visible) lies behind. The compass shows the A–P (anterior–posterior) axis orientation.

d. Dorsal surface of a left haltere at the same scale as the wing in **c**. Dashed lines indicate the homology of haltere subdivisions (labeled) with corresponding regions of the wing, as mapped by various methods. The haltere functions as an odd sort of gyroscope – one that vibrates instead of rotates – to sense body movements. The analogous device in a human is our inner ear: the fluid in our semicircular canals is the inertial mass [475].

e. A typical campaniform sensillum. This tiny device detects cuticular deformations [2517], as seen in top view (left) and cross section (right). Clusters of such sensilla lie at the base of the haltere [796], where they sense strains from pitch or yaw [542,2176]. None exist on the capitellum, which merely acts as a dead weight. Each sensillum (each ~3 μm in diameter) has a dome, a socket, a sensory neuron [1940], and a sheath. The sheath is omitted here [1121]. Six different types of sensilla have been documented on the fly wing and haltere [423]. The dendrite attaches to the dome via a tubular body (not shown) [2212]. When the dendrite is stretched, it causes the cell body to emit action potentials that travel down the axon (truncated here) to inform the central nervous system [651]. Redrawn from ref. [1940].

f. Nests of campaniform sensilla on the wing and haltere (dorsal only) compared side-by-side at higher magnification. Two homologous nests are shown: (1) a cluster of fused, type-5 sensilla near the base and (2) a more distal array of unfused, type-2 sensilla that are graded in size from proximal to distal (one row on the wing and six rows on the scabellum). Further homologies exist ventrally (not shown) [423]. The bottom arrow indicates an array on the pedicel that has no counterpart on the wing. It is about the same size as the group on the scabellum, but its sensilla are fused into transverse rows. The trellis-like patterns in the latter two haltere arrays may facilitate the monitoring of torque strains [1784].

... Similarly, the ventral radius bears **a field of sensilla which is almost identical to that found on the ventral scabellum**. The v.Rad.A sensilla are elliptical, low-profile sensilla bearing sockets (type 6). There are four and occasionally five sensilla in this field, arranged in a diamond pattern. The ventral scabellum bears five elliptical high-profile sensilla with sockets which share the same size range and pattern of distribution. In fact, it is only the difference in dome prominence which makes this field distinguishable from that of the wing. [423]

This parable of the humble haltere illustrates some rather profound principles concerning how evolution manipulates development:

Take-home lessons

1. **Serial homologs**, such as the wing and haltere, use the same developmental recipe in different ways. They are variations on a theme.
2. Evolution can reshape anatomy by rewiring gene circuitry.
3. *Hox* genes can reconfigure organs by gradually recruiting target genes, each of which tinkers with anatomy in a novel way.
4. Eons of rewiring can be reversed by a single **atavistic** mutation.

Anyone who has ever watched videos of dragonflies catching mosquitoes in midair knows that dragonflies are at least as adept at flying as dipterans [2450,2453], but how can this be if they have no halteres? In an astounding case of **convergence**, dragonflies use their head in the same way that flies use their halteres. The loosely attached head wobbles on the neck like a bobblehead doll, and its motions (pitch, yaw, and roll) are detected by trichoid sensilla where it rubs against the thorax [452].

How the fly tattooed its arms

Aside from their prowess as aerialists, flies are also notorious for being skilled troubadours [407,562]. *Drosophila melanogaster* males serenade potential mates by vibrating first one wing and then the other as they perform a ritualized dance [1386,2289]. The reputation of flies as singers and dancers – and party animals in general – was immortalized in William Blake's poem "The Fly," which was part of his collection, *Songs of Innocence and of Experience* (1794):

Little fly,
thy summer's play
my thoughtless hand
has brushed away.
Am not I
a fly like thee?
Or art not thou
a man like me?
For I dance
and drink and sing,

till some blind hand
shall brush my wing.
If thought is life
and strength & breath,
and the want
of thought is death,
then am I
a happy fly,
if I live,
or if I die.

Courtly manners notwithstanding, male flies have a disturbingly dark side to their personalities that should give us pause [20,2322]. Alcohol is a constant temptation on the fermenting fruits that they visit, and they occasionally overindulge. As with men, liquor enhances their sexual arousal but decreases their performance [1262], and rejected suitors tend to drown their sorrows in the dipteran equivalent of daiquiris and mai tais [1307,1871,2036]... giving new meaning to the term "bar flies."

Moreover, male flies become downright brutish when vying for the same female [1477]. If two rivals happen to meet, they rear up on their hindlegs and use their forelegs to box one another furiously [375], looking for all the world like love-crazed kangaroos [546,1033]. Later, the victorious male uses his forelegs again like arms – but this time tenderly – to caress the female during foreplay [920,1175]. As he strokes her, chemosensory bristles on his tarsi savor her aphrodisiac pheromones [1334,2199,2211].

Given that male flies sometimes behave like drunken sailors [2525], it seems apt that they bear tattoos on their arms. (*Disclaimer:* The author is aware that this fad has spread to sober landlubbers and doesn't think there is anything wrong with tattoos in general.) The forelegs of males bear pigmented insignia that might aid in mate selection [1191] or species recognition [1386,2227], in addition to performing grooming functions as explained below. The light-colored (yellow) bristles are arranged in transverse rows, while the dark (black) ones are aligned orthogonally in a structure called the sex comb.

In 2009 these decorations began to be analyzed methodically in terms of both their evolution and their development [84,85]. The foreleg begins as an ordinary midleg but veers off in a new direction under the guidance of the *Hox* gene *Sex combs reduced* (*Scr*) [970,2171]. Praying mantises jigger their forelegs even more drastically (also via *Scr*?) by turning them into an elaborate spring-loaded trap for prey capture [1518,1896]. Thus, mantises are the true centaurs of the insect world [1431] because their front legs are used as arms all the time, while flies only use them as arms occasionally, to box or flirt. Flies also differ from mantises insofar as the forelegs are sexually dimorphic: the sex combs are confined to males via the action of a separate gene belonging to the sex-determination hierarchy.

As with the haltere, the foreleg (first leg) cannot be understood in isolation. We must first delve into how the fly builds its midleg (second leg). The midleg serves a default role ventrally akin to the wing's role dorsally. It is thought to represent the ancestral state for all three pairs of legs [349,761,965]. Hence, this digression may also be useful in illuminating some aspects of the archetypal dipteran leg [608,1211,1792].

Like the wing disc (Fig. 2.2), the midleg disc (Fig. 2.4) anchors adult geometry to the (A/P) line separating posterior *en*-ON cells from anterior *en*-OFF cells:

1. En turns ON *hedgehog* (*hh*), which by default is OFF (Fig. 2.4c).
2. All P cells secrete Hh protein, but they cannot respond to it (Fig. 2.4d).
3. Hh diffuses over the A/P line a short distance into A territory (Fig. 2.4d).

Based on how the wing disc develops, we might expect all cells that receive Hh to turn ON *dpp*, but here in the leg disc, only the dorsal (D) cells do so (Fig. 2.4e) [136,539]. Ventral (V) cells instead turn ON *wg*. We do not yet know how V and D leg cells are primed to respond differently to Hh in this manner, but it undoubtedly depends on the D/V differentiation pathway [310,1544].

The next step also seems strange. If Dpp and Wg were to diffuse as randomly in the leg disc as Dpp does in the wing disc [2511], then their domains of influence should have fuzzy edges and overlap to a large extent [1261], but they don't. Instead, their dominions are sharply defined and perfectly complementary [974]. When *dpp* is disabled on the D side of the disc, a Wg stripe arises there instead, yielding a V/V symmetric anatomy (like a hand with palms on both sides), and when *wg* is disabled on the V side, the result is a reciprocal D/D anatomy (like a hand with two back sides and no palm) [974,975]. In the wing disc the *dpp* and *wg* stripes intersect without interfering, but in the leg disc Dpp suppresses *wg*, and Wg suppresses *dpp* [965]. Clearly, the gene circuitry for these two discs is quite different.

The only way to explain these phenotypes is to assume that Dpp and Wg travel in arc-shaped paths [963], but how can they do that? Conceivably, they are escorted by carrier molecules [1405], or they might move through cells by transcytosis [1194,2292,2514] instead of around them, with the direction being dictated by cell polarity [812]. Either way, cells would be informed of their distance from each reference line (*dpp*-ON or *wg*-ON stripe) as measured along arcs rather than along straight lines [458].

This hypothetical scenario is reminiscent of the old Polar Coordinate Model [271], which postulated polar, instead of Cartesian, coordinates: an angular (θ) one as just described, and a radial (r) one that tells cells their distance from the disc center. The Polar Coordinate Model fell out of favor because the regenerative phenomena that it tried to explain were shown to actually obey different rules [965], but some of its features are worth a fresh look in leg discs since a more popular model that was based on random diffusion [1261] was recently disproven [634].

Early in leg disc development, Dpp and Wg do overlap briefly at the disc center where they jointly activate the homeobox gene *Distal-less* (*Dll*) [311,635]. Then, Dll, in conjunction with Dpp and Wg, turns ON a fly version of the toolkit gene *Epidermal Growth Factor* (*EGF*) [724,1178]. Dll causes distal outgrowth wherever it is expressed, but it does not diffuse [416]. In contrast, the EGF **morphogen** (named "Vein" [636]) *does* diffuse and is thought to encode the postulated "*r*" coordinate, at least within the tarsus [309,636,725].

During metamorphosis, the leg disc telescopes out to become a cylinder. The disc's center becomes the leg's tip, and the disc's perimeter encircles the leg's base. Spokes of gene expression (θ) become longitudinal stripes (e.g., eight bristle rows) [962], and concentric annuli (*r*) become spaced rings (e.g., five tarsal joints) [271].

The midleg manifests an odd symmetry that now starts to make sense. The further a tarsal row is from the V midline, the fewer bristles it possesses, the greater the lengths of those bristles, and the wider the intervals between them (Figs. 2.4i and 2.5c). These trends reveal that rows at the same θ level of Dpp or Wg (row 1 = row 8, 2 = 7, 3 = 6, and 4 = 5) develop alike, regardless of whether they are on the A or P side of the A/P line. This programming must be at least 65 MY old [2062] because similar mirror-image trends are found on midlegs throughout the genus [958].

If the eight bristle rows are indeed "welded" to the Dpp and Wg sectorial gradients like the 12 hours of a clock are printed on its face, then this **morphological integration** should constitute a developmental **constraint**. Consistent with this prediction, the number of rows is not only constant from fly to fly but it is also constant throughout the genus despite a three-fold range in tarsal circumference [958]. Evidently, the links between the θ-coordinate genes *dpp* and *wg* on the one hand and the bristle-promoting genes *achaete* (*ac*) and *scute* (*sc*) on the other [1114] are hard to break, even when leg size increases tremendously. Land vertebrates exhibit a comparable constraint in our hands and feet, where five digits persist over a huge range of body size from mouse to elephant [731,852,2159]. Given that the eight bristle rows are hard-wired into the fly genome, the fly foreleg poses two formidable riddles:

Scr Puzzle #1. How did *Scr* manage to insert transverse rows into a pattern that appears to be extremely resistant to change?

Scr Puzzle #2. How did *Scr* manage to alter only the A side when the A and P sides appear to be locked into mirror symmetry?

The first puzzle was solved in 2007 [2043]. Unlike *Ubx*, which converts a wing into a gyroscope (T2 → T3) by "reaching down" into the circuitry (Fig. 2.3), *Scr* transforms a midleg into a foreleg (T2 → T1) by becoming part of the circuitry itself (Fig. 2.5) [1191,2171]. The following steps are involved, but they may not have evolved in this order:

1. *Scr* abdicated its disc-wide authority by reducing its background expression to a negligible level [1727,1820,1877].
2. *Scr* became enslaved (via a *cis*-enhancer?) to *wg*, so that *wg* only turned *Scr* ON at a θ level between rows 7 and 8 [126,2405].
3. *Scr* widened this interstripe by inducing lateral growth, presumably by conscripting a growth-stimulating target gene [970].
4. *Scr* evoked new bristles by preventing the default *ac* and *sc* expression from being turned OFF by Delta–Notch signaling [2043].
5. *Scr* forced these bristle cells to adhere to one another and align laterally by using the EGFR pathway as some sort of "glue" [966].
6. *Scr* endowed these new bristles with a lighter color (yellow vs. brown) by silencing the pigmentation gene *yellow* (*y*) [970].

What role, if any, did selection play in this process? Anyone who has ever watched a fly grooming itself knows that the transverse rows serve a behavioral function aside from any visual role they might play. Flies are as compulsive about cleanliness as cats. After a fly lands and steadies itself, it wipes its face with its forelegs [2069,2274], and

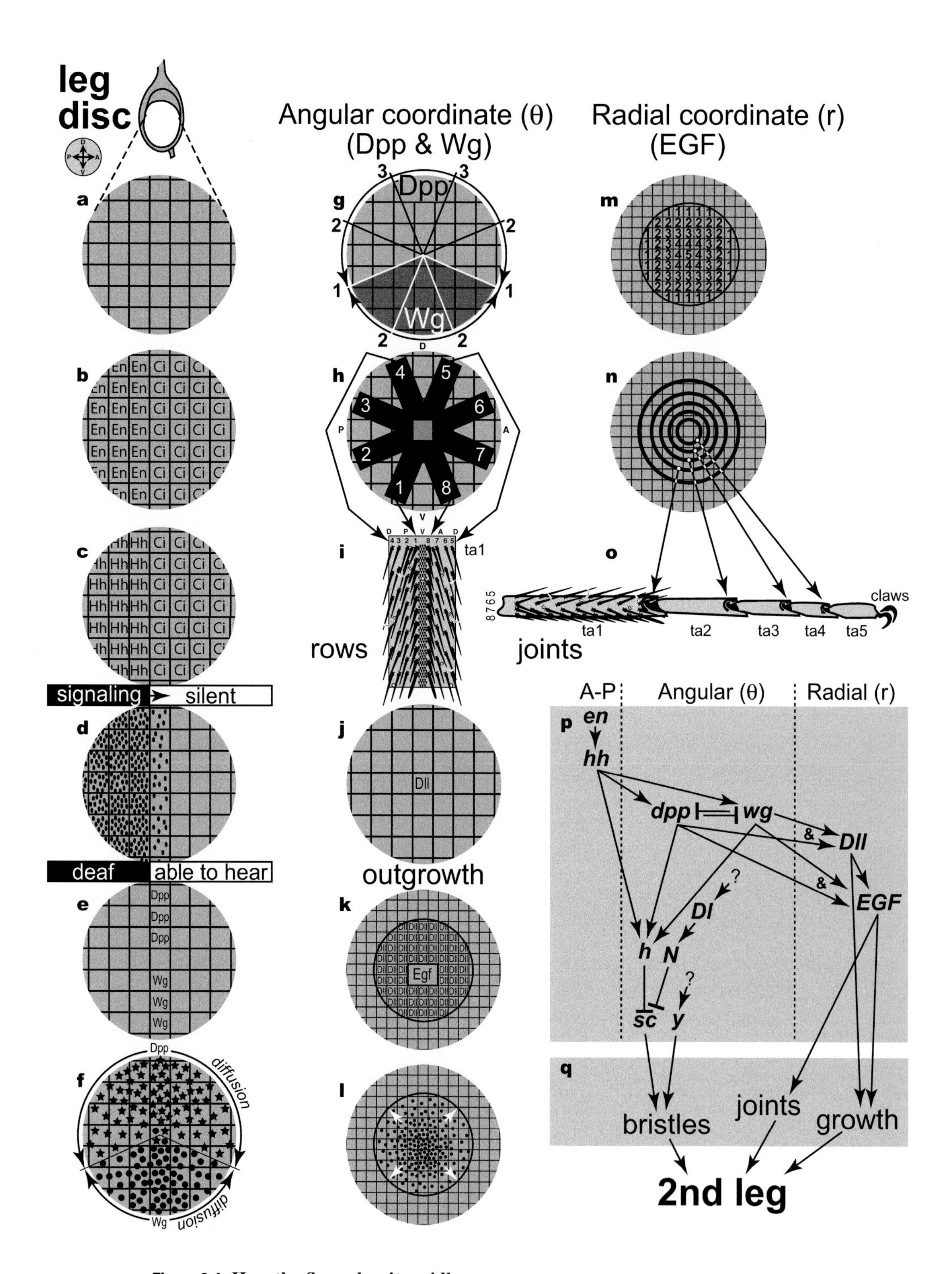

Figure 2.4 How the fly makes its midleg.

a–e. Stages of patterning along the anterior–posterior (A–P) axis. Modified from ref. [965]. (**a**) A right second-leg disc is depicted with its center removed, enlarged, and rendered as a circle. Squares represent cells, though real cells are less regular and more hexagonal. A mature disc actually has 1000s of cells, but fewer are shown because most events

Figure 2.4 (*cont.*)

(**b–j**) occur when it has only 10s or 100s of cells. The points of the compass (A, P, D, V) indicate cardinal directions (anterior, posterior, dorsal, ventral). (**b**) Posterior cells express Engrailed (En), while anterior ones express Cubitus interruptus (Ci). Each of these proteins is a transcription factor that is made by a **selector gene** (*en* or *ci*). (**c**) En turns ON *hedgehog*, which encodes a **morphogen** (Hh). (**d**) Hh proteins (small ovals) are made throughout the P region and cross into A territory, where Ci transduces the Hh signal. Because P cells lack Ci, they are refractory ("deaf") to their own signal, so they don't respond by making Dpp. A cells can hear the signal, but the signal only diffuses a short way into their territory, so that the responding cells occupy a narrow stripe along the A/P boundary. (**e**) Hh turns ON *dpp* dorsally and *wg* ventrally. Each of these target genes makes a longer-range **morphogen** (Dpp or Wg). We don't know what predisposes the two halves of the disc to respond differently to Hh.

f–i. Stages of patterning around the circumference (θ coordinate). Adapted from refs. [636,963,965,975]. Other insects apparently deviate from this scheme insofar as they do not use Dpp [1105,1648] or Wg [57,1669] for these roles. (**f**) Dpp (stars) or Wg (solid circles) proteins are secreted by dorsal or ventral cells, respectively. Dpp and Wg diffuse around the disc in arcs that meet at lines. These lines later become bristle rows 2 and 7 (**h**). The concentrations of Dpp and Wg diminish as they approach these lines, thus creating two stable, mirror-image gradients of different span (Dpp > Wg). (**g**) Cells assess their angular distance from the A/P line (θ coordinate) by the amount of Dpp or Wg (numbers around the periphery). Although P and A cells share the same coordinates (Dpp or Wg), they can interpret these values differently via their *en* or *ci* **selector genes** respectively. (**h**) Bristle rows (black bars) are established at specific angles. These angles are determined by the concentrations of Dpp and Wg, though bristles per se do not form until after disc eversion. (**i**) During metamorphosis the circle becomes a cone and then a cylinder, with the center of the circle becoming the tip of the leg. Hence, in the adult these rows run along the length of the leg (**o**). Shown here are rows (numbered) on the basitarsal segment (ta1). That segment is drawn as if it were slit along its dorsal midline and flattened into a rectangle. For convenience relative to **h**, the basitarsus (here and in **o**) is from a left leg, despite schemata in **b–h**, which are right discs.

j–o. Stages of patterning from center to periphery (radial coordinate). Adapted from refs. [309,636,724,1178]. (**j**) Cells in the center of the disc that are exposed to both Dpp and Wg (**f**) express Dll (Distal-less) – a (non-diffusing) homeodomain protein. (**k**) Dll causes outgrowth wherever it is expressed (denoted by smaller cells in greater numbers) [417,1714], and it turns ON the fly's *Epidermal Growth Factor* (*EGF*) homolog (*vein*) at the disc center [724,1178]. (**l**) EGF proteins (dots) are secreted and diffuse in all directions, eventually creating a stable gradient [309,636,725]. (**m**) Cells assess their distance from the center (radial coordinate) by the amount of EGF [1598]. Numbers denote EGF levels. (**n**) Cells at particular EGF thresholds form tarsal joints (rings). (**o**) Tarsus of a left leg, including ball-and-socket joints between segments ta1–ta5 [972]. For clarity, bristles are only shown for the basitarsus (ta1).

p. Circuitry for midleg patterning. Genes first establish the A–P axis, then angular (θ) and radial (*r*) coordinates. Arrows indicate activation of one gene by another; T-bars denote inhibition. Ampersands show where Dpp and Wg act jointly; *dpp* and *wg* mutually inhibit one another to maintain their respective sectors. Stripes of *hairy* (*h*) expression [346] are induced by *hh*, *dpp*, and *wg* [948,1222]. The *h* stripes inhibit the bristle-promoting genes *scute* (*sc*) and *achaete* (*ac*, not shown) [1114,1691], which are constitutively ON [2043]. The *yellow* (*y*) gene causes bristle pigmentation. We don't yet know the upstream regulators for *y* or *Delta* (*Dl*) [1114,1117], which encodes the signal received by Notch (N). Inputs to *Scr*'s radial coordinate are known but omitted [1191,1820]. For other omitted genes see refs. [7,459,1179]. Distilled from refs. [636,965,1114].

q. Features of leg development that depend on the circuitry in **p**.

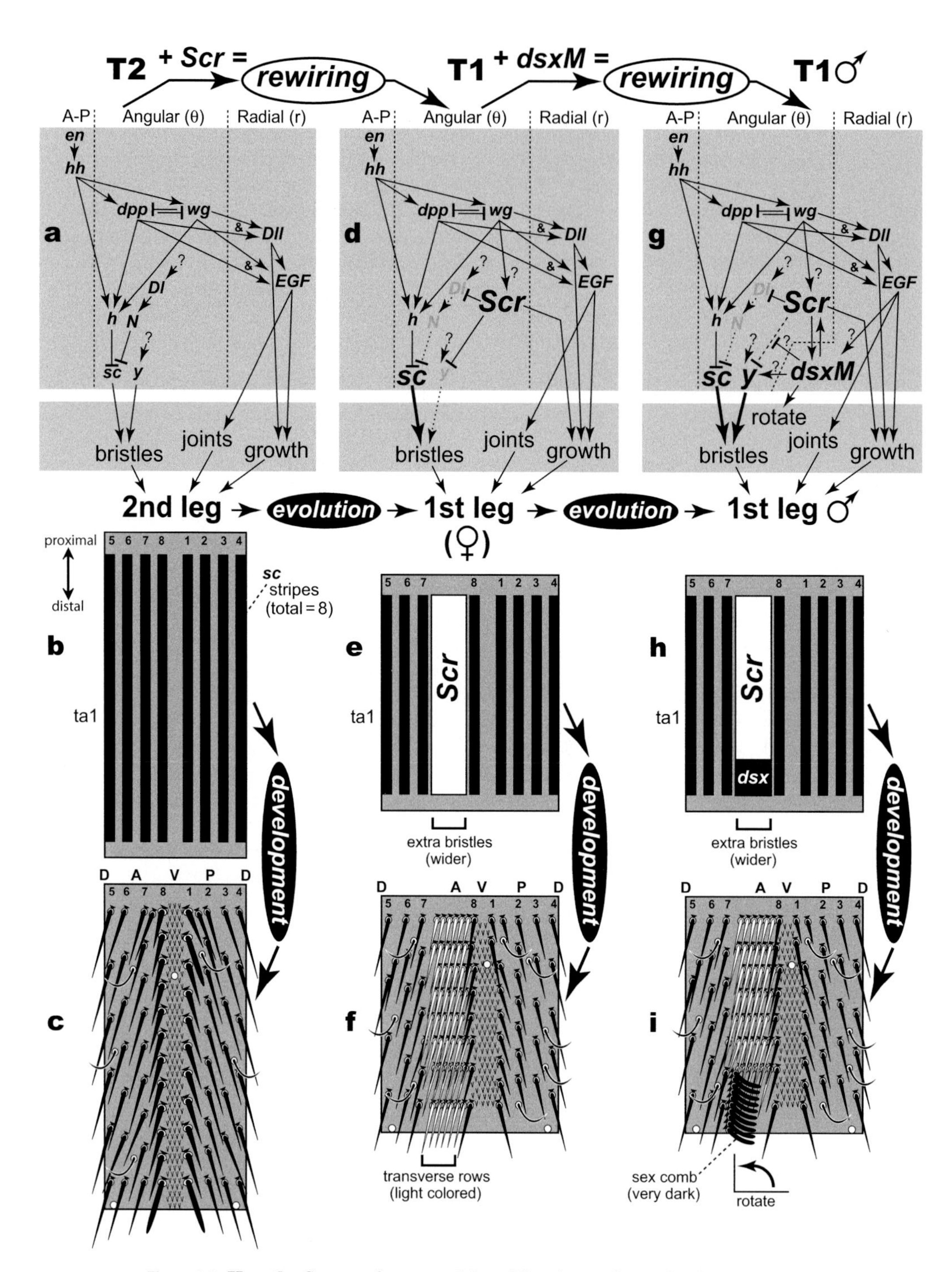

Figure 2.5 **How the fly transforms a plain midleg into a fancy foreleg.**

a–c. Bristle pattern formation on the *D. melanogaster* midleg. The midlegs are thought to be closest to the "ground state" for all six legs in ancestral flies [349,608,761].

a. Circuitry that evokes bristles and joints on the midleg (from Fig. 2.4). T2 denotes the second thoracic body segment.

the tarsal region that makes contact is rows 7 and 8. What better place to put a brush? Indeed, the transverse rows not only look like a hair brush but function like one too. After collecting dust from its eyes, the fly rubs the transverse rows of its forelegs together to rid itself of the dust altogether. Interestingly, the third legs separately evolved (via *Ubx*) transverse rows between rows 1 and 2 [965,2043] – the ideal spot for a brush to clean the wings.

Figure 2.5 *(cont.)*

b. Stripes (numbered) of *scute* (*sc*) expression on a right midleg basitarsus (ta1) of a pupa. Only those cells that express *sc* can make bristles. The basitarsus is a cylinder but is drawn here (and in **c**, **e**, **f**, **h**, and **i**) as a rectangle (slit along the dorsal midline and unrolled). Foreleg basitarsi are wider due to *Scr*, but we don't yet know why they are shorter. Bristle pattern maps are from ref. [965].

c. Basitarsus (panoramic view) of a right adult midleg. Males and females exhibit this same pattern. Rows 1 and 8 are separated by a wide gap that is full of hairs (Vs) – one per cell. (See refs. [1114,2043] for why this interstripe is so strange.) Tiny triangles above bristle sockets denote thick, dark hairs called bracts. Bracts are induced by bristles [521,964]. There are five curved (chemosensory) bristles which lack bracts, plus three campaniform sensilla (white circles).

d–f. Bristle pattern formation on the foreleg of a *D. melanogaster* female. The female foreleg is thought to retain the original pattern of both sexes before males acquired a sex comb.

d. Circuitry that elicits bristles and joints on a female foreleg. Arrows denote activation; T-barred lines denote inhibition. Faded letters signify genes inhibited by *Scr*. Dotted arrows or T-bars are disabled links. T1 denotes the first thoracic body segment. The T1 *Hox* gene *Scr* (*Sex combs reduced*) evidently acquired a new *wg*-responsive *cis*-enhancer. This supposed enhancer allowed *Scr* to later insert bristles between *sc* rows 7 and 8 by (1) stimulating growth that widened the segment and (2) activating *sc* (larger font) indirectly by blocking *Dl–N* (*Delta–Notch*) signaling. The new bristles were lighter in color due to an inhibitory link that *Scr* acquired to the *y* (*yellow*) gene. Effects of *Scr* were verified by disabling it in T1 or by ectopically expressing it in T2 [970].

e. Stripes of *scute* (*sc*) expression on a right foreleg basitarsus (ta1) of a female pupa. Compared to a midleg (**b**), the foreleg has an extra *sc* stripe (white) due to *Scr* being expressed here.

f. Basitarsus (panoramic view) of a right female foreleg. The ninth *sc* stripe (**e**) behaves differently: its bristle cells organize themselves into rows that run transversely [966]. Similar rows exist on the honeybee tibia [18] and the butterfly wing [2005].

g–i. Bristle pattern formation on the foreleg of a *D. melanogaster* male.

g. Circuitry that induces bristles and joints on a male foreleg. Amalgamated from refs. [970,1114,2043,2171]. Symbols as in **d**. The sex-determination gene *dsx* (*doublesex*) evidently acquired a new *cis*-enhancer (responsive to *EGF* and *Scr*?) that caused it to be expressed distally within the *Scr* area [1870,2171]. *M* denotes a male-specific mRNA (*dsxM*) from alternative splicing. Additional links allowed *dsx* to convert the distal transverse row into a sex comb by (1) overriding *Scr*'s inhibition of *y* to darken the bristles and (2) causing the row to rotate. Temporal control is as critical as spatial control but is less well understood [162,356,970]. Not shown is an inferred inhibition of *Scr* by *en* on the posterior side (see text).

h. The gene *dsx* is expressed at the distal end of the *Scr* area on the basitarsus (ta1).

i. Basitarsus (panoramic view) of a right male foreleg. The transverse row in the *dsx* zone (**h**) behaves oddly insofar as (1) it rotates 90° [84,973,2217] and (2) its bristles become thicker, blunter, darker, and more curved than ordinary bristles. We don't yet know which target genes of *dsx* control these events [1191]. The adornments on *D. melanogaster* legs are actually mild compared to the gaudy forelegs of other insects [347,2069].

Any ancestral fly lucky enough to be equipped with such brushes could have lightened its ballast for the next takeoff and thus could have outperformed its conspecifics. Even a ragged brush would have helped a bit, so fortuitous mutations that implemented only steps (2) and (4) could have given their bearers a leg up, so to speak, on their competitors. For that reason, any mutations that added bristles near rows 7 and 8 on the foreleg would have spread through the population. *Scr* happened to mediate this increase by turning ON there [1877], promoting growth, and spawning bristles. The next upgrade might have occurred when mutations forced the extra bristle cells to stick to one another side-by-side to make transverse rows [965,966]. Over time, selection would have rewarded many such refinements that enhanced grooming efficiency.

In summary, the solution to *Scr* Puzzle #1 seems to be that *Scr* circumvented the **constraint** of row constancy by shoehorning a whole new **module** (transverse rows) into the old framework (eight longitudinal rows) without changing the framework itself. The details of the regulatory linkages and the cellular mechanisms remain to be worked out.

Scr Puzzle #2 (how *Scr* overcame A/P mirror symmetry) was solved in 2011, at least to a first approximation [2171]. Before evaluating the evidence we must first examine how *Scr*, along with *dsx*, creates the sex comb – the structure that gave the gene *Scr* its name (*Sex combs reduced*).

The sex comb is a row of peculiar (thick, dark, blunt, and curved) bristles found only in males [1191]. It gets its name from the resemblance of its bristles to the teeth of a pocket comb for combing one's hair. It evolved ~62 MY ago [80,1193]. In *D. melanogaster*, the sex comb starts out as an ordinary transverse row but then pivots 90° to a longitudinal orientation [84,973,2217]. The comb has been shown to enhance mating success [1617], but we don't know exactly how it does so [1191].

Two genes must be expressed at the site of the sex comb for it to develop normally. One is *Scr* [1877]. The other is *doublesex* (*dsx*) [2171] – another charter member of the bilaterian toolkit (cf. Table 1.2) [1404,2516]. Amazingly, virtually all bilaterians (from flies to humans) use the very same sex-determination gene (*dsx*) to distinguish their sexually dimorphic structures [1192]. In fruit flies, the *dsx* transcript is spliced differently in males (*dsxM*) versus females (*dsxF*) [289,391].

When *dsxM* expression is artificially forced to expand from the sex comb spot to the entire *Scr*-ON territory between rows 7 and 8, all of the bristles in the transverse rows now resemble those in the sex comb, suggesting that *Scr* and *dsx* are sufficient to endow bristle cells with this identity. However, none of the proximal rows rotates, so some third factor besides *Scr* and *dsx* must be needed to drive rotation [2171].

Did *dsxM* mold a sex-comb bristle from an ordinary bristle? Probably not. Bristles shaped like those in the sex comb already existed on the clasping organs of male genitalia, so *Scr* and *dsx* might have elicited this shape by conjuring an old (genital) subcircuit at a new location (foreleg) [1193]. If so, then this **novelty** would have arisen by both **co-option** (new use of an old module) and **heterotopy** (new site for an old structure). After the sex comb arose, *dsxM* must have darkened its bristles by not only (1) overriding *Scr*'s inhibition of *y*, but also (2) goading the *y* gene to an even higher level of expression than in ordinary (brown) bristles [2227].

The best clue we have regarding *Scr* Puzzle #2 is the *engrailed*[1] (*en*[1]) mutant allele. Adult *en*[1] males have a typical sex comb between rows 7 and 8, but they also have an extra sex comb between rows 1 and 2. Strikingly, the second comb rotates in mirror-symmetry to the primary one [2216]. The simplest interpretation of this phenotype is that the wild-type *en* allele normally prevents posterior cells from making a comb by inhibiting *Scr* and *dsx*. When *en* malfunctions, *Scr* and *dsx* turn ON at the same Wg level (θ value) on both sides of the segment (A and P). Evidently, when *Scr* yoked itself to being activated by *wg*, it also allowed itself to be inhibited by *en*. If we can document this linkage in *Drosophila* and trace its origin in the dipteran clade, then the puzzle would be solved.

Interestingly, *en* seems to be playing a comparable role in the wing, because this same allele (*en*[1]) causes the P half of the wing to make veins and marginal bristles typical of the A half [744]. If so, then the wild-type *en* gene must be encoding an abstract (P vs. A) identity regardless of which physical structures (leg or wing) happen to occupy the P region [1256]. Historically, the structure-independent properties of *en*[1] provided our first clue that cells operate like computers in calculating fates [741]. Many other genes have since been found that also act as abstract binary switches [961,965,967]. This conclusion is listed below, along with a few other generalizations shown by this case study:

Take-home lessons

1. Genes (e.g., *en*) can function as abstract binary switches. By acting in concert, multiple such binary switches can encode an exponential number (2^n) of identity states [1126].
2. *Hox* genes can reshape the organs where they are expressed not only by acquiring target genes but also by becoming target genes themselves.
3. Sexual dimorphism resembles **serial homology** insofar as a single gene (*dsx* instead of a *Hox* gene) can rewire circuitry to reshape anatomy [1160].
4. Patterns can be elaborated incrementally by evolution (e.g., transverse rows, then sex comb) to become richly intricate.

Another clue may hold the key to deciphering the area code (r, θ) for the sex comb: when the amounts of Dpp and Wg are experimentally titrated to parity in the tarsus (overcoming their mutual antagonism), the sex comb expands to a nearly complete 360° ring around the leg circumference – as if all the points of a compass were being forced to read "south" (cf. Fig. 2e in ref. [975] and Fig. 2b in ref. [1544]). So far, however, no one has figured out what this anomaly is trying to tell us about anatomy.

It also remains unclear how the *en*[1] mutation mimics the sex comb's (r, θ) area code on the P side. If *en*[1] makes a P comb by unleashing *Scr*, then why doesn't it also evoke transverse rows? (It doesn't [650,2216]!) Equally puzzling is why no other *en* alleles make an extra comb [586,885,1256]. Part of the answer to the latter question may lie in the gene *invected*, a paralog (i.e., related duplicate) of *en* that acts cooperatively with *en* to specify compartment identity [891,2050,2051]. Perhaps the *en*^{1} mutation disables both paralogs, while other mutant *en* alleles do not.

The genus *Drosophila* offers a rich cornucopia of sex combs [126,1820]. For example, some species resemble *D. melanogaster* in having a comb only on the first tarsal segment (ta1), while others have combs on ta2 and ta3 as well, implying that tarsal segments are **serial homologs** of one another [1191,1820]. The combs of many species rotate, while those in some species do not, in which case the sex-comb bristles tend to look more like ordinary bristles. Nevertheless, all combs rely on *Scr* and *dsxM*, regardless of their geometry or polarity [126,1191,2171].

The "tattoo" property of these patterns remains enigmatic. Why should sex combs be black and transverse rows be yellow, while ordinary bristles are brown? As every fly geneticist knows, courtship and mating occur just as avidly when males and females are mixed in the dark as in the light, so visual insignia seem irrelevant.

Historically, the study of sex combs figured prominently in fostering the field of developmental genetics [2109,2218]. During the past decade this subfield has blossomed to become one of the most productive sectors on the evo-devo frontier (Fig. 2.6) [1191]. Among the surprises harvested from recent investigations are the following [86,1191,2170], all of which teach us the same lesson: we cannot infer process from pattern alone.

1. Sex combs that look as if they have rotated may arise instead via interdigitation of cells from adjacent bristle rows (Fig. 2.6e, g)!
2. For features such as sex-comb orientation, species can use drastically different cellular mechanisms to reach the same outcome (Fig. 2.6c vs. g)!
3. The ta1 and ta2 tarsal segments of the very same species (*D. rhopaloa*) make their respective combs via different routes (Fig. 2.6c, f)!
4. Distantly related species have occasionally adopted the very same mechanism independently by evolutionary **convergence** (Fig. 2.6d)!

Few **novelties** in any other animal rival the sex comb's potential as a model system for dissecting the gene circuitry underlying anatomy at the cellular level [84,1190]. Combs therefore offer a playground for researchers to explore how *Hox* genes **individuate** body segments and how sex-determination genes enforce dimorphisms. Finally, the sex comb's appeal is enhanced by its malleability: few other traits are as modifiable under pressure from the fickle fads of sexual selection [23,1820,2170]. The only other structures that rival the comb's **evolvabilty** are the penis and surrounding parts of the male's genitalia [1026,1193,2091].

Why the fly twirls his penis

Animals do many silly things (e.g., nutty courtship rituals [2375]), but flies do something so crazy that it boggles the mind: the genitalia of male flies rotate through an angle of 360° during metamorphosis (Fig. 2.7) [793]! This pointless pirouette has no net effect on any external structures, and its only effect internally is to wrap the ejaculatory duct around the rectal tube (Fig. 2.8c). Might that inner loop serve some function? No, because flies that lack this twist are fully fertile [2151]. So why do the genitalia go to all this trouble for nothing?

In 2010 this mystery was solved by new findings from an evo-devo investigation worthy of Sherlock Holmes himself [2151]. The answer shows how ridiculously inefficient evolution can be from an engineering perspective. The following scenario was deduced from the new data, and it is buttressed by corroborating evidence from molecular systematics, comparative anatomy, and developmental genetics.

Originally, flies mated doggy-style, with males mounting females from the rear [1422], and their penis angle fit this orientation nicely (Fig. 2.7b). However, mating postures are notoriously labile in insects [1026,1044], and one descendant branch – the muscomorphans – changed to a tail-to-tail style, whereupon males had to twist their abdomens awkwardly to fit their penis into the vagina. To alleviate this torque, a mechanical contraption evolved which rotated the male genitalia 180° during development. This device was dubbed the "L/R module" because it imposed left–right asymmetry on a symmetric rear end [461].

Subsequently, for some unknown reason, one offshoot of muscomorphans – the cyclorrhaphans – returned to the dog-like mating posture [1422], forcing love-struck males to contort their abdomen into yet another tantric pretzel. At this point, the simplest way to restore the old penis angle would have been for **atavistic** mutations to abort the rotation before it even began (i.e., stop it at 0°). Instead, mutations arose which duplicated the module, adding another 180° for a total of 360° [2151]. This iterative solution may be as effective as the atavistic alternative in solving the angle problem, but it is patently inefficient because it wastes energy needlessly to (1) build a new module and (2) rotate it.

Thus, *Drosophila* (a cyclorrhaphan) currently displays two phases of rotation, each of which is controlled by a separate L/R module in its eighth abdominal segment (A8). First, the posterior ring (A8p) rotates, turning the genitalia 180° (Fig. 2.8b), and then the anterior one (A8a) pivots 180°, restoring the genitalia to their original orientation (Fig. 2.8c). In this way, the mechanism "covers its tracks" externally but leaves incriminating evidence internally in its asymmetric twisting of the sperm duct around the rectal terminus.

For this apparatus to do its job, the A8p tumbler must detach from the A8a tumbler during phase 1, and the A8a ring must detach from the A7 segment in phase 2. These slippages are mediated by zones of cell death [3], one zone per module, that function as a clutch [1828] or brake release [1216]. Moreover, A8a must reattach to A8p in phase 2 to engage the genitalia as it turns. How this rejoining occurs is unknown. Overall, this gadget operates like the interleaved cams of a combination padlock, which are aptly reminiscent of a medieval chastity belt.

Because this clockwork was discovered so recently (2010) [2151], it is not surprising that we are still fumbling with how its gears mesh. Even so, the basics of the system that we have been able to grasp thus far afford a few insights into how evolution works in general:

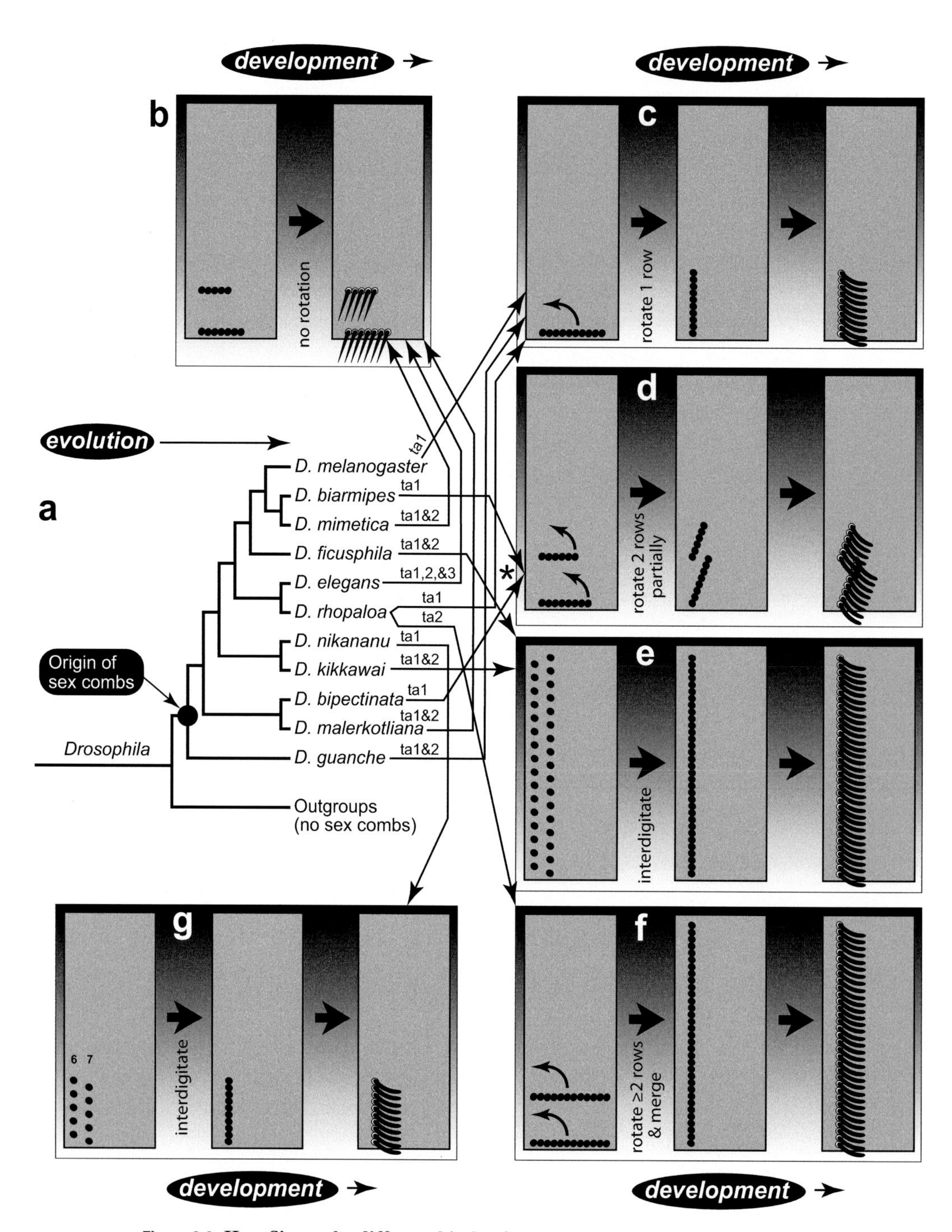

Figure 2.6 How flies make different kinds of sex combs.

a. Phylogeny of some *Drosophila* species in the *melanogaster* or *obscura* groups (clades with sex combs) of the subgenus *Sophophora* [1191,1193]. Sex combs are also found in a separate fly genus (*Lordiphosa*) [1191]. Arrows connect species (**a**) to the mode of development of their sex combs on one or more tarsal segments of the male foreleg

Take-home lessons

1. Wheels really do exist in animals, despite expectations to the contrary [504,833,838] – proving Haldane's point that "[evolution] is not only queerer than we suppose, but queerer than we *can* suppose" [902].
2. Evolution is always opportunistic, often inefficient, and sometimes counterintuitive – e.g., asinine 360° twirls that tie your guts in knots.
3. Evolution is an amateur tinkerer, not a professional engineer [1077,2410], so its contraptions can become as cumbersome as Rube Goldberg machines.
4. Evolution is myopic. It rewards any device that fixes a current problem, even when that solution causes a new problem later – a pitfall termed the "bislagiatt" trap ("but it seemed like a good idea at the time") [967].
5. Mutations are not made-to-order. The response to selection pressure (e.g., to reorient the penis) will be mediated by the first helpful mutations that occur and spread, not necessarily by the best (most optimal) ones.
6. Body segments are iconic modules, but it is also possible for modules to arise serially inside segments – in this case, the A8a and A8p halves of A8.

Figure 2.6 (*cont.*)

(**b–g**). Labels on arrows denote which tarsal segment(s) obey which mode: first (ta1), second (ta2), or third (ta3). Based on ref. [2170] with data added from refs. [85,126,1191,1193,1820].

b–g. Cellular mechanisms of sex comb development on tarsal segments (ta1, ta2, or ta3) of right forelegs, as documented in various species. Black circles denote bristle cells. Curved arrows show the path of rotation. Bristles other than those in the sex comb are omitted for clarity, as are bracts.

b. In species whose combs do not rotate, the sex comb bristles are shaped more like ordinary bristles [126].

c. The style of rotation for *D. melanogaster* is also seen in distantly related species. N.B.: This diagram is correct for the comb on ta1 of *D. guanche*, but its ta2 comb spans ~70% of the segment [1191,2170].

d. Rotation of more than one transverse row. Asterisk denotes a likely case of convergent evolution [2170]; cf. the adoption of the same pathway by distant relatives *D. melanogaster* and *D. guanche* (**c**). Proof of **convergence** requires knowing the ancestral state, which has not yet been ascertained [2170]. Unlike *D. biarmipes*, *D. bipectinata* has a few sex-comb bristles on ta2 (not shown) [1191,1766].

e. Interdigitation of bristle cells in rows 6 and 7, yielding a comb that spans the segment [85]. N.B.: There is a row of sparse sex-comb bristles (not shown) parallel to the main comb in *D. ficusphila* [2170].

f. Merging of two rotating rows. N.B.: The ta2 comb of *D. rhopaloa* actually spans the entire segment [2170]. The ta1 comb in *Lordiphosa magnipectinata* takes this theme to an extreme (not shown): it arises from the rotation and merger of ~6 transverse rows [86].

g. Same mechanism as in **e**, but confined to distal region. Interdigitation of bristle cells in rows 6 and 7 (labeled) produces a comb that looks like the one in *D. melanogaster* but is formed entirely differently [85,2170].

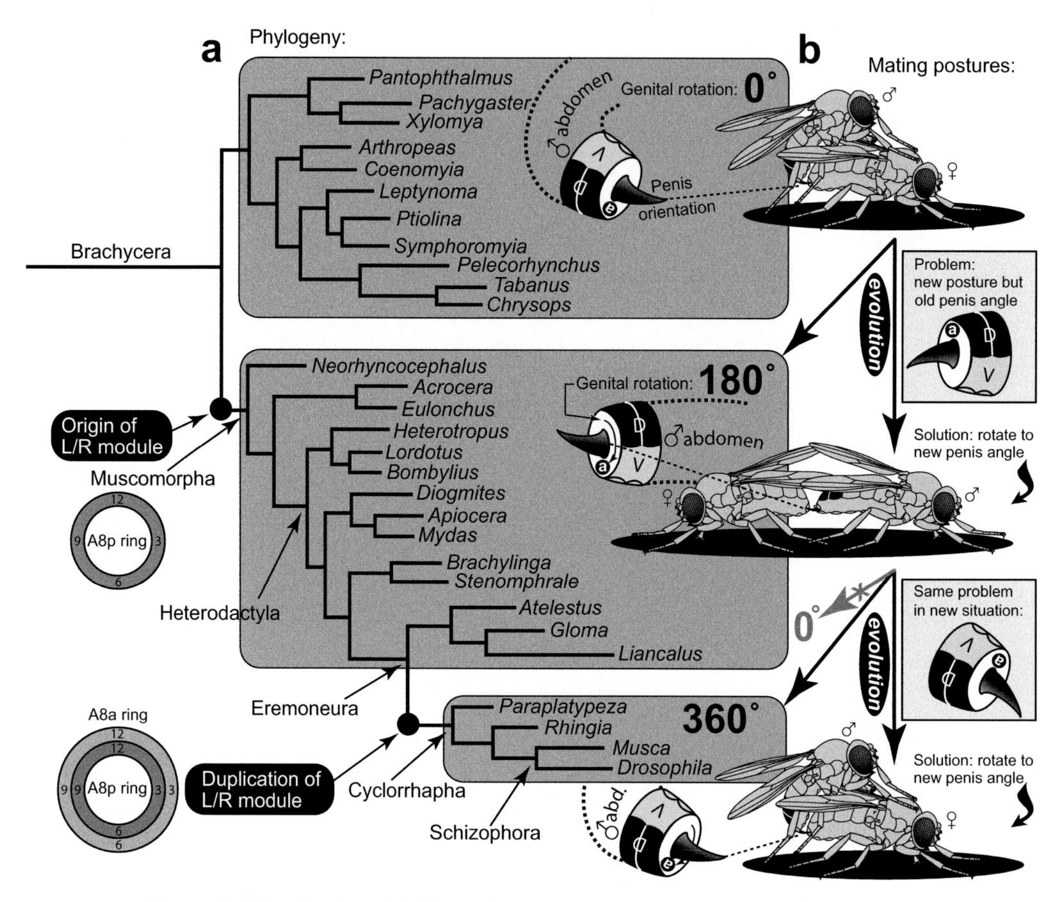

Figure 2.7 **Why the fly twirls his penis.**

a. Phylogeny of higher dipterans based on 28S ribosomal DNA sequences. Brachycerans arose ~210 MY ago as a suborder of dipterans, which originated ~250 MY ago [880,2387]. Italicized names on branches are genera. Names along the left edge (Muscomorpha, etc.) are nested clades that lead to *Drosophila* (Schizophora, ~80 MY ago). Genera fall into three groups based on the extent of genital rotation in males: basal brachycerans (upper box), where the genitalia do not rotate (0°), muscomorphans (middle box), where the genitalia turn 180°, and cyclorrhaphans (lower box), where the genitalia turn 360° [2151]. Black circles denote the times when the L/R modules that cause these rotations probably evolved. A8p and A8a are adjacent rings of tissue in the eighth abdominal segment. See Fig. 2.8 for how the module is thought to operate. Redrawn from refs. [2387] and [2151] (Supplementary Fig. S3) and vetted by Stéphane Noselli. Penis angles (conjectural) are based on genital rotations (documented). Male genitalia can evolve amazingly quickly in *Drosophila* [1026,1193,2091].

b. Mating postures (on black ovals) typical of the three groups in **a** [1422], with transitional stages sketched in boxes at the right. To better visualize the genital region, the hind end of each male is redrawn larger at the left. The penis is depicted as a thorn, and the anus is denoted by an "a" to facilitate comparison with Fig. 2.8. When muscomorphans evolved (middle panel), they started mating tail-to-tail for some reason, but could only do so by awkwardly contorting the abdomen during intercourse (not shown). Eventually, mutations crafted an L/R module, which reoriented the penis (to point dorsally instead of ventrally) by rotating the genitalia 180° during metamorphosis. When cyclorrhaphans (bottom panel)

One obvious mystery remains. Where did the L/R module come from in the first place? We don't yet know, but there is one intriguing clue: the module uses the same molecular motor as other structures that exhibit left–right asymmetry during development [461,944]. Mutations in *myoID*, the gene that encodes the myosin protein MyoID, cause a pervasive syndrome of mirror-image reversals (situs inversus) [1034,2092]: (1) the genitalia rotate counterclockwise instead of clockwise, (2) the testes twist in spirals that are counterclockwise instead of clockwise, and (3) the midgut and hindgut follow paths that are left–right reversed [1750,2093].

Myosin is best known for its role in muscles, where it assembles into filaments that use adenosine triphosphate (ATP) as fuel to ratchet along actin fibers to mediate muscle contraction. In contrast, MyoID is an unconventional myosin [1541] that does not form filaments [2462]. Instead, MyoID motors may be crawling along polarized actin fibers [140] toward intercellular adherens junctions [2093], where they could cause chiral changes in cell shape or cell movement [2175]. Those alterations at the cellular level could collectively culminate in visible asymmetries at the organ level [461,1024].

Strangely, flies whose *myoID* gene has been disabled show no ill effects on viability or fertility (except that males with partly rotated genitalia are sterile), despite a mirror-image reversal of their internal organs [2093]. Humans are likewise only mildly inconvenienced by situs inversus [2095]. These odd facts beg the haunting question: Are there species of dipterans (or families of humans) out there, yet undiscovered, living a secret life of reversed asymmetry? Indeed, Lewis Carroll's *Through the Looking-Glass* was based on just such a conceit [333].

Another lingering riddle concerns what sinister device may be taking over in the absence of MyoID [461]. How did this default mechanism, whatever it is, arise? Part of this riddle was solved in 2013: the very same *Hox* gene that governs genital identity – *Abdominal-B* (*Abd-B*) [460] – was found to also be a key upstream regulator of *myoID*. Reducing *Abd-B* function to a mild extent prevents *myoID* activation. Curiously, it also prevents any sort of rotation (dextral or sinistral), indicating that *Abd-B* is critical for the operation of the *entire* apparatus – including *both* the normal *and* the default components.

When *Abd-B* is completely disabled (not just partially inhibited), the genitalia are **homeotically** replaced by a leg [638]! It seems strange to think of fly genitals as a leg in disguise [418,1484], but the genital and leg discs turn out to be governed by the same genetic circuits [83,637,822]. Astoundingly, comparable data suggest that the mammalian penis also evolved as an altered appendage [410,2472] by **co-option** and **heterotopy** of a genetic cassette [1304].

Figure 2.7 *(cont.)*

returned to the posture of ancestral brachycerans (upper box), they could have easily restored the original penis angle (to point ventrally again) by simply inactivating the L/R module to yield a 0° rotation (X'd gray arrow), but they duplicated the module instead – extending the rotation by an extra 180° [2151]. For convenience, all schematics depict *Drosophila*, where the male's last two segments (D, dorsal; V, ventral) are black. (Indeed, "*melanogaster*" means black abdomen.) Modified from ref. [1422] and approved by Stéphane Noselli. An equally twisted tale of body contortion – without the kama sutra – was recently reported in echinoderms [2073].

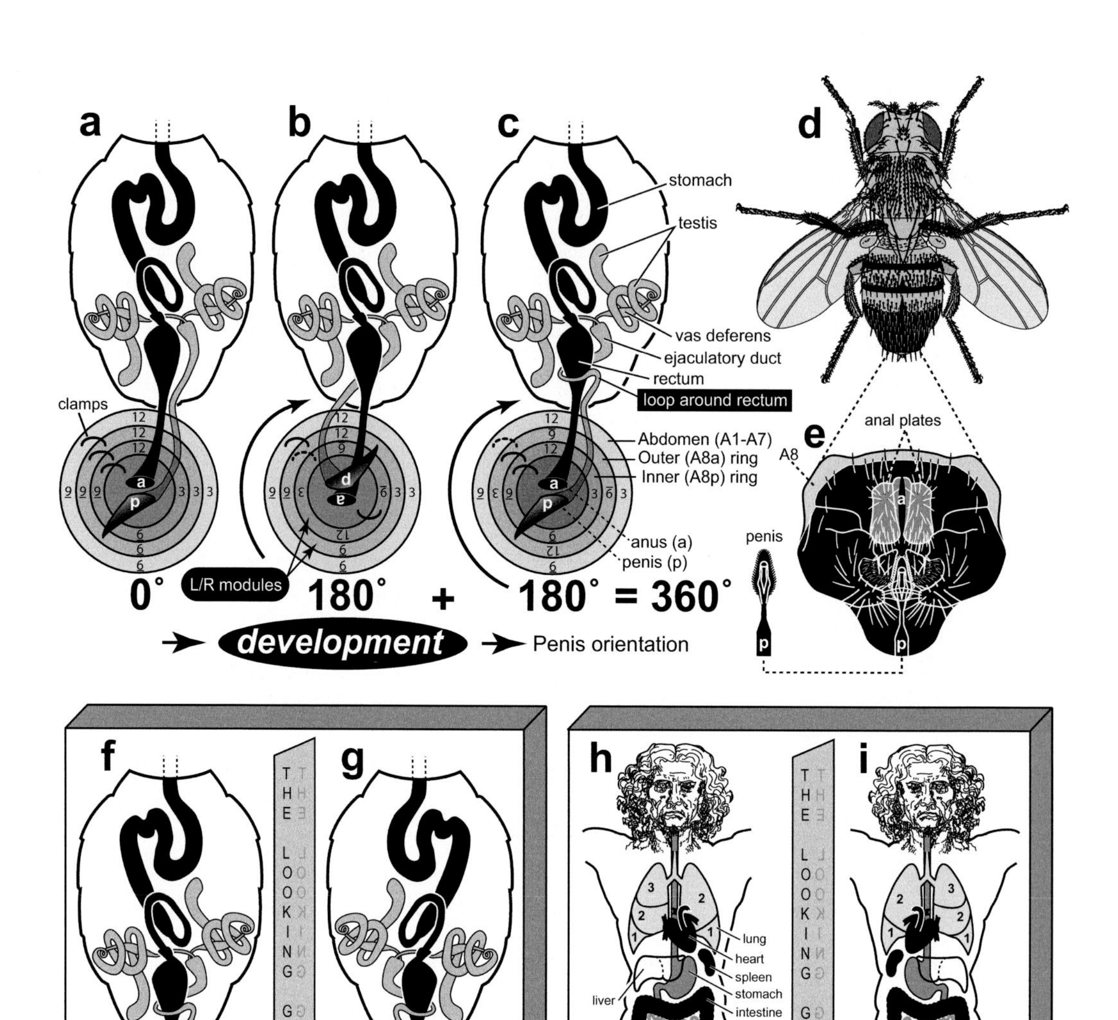

Figure 2.8 How the fly twirls his penis.

a–c. Genital rotation in *Drosophila* [2151]. The machinery works like a combination padlock. Contents of a male abdomen (labeled in **c**) are drawn at successive stages of development. The paragonia and sperm pump are omitted for clarity. Black organs are digestive; gray organs are reproductive. Food is digested in the stomach, and waste is expelled via the rectum to the anus (black oval "a"). Sperm are made by spiral testes and exit via vasa into an ejaculatory duct leading to the penis (simplified as the gray cone labeled "p"). The "archery targets" are schematics of **e** (the fly's posterior as seen from its rear), with dorsal up and ventral down. Abdominal segment 8 (A8) has two rings – one anterior (A8a) and one posterior (A8p). These rings are the L/R modules, which can rotate separately. Segments A1–A7 (schematized as outermost ring) do not rotate, and A7 degenerates in males [678]. The genitalia and anal structures (drawn under A8 in **e**) rotate with A8p because they are fixed to it [1936]. Attachment of adjacent rings is denoted by a solid arc

One of the most intriguing cases of co-option and heterotopy in any animal system concerns how butterflies got their eyespots. Depending on which theory turns out to be right, eyespots are either collapsed legs in the middle of the wing blade (disguised as archery targets) or pigment stripes displaced from the wing margin (disguised as concentric rings). That debate is presented in the next chapter.

Figure 2.8 (*cont.*)

("clamp") and detachment is denoted by a dashed arc. Detachment of rings from one another (permitting rotation) involves apoptosis [1216]. Adapted from ref. [2151].

a. Initial orientations of A1–A7, A8a, and A8p are marked as clock faces. Digestive and reproductive tracts have been graphically elongated to reach the "bull's eye" containing "a" (anus) and "p" (penis).

b. A8p (inner L/R module) detaches from A8a and rotates 180° clockwise, inverting "a" and "p" and twisting the ejaculatory duct halfway around the rectum.

c. A8a (outer L/R module) detaches from A1–7 but reattaches to A8p. It then rotates 180° clockwise, restoring "a" and "p" to their original geometry and winding the duct so that it loops around the rectum by a full turn. This fascinating "lock" offers a new and powerful system with which to unlock the secrets of L/R asymmetry in animals more generally.

d. *D. melanogaster* male, showing the abdomen in top view.

e. Posterior end of the abdomen as seen from the rear, with anus (a) and penis (p) marked. For clarity, the penis has been redrawn to one side.

f–i. Reversals of organ asymmetry (situs inversus) in flies and humans. Adapted from refs. [967,2093].

f. Abdominal organs in a wild-type (normal) fly. The shaded circle schematically represents the genital plate (**e**).

g. Mirror-image abdominal organs in a mutant fly that lacks a functional Class I Type D myosin motor protein due to (1) a null mutation in the *myoID* gene or (2) RNA interference with the *myoID* gene [1034,2092]. The hindgut has a mirror-image shape, the testes coil counterclockwise instead of clockwise, and the sperm duct likewise loops counterclockwise around the rectum. There is no visible change in the orientation of either the anus or penis.

h. Thoracic and abdominal organs in a normal human.

i. Mirror-image organs in a person exhibiting primary ciliary dyskinesia, also known as Kartagener syndrome [887]. The heart and spleen are on the right, the liver is on the left, the lobes of the lung are reversed, and the colon turns counterclockwise. In general, left–right asymmetries in vertebrates appear to require microtubules, while those in insects require actin, but this distinction may mask a more deeply conserved mechanism that was inherited from our common ancestor [2093].

3 The butterfly

How the butterfly got its scales

Butterflies and moths are lepidopterans. The name means scaly wings. Under a microscope their wings resemble shingled roofs [335,1636]. What seem to be solid areas of color to our naked eyes are actually tiled mosaics of overlapping scales [88], each of which has a uniform color [1087,1513,1630]. There are on the order of a million scales per wing, approaching the megapixel capacity of a cheap digital camera.

Why scales? We don't know their original function [880,1901], but they certainly come in handy when their bearer is caught in a spider web. Scales detach easily from their sockets, so the struggling captive can often wriggle free, leaving its scales behind on the web like a dusty fingerprint, or "wingprint" as it were [602].

Butterfly scales are **homologous** to fly bristles [722,2510], and bristles pre-date scales [880], so our evo-devo quest for the origin of scales leads to a question of how a conical bristle became a flat scale. Bristles are erected on a scaffold of microfilaments and microtubules [197,888], but we know too little about the trestlework of scale anatomy to guess how the cytoskeleton was reshaped during the transitional period [774]. Even so, we can glean a few clues about scale origins from what is known about the cellular and genetic basis of scale development.

Scales and bristles are both variations on a theme that pervades all insect sensilla [1230]. Those sensilla (miniature sense organs) are assembled from small cohorts of cells that descend from a single cell – the sensory organ precursor (SOP). Regardless of whether the sensillum detects touch, smell, or taste, its SOP undergoes a fixed number of cell divisions, each of which has a fixed orientation [1470,1654]. Stereotyped mitoses are common in embryonic cleavage but unusual thereafter [490]. The SOP subroutine constitutes a small-scale **module** in the arthropod genome.

Based on comparative studies, the primordial cell lineage is thought to have produced five descendant cells via three mitoses (Fig. 3.1) [1230]. Bristles reduce the cohort to four via a cellular suicide after the second mitosis (Fig. 3.1l) [660], while scales reduce it to two via a suicide after the first mitosis (Fig. 3.1r) [1254]. Hence, scales lost their ability to act as sensilla because they aborted their neuron [722]. (Exceptional scales along wing veins manage to keep their neurons by averting suicide [722].)

Aside from sharing a pedigree for differentiation, scales and bristles rely on the same toolkit genes for their spatial distributions. The genes *achaete* (*ac*) and *scute* (*sc*) endow groups of cells with the potential to make bristles [743], and an *ac–sc* homolog (*ASH*)

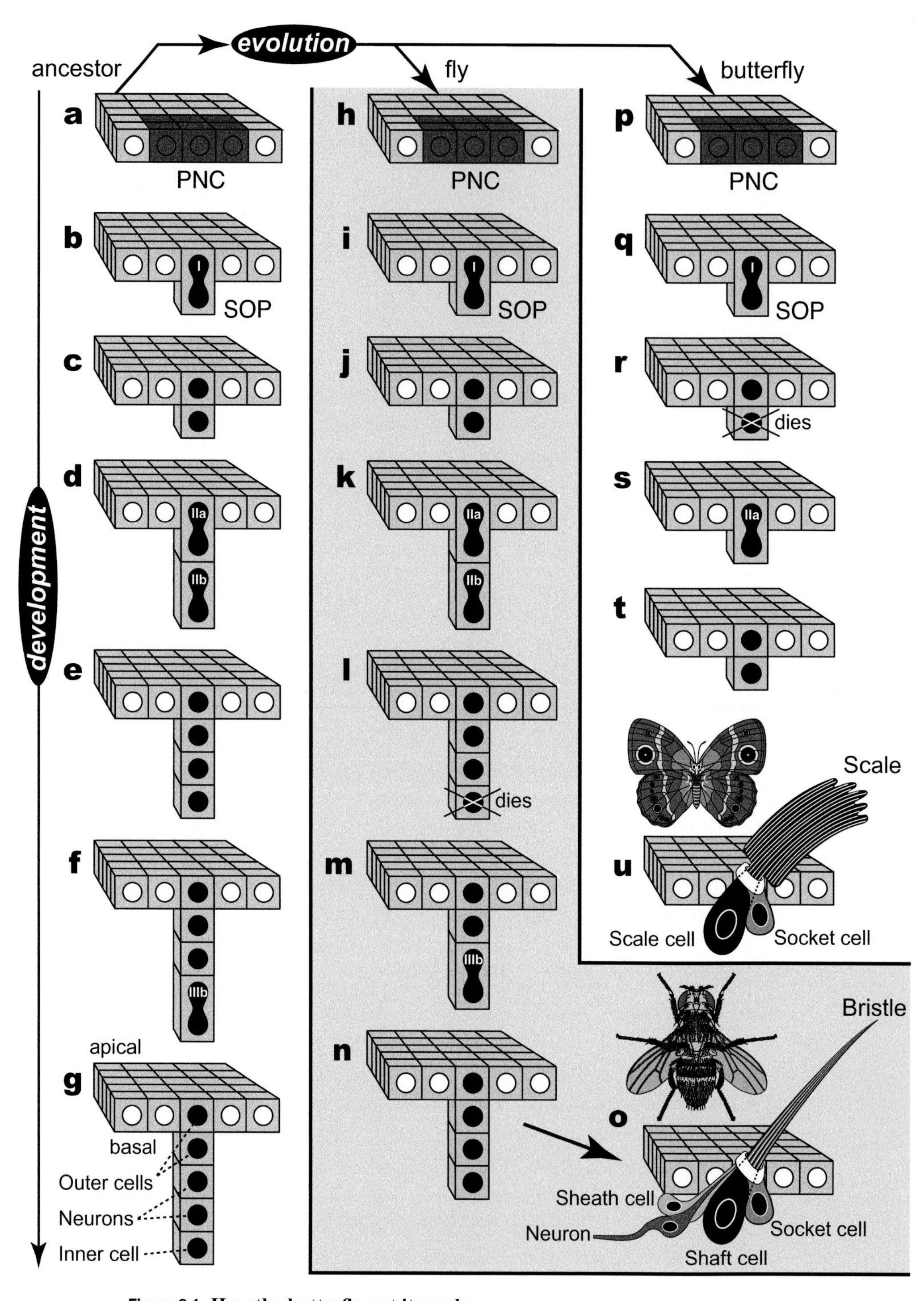

Figure 3.1 How the butterfly got its scales.

plays the same role for scales [1600]. Within these "proneural clusters" (PNCs), only one of the would-be bristle or scale cells usually goes on to become a SOP [357,2005]. The remaining cells are blocked by a signal that is transduced by the Notch pathway [1229,1832].

Unlike fly wings, which only have bristles along their outer edge [965], and dragonfly wings, which lack bristles entirely (the ancestral condition?), butterfly wings sprout a confluent lawn of scales over their entire surface. How *ASH* engenders this lawn is not yet known [722,1832].

When scale SOPs first appear during development, they are typically scattered randomly [1596]. In primitive butterflies they remain so [1633], but in more advanced species they align along the anterior–posterior (A–P) axis [1595,2005]. Curiously, in still other species, A–P rows exist from the very start of scale development without the need for any corrective movements [1628]. This diversity is reminiscent of the varied styles of sex-comb etiologies (Fig. 2.6).

Within each row the SOPs make two types of scales that elongate to different extents [2483]: (1) short "ground" scales that form an understory and (2) long "cover" scales that make a canopy [774]. The two types alternate [388] – ground, cover, ground, cover, and so on. Having two layers of scales allows butterflies to escape at least two spider

---

Figure 3.1 (*cont.*)

a–g. Development of an archetypal insect sensillum, as deduced from sensilla types among insect orders. Adapted from ref. [1230].

a. Square piece of insect skin, which is characteristically one cell thick [1848]. Cubes represent cells, though cells are actually packed hexagonally. Circles are nuclei. A proneural cluster (PNC) of six cells is shaded. Any of these cells can form a sensillum because they express the proneural genes *achaete* or *scute*.

b–c. One cell within the PNC becomes a sensory organ precursor (SOP). The SOP (black nucleus) undergoes the first (I) of several mitoses, denoted by nuclear fission (dumbbell shape), to yield two daughters.

d–e. Each daughter (IIa and IIb) divides once, to make a total of four granddaughters.

f–g. One granddaughter (IIIb) divides, bringing the total number of descendants to five. Two (outer) cells make a transducer of some sort (not shown) on the skin surface, while the other three remain submerged (inner) as neurons or supporting cells. N.B.: For convenience, all mitoses are drawn along the apical–basal axis (labeled), but mitoses in **b** and **i** actually occur in the plane of the epidermis [1230].

h–o. Development of a fly bristle (cf. neuroblasts [367]). The canonical lineage (**a–g**) is obeyed, except that one granddaughter cell dies (**l**). Outer cells enlarge by endoreplication (i.e., repeated replication of DNA sans nuclear division) [687,1932] and make the bristle shaft and socket, while inner ones become a neuron plus a sheath cell (**o**). The shaft elongates via microfilaments [888] and microtubules [197] while being braced by a cuticular sheath [1584]. Based on refs. [960,965].

p–u. Development of a butterfly scale, which resembles a flat bristle [774]. The standard lineage (**a–g**) is obeyed, except that one daughter dies, aborting the IIb lineage, leaving only outer (IIa) cells [722]. As in the bristle (**o**), one outer cell makes a socket, whereas the other makes a scale (**u**), and the larger the scale, the more rounds of endoreplication it must undergo [1207]. Scales are thought to have evolved from bristles, rather than the other way around [774]. Scales on the wing point distally [1757]. Modified from refs. [1254,1470].

webs in their lifetime before their wing surfaces become too denuded for this Houdini trick to work.

This vignette of scales versus bristles has served to illustrate some broader evo-devo principles:

> **Take-home lessons**
>
> 1. Development often proceeds from a crude first approximation (the PNC and scattered sensilla) to a precise final state (the SOP and aligned rows).
> 2. Evolution often creates diverse structures (scales vs. bristles) by tinkering with the same regulatory **module** (SOP subroutine).
> 3. Cell death is often used to prune the products of subroutines, presumably because this is genetically easier than rewriting the recipe itself.

How the butterfly got its spots

The ornate wings of butterflies have delighted humans from time immemorial [620]. Collectively, they comprise a fluttering gallery of ~14,500 tiny works of art [880]. Most are collages of stripes, spots, ripples, or other motifs in the pointillist genre [150,335,1633]. For any given species the layout of the constituent elements is so distinctive that its members can recognize one another as easily as a clan of Scots wearing the same tartan kilt [456,1680].

How do these designs develop? And how did they evolve? Here we encounter one of the richest subfields on the evo-devo frontier [153,1190]. Its potential is just starting to be tapped by molecular approaches [157,1537]. Among its advantages are (1) its Euclidean simplicity (the surface is planar and only one cell thick), (2) the absence of tissues that could confound the analysis (dermis, etc.), and (3) the dazzling diversity of the designs themselves, which offer endless puzzles for inquisitive explorers [229,335].

Much of the research on butterfly wings has dealt with eyespots, because they are so conspicuous [41]. Moreover, they are precise: unlike leopard spots, which are highly variable from individual to individual, butterfly eyespots are relatively fixed in position within any given species (cf. Fig. 5.3) [1635].

Eyespots are round dabs of color surrounded by contrasting concentric rings [1681]. They are so named because they resemble vertebrate eyes [1528]. This illusion is often enhanced by white scales near the center that mimic the reflection of a light source by a dark pupil [2115].

Large eyespots can startle would-be predators by being uncovered suddenly [1172,2259,2260], though the element of surprise is not essential for scaring off birds [2117]. Small eyespots can deflect an attack away from the actual head by acting as decoys [1682,2258,2454]. Either way, the intended victim can get away relatively unharmed [1173,2192] (cf. butterflyfish [1135]).

Eyespots therefore help butterflies escape birds just as scales help them escape spiders. However, the first eyespots were likely less noticeable and may have served a different role such as mate choice [1797] or species identification [51,1868]. Alternatively, like many other animal traits, eyespots may have initially served no function at all [860].

The center of a developing eyespot is called its focus, akin to the bull's eye of an archery target (Fig. 3.2b). The focus is necessary and sufficient for induction of the entire eyespot because (1) killing the cells of a focus can abort the whole eyespot [704], and (2) transplanting a focus to a different part of the developing wing disc can elicit a new eyespot from surrounding tissue [705,1629], provided that the tissue is competent to respond [154,230,1538]. Apparently, each focus is the reference point for a radial coordinate system that acts separately from the Cartesian coordinate system of the wing as a whole.

A cocktail of transcription factors has been found in the focus [267]. Among them are the homeodomain proteins Distal-less (Dll) [335,2035,2351], Engrailed (En) [153], and Antennapedia (Antp) [352,1921], but the role of these proteins, if any, in erecting the coordinate system is unknown.

In 1978 Fred Nijhout, a pioneer of this field, proposed that the focus emits a diffusible **morphogen** whose concentration declines with distance [1627]. The morphogen's range would have to be on the order of ~100 cell diameters for it to reach the spot's perimeter [703,1636]. Such a conical gradient could theoretically elicit colored rings at different radii [267,1530] by activating different pigment genes at different levels [123,350,1629], just as a hilltop is encircled by contour lines in a topographic map (Fig. 3.2c). Indeed, adjacent eyespots fuse together exactly as predicted by this model [1533].

Ever since the idea of a focal morphogen was put forth, the Holy Grail for evo-devotees has been the identity of the morphogen. The Quest may now be at its end, though there have been a few red herrings along the way.

Hedgehog (Hh) became a morphogen candidate in 1999 when *hh* transcription was found to be associated with eyespot sites in *Precis coenia* [1143]. However, *hh* is transcribed adjacent to each focus (anteriorly and posteriorly), rather than inside it. Hh is evidently diffusing from these flanking sites into the focus, where it activates its receptor. This centripetal polarity runs counter to the centrifugal diffusion predicted by Nijhout's model, so Hh is probably not the morphogen per se [2222].

A 2006 study of *B. anynana* showed the presence of Wingless (Wg) in the focus (Fig. 3.2h) [1534]. Earlier probes had failed to detect Wg there because they had not looked during the brief period after pupation ("AP") when it is expressed. Reassuringly, this period (10–16 h AP) overlaps the window of time (1–12 h AP) when cautery of the focus aborts the entire eyespot [704].

In 2010 three mutations with seemingly different phenotypes (affecting spot size, spot color, or ring width) were found to be alleles of the very same gene – a gene that affects the Wg pathway [1919]. Also in that same year Wg was shown to be the inducer for pigment spots in fly wings [2369]. Both of these clues suggest that Wg is the eyespot morphogen.

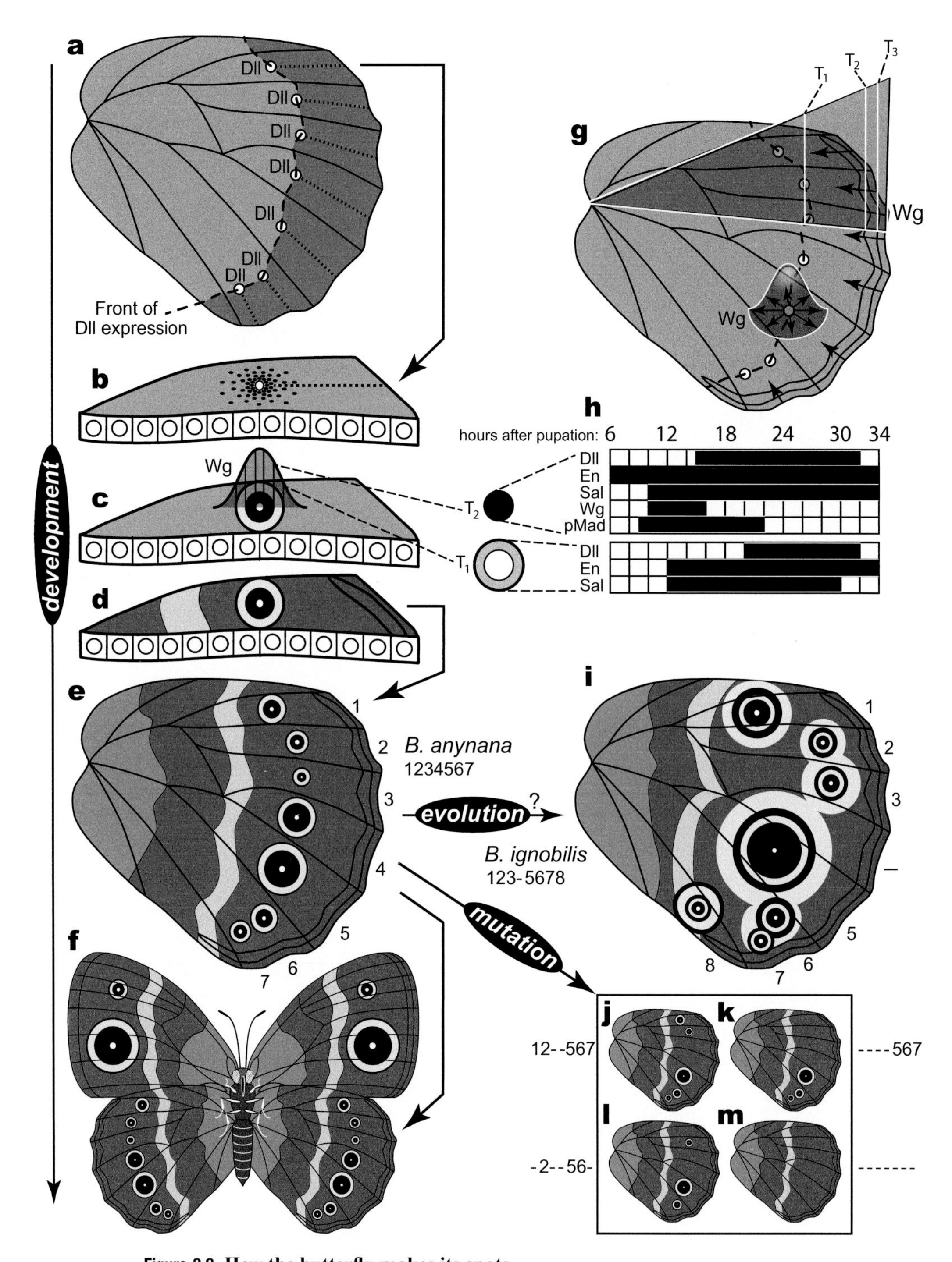

Figure 3.2 How the butterfly makes its spots.

a–f. Development of the ventral surface of the left hindwing of the African squinting bush brown butterfly *Bicyclus anynana*. Adapted from refs. [231,1634]. All panels are oriented in conformity with **f**. (**a**) Developing hindwing of a mature caterpillar, showing future veins

How did eyespots arise evolutionarily? Four theories have been proposed (Fig. 3.3). They are discussed below in historical order. For convenience, each is given a name that befits its central claim.

The "Flat Leg Model" (Fig. 3.3a, b) [1921] was put forward in 1994 by Sean Carroll and coworkers [343]. They conjectured that eyespots are **heterotopic** replicas of leg

Figure 3.2 (*cont.*)

(solid lines) and places where Dll (Distal-less) is expressed, including (1) distal zone (dark gray) up to the transverse dashed line, (2) intervein stripes (dotted lines), and (3) future foci (open circles) [335,708]. Veins are a universal feature of insect wings [508] that serve as conduits for nerves, tracheae, and blood, and as rigid struts to brace the wing during flight [2450]. (**b**) Magnified sector from **a**, where the most anterior eyespot will develop. Its focus emits a **morphogen** (tiny ovals) that diffuses away in all directions. The morphogen is probably Wingless (Wg) [1534]. Squares are epidermal cells (size grossly exaggerated) that form a monolayer; inner circles denote nuclei. (**c**) Concentration profile (hill-shaped gradient) of the focal morphogen (presumed to be Wg) plotted above a future eyespot. A high threshold (T_2) leads to the dark spot and a low threshold (T_1) to the light annulus [150]. (**d**) Final eyespot, plus surrounding markings. (**e**) Adult hindwing. The presence of two eyespots – 6 and 7 – in a single intervein sector seems to violate the rule that eyespots arise midway between veins, but, in fact, an ephemeral vein does exist in the larva between these future spots (A. Monteiro, personal communication). (**f**) Wet-season *B. anynana* female from below [231]. Blending of fore- and hindwing stripes (known as Oudeman's Principle [1633,1775]) helps camouflage the animal's outlines like a tiger's stripes [809,1480]. Nymphalid forelegs are too short to be used for walking and hence are deemed to be **vestiges** [2436]. Traced from ref. [335] with venation after ref. [1868].

g. Types of Wg gradients. Arrows indicate diffusion. Wg diffuses (1) proximally from the distal margin at an early stage to form a triangular gradient [343] and (2) radially from each focus (only one is detailed) at a later stage to form conical gradients (**c**) [1534]. As in flies, Wg specifies a *y* coordinate along the D–V axis (Fig. 2.2) [343], but butterflies do not appear to use Dpp along the A–P axis. Their A–P **morphogen** is unknown. The triangle is a 2D slice through a 3D ridge. Stripes E_2 and E_3 (Fig. 3.3j) are specified at levels T_3 and T_2 (solid lines) and gene *B* (Figs. 3.3g and 3.4a) is turned ON at T_1 (dashed line; edge of Dll front in **a**). Analogous thresholds elicit circles in eyespots (**c**) plus gene transcription (**h**).

h. Times when the listed proteins are expressed in the center of eyespots (black circle) or the outer scale-forming region (pale ring). N.B.: For each protein, expression is shown as a bar stretching from first detection to time of fading or to last time point taken (34 h after pupation), but start times are limited by sampling interval, and protein levels vary within respective time periods. Gene transcription obeys a different order (e.g., *sal* before *en*) [1681]. Abbreviations: Dll (Distal-less), En (Engrailed), Sal (Spalt), Wg (Wingless), and pMad (phosphorylated form of the Mad transcription factor, which acts in the Dpp pathway). From ref. [1534].

i. Hindwing (ventral surface) of *B. ignobilis*. Note (1) larger spots, (2) extra rings, (3) fusions of outer rings, (4) expansion of outer rings beyond bounding veins as on the forewing of *B. anynana*, (5) shifts of spots along proximal–distal axis, causing misalignment, (6) absence of spot 4, and (7) presence of a new spot 8 posterior to spot 7. N.B.: The arrow does not imply that *B. ignobilis* evolved from *B. anynana*, merely that it evolved from a species whose eyespot array was as orderly as that of *B. anynana* [42]. Redrawn from ref. [153].

j–m. Phenotypes obtained after x-ray mutagenesis. Mutant hindwings either lose spots 3 and 4 (**j**), spots 1–4 (**k**), spots 1, 3, 4, and 7 (**l**), or all spots (**m**). From ref. [1536].

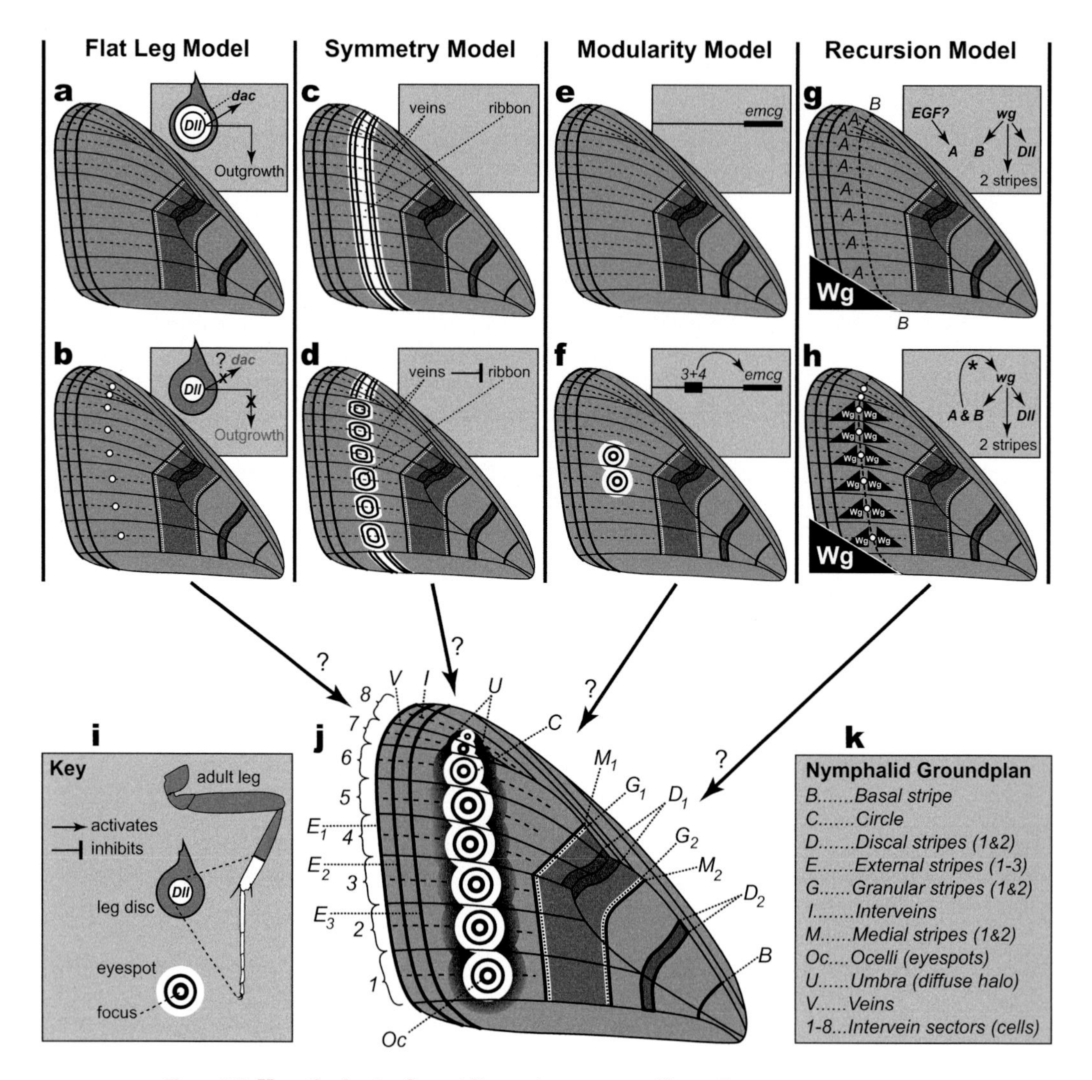

Figure 3.3 How the butterfly got its spots: a menu of hypotheses.

a–h. Alternative hypotheses for how eyespots originated in the progenitor of the Nymphalidae, using Schwanwitsch's prototype of the dorsal forewing (**j**) as a basis for comparison.

a, b. Flat Leg Model [343]. The teardrop shape in the gray box denotes a leg disc, where expression of two genes is indicated (**a**). *Distal-less* (*Dll*) is transcribed in the central area that grows out to become the leg tip (**i**), while *dachshund* (*dac*) is transcribed around it [792]. A mutation accidentally turned ON *Dll* (but not *dac*?) at certain sites (white dots) in the wing blade that became eyespot foci (**b**). Outgrowth was simultaneously blocked somehow (X).

c, d. Symmetry Model [1635]. The ancestor (**c**) is assumed to have had a distal ribbon (white band presaging U in **j**) like its central ribbon (MGDGM in **j**). A mutation caused veins to suppress ribbon identity (T-bar in gray box), so the ribbon was sliced into islands that became eyespots (**d**).

discs. The analogy is based on their discovery that *Distal-less* (*Dll*) is transcribed in eyespot foci, akin to its central expression in the developing legs of flies [792], butterflies [1714,2340], and other insects [8,58,2173]. From this similarity the authors inferred that mutations **co-opted** *Dll* for this new role, but they offered no explanation for (1) how the mutations targeted *Dll* to sites near the wing margin, nor (2) why eyespots are not bumps [613,821,1715], given that *Dll* causes appendage outgrowth in insects [417,1145,1714]. Other genes are expressed in rings around the *Dll* bull's eye in leg discs [1178,1534], but none of these "leg genes" (e.g., *dachshund*) has yet been found to be transcribed in eyespots [1535].

The "Symmetry Model" (Fig. 3.3c, d) was formulated in 2001 by Fred Nijhout [1635]. He argued that a symmetric ribbon of pigment bands was sliced into pieces by the wing veins to form a chain of eyespots. This model is based on a similar kind of fragmentation that is commonly observed where wing veins chop the "central symmetry system" (bounded by M_1 and M_2 in Fig. 3.3j) into sections that can then shift relative to one another along the proximal–distal axis [1627]. Each of the resulting pieces would have automatically acquired a focus if (1) ribbons use the same morphogen (Wg) along their line of symmetry as eyespots use at their focus and (2) veins suppress *wg* via a

Figure 3.3 (*cont.*)

e, f. **Modularity Model** [1528]. The ancestor (**e**) is supposed to have lacked eyespots due to a hypothetical *eyespot master control gene* (*emcg*) being constitutively turned OFF (or not having been co-opted yet to be ON here). Mutations turned *emcg* ON at particular sites (e.g., spots 3 and 4) by adding *cis*-enhancers (**f**). More enhancers were added sporadically in certain butterfly clades until the full set of eyespots was attained.

g, h. **Recursion Model** (see Fig. 3.4 for details) [971]. The ancestor (**g**) is postulated to have expressed a gene *A* midway between adjacent veins and a gene *B* parallel to the wing margin. Lines of *A* might be dictated by gradients of EGF (Epidermal Growth Factor) that emanate from veins (Fig. 3.4b) [965] or by Dpp [1405]. The Wg gradient is only drawn up to the line (dashed) where Wg turns ON *B*, but it actually spans the wing (Fig. 3.2g). A random mutation (asterisk in box) caused *wg* to be turned ON wherever *A* and *B* are both ON, so cells at intersections of lines *A* and *B* now act as if they belong to the margin (**h**), and the resulting foci (white dots) secrete Wg (triangles are gradients; cf. Fig. 3.2c, g), turn ON *Dll*, and elicit stripes that take the form of circles. The model is named for this reiteration (recursion) of patterning steps.

i. Key. Gene interactions are stimulatory (arrow) or inhibitory (T-bar). The center of an eyespot is its focus. The center (white circle) of a leg disc (teardrop) expresses *Dll*, which causes the disc to telescope out to form a cylinder (adult leg) [311].

j, k. Nymphalid groundplan deduced by B. N. Schwanwitsch (1924) [1983]. Dorsal forewing. Hindwing is similar but D_1 stripes are longer, D_2 stripes are shorter, and there is one less eyespot. The groundplan accounts for most of the diversity in this family [1633,1635]. The resemblance of eyespot rings to border stripes (E_2 and E_3) inspired the Recursion Model (**g, h**). The inner circle of each eyespot has a thinner line here than in Schwanwitsch's scheme but is similar to Süffert's scheme. Both plans are amalgams and may not reflect the ancestral state [1983]. D_1 stripes are flanked by G and M stripes in mirror image, so the area between M_1 and M_2 is called the central symmetry system [1633]. Redrawn from ref. [1985] with superscripts changed to subscripts, but numbering retained (rear to front) even though modern convention is reversed (front to rear) [1536].

diffusible inhibitor so as to confine the source of Wg to a point (focus) midway between them. The first supposition finds support in the fact that Wg is indeed expressed along the D_1 zone [1393]. However, this zone only extends one-third of the way along the line of symmetry. The second tenet, which envisions a ridge of Wg being carved into a chain of mountain peaks [1635], is certainly plausible, but it lacks any corroboration at present.

The "Modularity Model" (Fig. 3.3e, f) was advanced in 2008 by Antónia Monteiro, another patterning pioneer [1528]. She postulated that eyespots emerged sequentially one by one, rather than jointly as a whole chain, via *cis*-regulatory mutations in a **master gene** for eyespot initiation. This model is based on an analogy with how the gene *even-skipped* (*eve*) is controlled by stripe-specific enhancers in fly embryos [2404]. In the case of *eve*, certain enhancers regulate more than one stripe [928], and the same might be true for eyespots because single mutations can delete multiple eyespots simultaneously (Fig. 3.2j–m) [1536]. When such mutations are eventually mapped, they may lead us to the master gene, which might be *wg*. A similar mutagenic approach succeeded in dissecting the headquarters for bristle patterning in *Drosophila* [807]. Like the Flat Leg Model before it, the Modularity Model fails to explain why eyespots are arranged in a chain.

The "Recursion Model" (Fig. 3.3g, h) was devised in 2013 to solve this riddle of eyespot alignment [971]. Instead of comparing eyespots to legs or color bands or embryo stripes, this model sees eyespots as tiny versions of the wing edge. It is based on a subtle clue embedded in Schwanwitsch's 1924 groundplan for nymphalids (Fig. 3.3j) [1983,1985]: the number and spacing of concentric circles in large eyespots echoes the number and spacing of border stripes (E_2 and E_3). Since the margin and the focus both use Wg as a morphogen and Dll as a transcription factor [343,1534], they could both be evoking the same pigment contours. Those contours would become stripes or circles depending upon whether the morphogen source is a line (margin) or a point (focus), just as the ripples in a pond obey the geometry of their source: boat wake (lines) or pebble splash (circles).

How did clusters of cells in the wing blade come to act as if they were at the wing margin? According to the Recursion Model, they were incited to do so by a mutation that happened to activate *wg* in response to input from two genes, here called *A* and *B* (i.e., *A* & *B* → *wg*). This scenario assumes that (1) *A* was expressed midway between the veins, (2) *B* was expressed at a certain distance from the wing edge, and (3) the mutation turned *wg* ON wherever these two lines intersected. These tenets are explained more fully below:

1. Each vein emits a **morphogen** – perhaps a variant of EGF as in flies [1920] – that diffuses at least halfway to the nearest vein (Fig. 3.4b), turning ON gene *A* at a certain level. Reassuringly, some butterflies have stripes midway between their veins (Fig. 3.5a) [1632,1985], though gene *A* must encode a transcription factor for this model to work, not a pigment.
2. The wing margin emits Wg [343,1534,2351], which then turns ON gene *B* at a certain distance (T_1 in Fig. 3.4a) beyond E_2 and E_3 (Fig. 3.3j). Expression of gene *B*, like that

of gene *A*, need not be manifest by any overt pigment, though it is comforting that one *Prepona* species does have a stripe that coincides with its eyespots (Fig. 9b in [1986]), and other nymphalids have a marginal pigment zone whose inner edge traces this same contour (hindwing in Fig. 3.5b).

3. A random mutation accidentally caused genes *A* and *B* to jointly activate the circuit that encodes wing margin identity (Fig. 3.4c), thereby evoking a chain of eyespots that instantly appeared along the line of *B* expression (Fig. 3.5c). This mutant became the ancestor of all eyespot-bearing butterflies – the nymphalids and related families.

Step (3) was theoretically the key event. The resulting genetic link is recursive insofar as it "reboots" a subroutine that was used earlier in development (at the margin). Cells midway between the veins are thereby duped, so to speak, into thinking that they are at the margin, and they behave accordingly by emitting Wg, activating *Dll*, and making "ripples" that become circles. If this argument is correct, then the reason *Dll* is expressed in foci is not because it is conjured by a leg circuit (*wg* & *dpp* → *Dll*) [792] as the Flat Leg Model claims, but rather because it is conjured by a wing circuit (*wg*→ *Dll*) [1608,2498]. Dll is only detectable in eyespots 7 hours after Wg (Fig. 3.2h) [1534], consistent with *Dll*'s rank as a downstream target of *wg*.

The genomic rewiring that is being envisioned here entails both **co-option** and **heterotopy** [75]. Co-option is the conscription of an old circuit (wing margin) for a new function (eyespot patterning) [2228], while heterotopy is the reuse of an old circuit (wing margin) at a new location (intervein midpoints) [75]. It also relies on **saltation**, a discontinuous alteration of the phenotype that is usually exemplified by **homeosis** [139], though homeosis is probably not involved here. Initially there were no eyespots. Then, *voilà*! A mutation occurred, and eyespots appeared on the wings of the offspring. Simultaneously, this mutation transformed a global, Cartesian (x, y) coordinate system (Fig. 2.2) into a local, radial one (Fig. 3.2g).

As far-fetched as the Recursion Model may sound, its premises are supported. The notion that *veins* cause eyespots (x coordinate) is buttressed not only by the fact that eyespots lie midway between them [1627,1631,1635], but also by the finding that extra veins lead to extra eyespots [1920]. The idea that the *margin* causes eyespots (y coordinate) explains why eyespots arise at a constant distance from the wing edge [42,1171], even when that edge is shaped oddly. Finally, when an eyespot shifts proximally, the inner marginal stripe in that same sector (E_3 in Fig. 3.3j) bends as if to follow it (Fig. 3.5d).

One virtue of this model is that it is testable directly. We should be able to pinpoint the predicted *cis*-enhancer that turns *wg* ON in foci by (1) coupling pieces of DNA from the *wg* locus to reporter genes [429,698,1380] and (2) seeing where these constructs are expressed in host butterflies [1537,1816]. A similar approach worked well in mammals [2040]. The eyespot enhancer should be present in nymphalids and other spotted families, but not in groups that lack eyespots. Gene *A* might be *Dll*, since *Dll* is expressed midway between veins before it intensifies in foci (Fig. 3.2b) [335,708]. If so, then we might be able to find the *wg* enhancer by probing for Dll binding sites in the *wg* region. There are no obvious candidates for gene *B*.

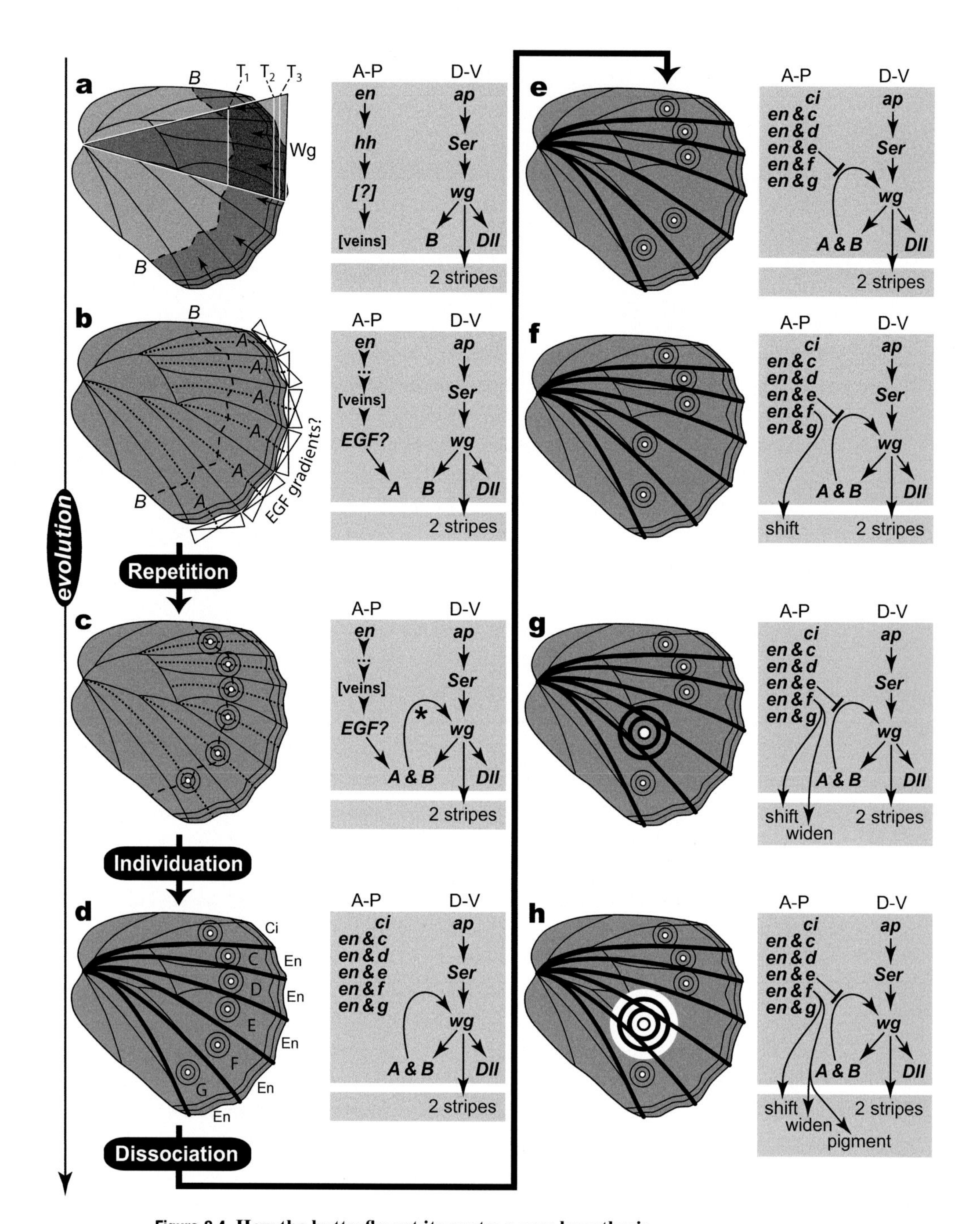

Figure 3.4 How the butterfly got its spots: a new hypothesis.

a–c. Recursion Hypothesis for the origin of eyespots [971]. For ease of comparison with Fig. 3.2, stages of evolution are drawn onto a ventral left hindwing of *B. anynana*. Gray boxes contain gene circuits (above) acting along the anterior–posterior (A–P) or dorsal–ventral (D–V) axis, and events at the tissue level (below). Arrows denote activation; T-bars inhibition.

According to the Recursion Model, the first species to acquire eyespots would have had an eyespot in every intervein sector [971], but many modern Nymphalids do not [41,1171]. How did certain lineages lose eyespots at various locations?

This sort of riddle is familiar to evo-devotees as Williston's Law [339,1485,2418]. It applies to any pattern that begins as serially repeated elements spaced at uniform intervals [961].

The solution to the riddle is **individuation** [1857,2305]: genes assign different "area codes" to members of the series [154,156,1528]. Once the individual elements acquire separate identities via these codes, they can then diverge evolutionarily – a process termed **dissociation** [1599,1809]. Individuation is covert (a crosslink in the genome), whereas dissociation is overt (a visible change in anatomy). Individuation gives **meristic** patterns a reservoir of plasticity [231,1634] that can be exploited at a later time [42,156].

Figure 3.4 (*cont.*)

a. Specification of cellular positions along (1) the A–P axis by *engrailed* (*en*) and *hedgehog* (*hh*) acting via an unknown (*dpp*-like?) agent [2222] and (2) the D–V axis (= proximal–distal axis in adult) by *apterous* (*ap*) and *Serrate* (*Ser*) acting via *wingless* (*wg*). Wg is secreted at the margin, as in flies (Fig. 2.2), and diffuses proximally to establish a concentration gradient (triangle). Based on what happens in flies [1608], *Distal-less* (*Dll*) is likely activated above a Wg threshold T_1 (Fig. 3.2a). Pigment genes turn ON at levels T_2 and T_3 to make stripes E_3 and E_2 (Fig. 3.3j). Imaginary gene *B* turns ON at level T_1 (dashed line). Genes are the same as in flies (Fig. 2.2), except that an unknown morphogen "[?]" governs the A–P axis instead of Dpp [343,1143].

b. Metameric control of the A–P axis by a **morphogen** that diffuses from each vein (triangles) [508]. Imaginary gene *A* (*Dll?*) turns ON (dotted lines) at a certain level (cf. Dll "lollipops" in Fig. 3.2a). Based on fly circuitry [965], the vein morphogen may be EGF (Epidermal Growth Factor).

c. A chance mutation (asterisk) forced *A* and *B* to jointly activate *wg*, rebooting its subcircuit. Hence, a recursive loop was forged, "duping" cells at *A/B* intersections into "thinking" that they lie at the margin. Foci (white circles) thus (1) secrete Wg, (2) activate *Dll*, and (3) elicit two contours, yielding eyespots. Contours are circular (vs. linear) because they emanate (like ripples in a pond) from a point (focus) instead of a line (margin).

d. Actual (Ci and En) [1143,1534] and hypothetical (C–G) [1536] sectors of the wing blade, though En spans the whole posterior compartment, which comprises most of the wing blade [2222]. The eyespots still look alike, but they have the potential to diverge anatomically because each of them expresses a unique transcription factor (Ci) or combination thereof (En with C, D, E, F, or G). Thus, the spots undergo **individuation** by being given unique "area codes." Adapted from ref. [1536].

e–h. Types of **dissociation** that may have occurred in the evolution of the *B. ignobilis* hindwing (Fig. 3.2i). Dissociation is the conversion of covert genetic differences into overt anatomical differences. Gray boxes depict possible rewiring events (above) that could explain eyespot alterations (below). Further changes would be needed to make this eyespot look like the real one. Removal of the fourth eyespot was presumably a co-requisite for enlargement of the fifth eyespot because otherwise they would have fused. The proximal shift of the fifth eyespot may have likewise "made room" for it to enlarge. Eyespot shifts alone (sans growth) may be aimed at a butterfly audience (vs. bird predators) where they function in mate choice [51,1797] or species recognition [1463,1716]. If so, then such shifts would literally exemplify the process of **character displacement** [986,1676,1751], which permits pre-mating reproductive isolation [2359].

Figure 3.5 How butterflies decorate and sculpt their wings. Images of these six species in the family Nymphalidae are from the lepidopteran database (http://www.lepdata.org; ©2012 by W. H. Piel, L. F. Gall, J. O. Oliver, A. Monteiro, and the Yale Peabody Museum). Only in 2011 did we discover that "transparent" fly wings are actually just as garish as butterfly wings [2027]! Their iridescent magnificence is only visible against a dark background. *For color plate see color section.*

a. *Pseudacraea dolomena*, ventral surface. As with many species (e.g., the monarch butterfly), the veins are pigmented [2128], but what is unusual here is that pigment stripes also appear midway between the veins. These intervein stripes suggest that the veins are activating pigment genes at a distance, perhaps via the morphogen EGF (Fig. 3.4b). According to the Recursion Model (Fig. 3.3g, h), they can also activate a hypothetical gene *A* (Fig. 3.4b) that may encode a transcription factor.

b. *Pseudacraea eurytus*, ventral surface. The hindwing has a broad band of pigment at its margin that is superimposed on the stripe array of *P. dolomena* (**a**). This band resembles the zone where *Distal-less* is expressed (Fig. 3.2a) before its expression recedes to the foci (Fig. 3.2c). The band's inner edge traces the same line as the eyespots in most nymphalids (e.g., **c**), so it may be revealing a contour line in the wing's coordinate system where the

One appealing scheme for assigning eyespot area codes, which was proposed by Monteiro *et al.*, is depicted in Figure 3.4d [1536]. It overlays half a dozen transcription factors – some known and some conjectured – onto the eyespot series so as to endow the eyespots with unique identities.

Figure 3.4e–h shows how subsequent dissociation might work, including the loss of particular eyespots. For example, the eyespot in sector E would vanish if a mutation inserted an enhancer near *wg* that blocks *wg* transcription when it is bound by both En and E (Fig. 3.4e). Indeed, another way to test the Recursion Model would be to look for such enhancers at the *wg* locus in species that possess only a few eyespots. All those enhancers should be inhibitory [971], in contrast to the enhancers expected by the Modularity Model, most of which should be stimulatory. Interestingly, a similar dilemma concerns our own species [968]: did we evolve our present distribution of hair (on our heads, armpits, etc.) by evolving large hairless regions from a previously all-hairy state (via negative enhancers) or by addition of hairy patches to a previously all-naked state (via positive enhancers)? No one knows.

Based on the opposite predictions of the Recursion and Modularity Models, one way to test their relative merits would be to examine nymphalid phylogeny. Are lineages more consistent with a spotless ancestor whose descendants gained eyespots stepwise

---

Figure 3.5 (*cont.*)

hypothetical gene *B* turns ON (Fig. 3.4a). According to the Recursion Model (Fig. 3.3g, h), eyespot foci are induced to emerge wherever the *A* and *B* lines intersect because the transcription factors A and B jointly "reboot" wing margin identity.

c. *Cyrestis acilia*, ventral surface. The chain of eyespots on each wing corresponds to where intervein stripes intersect the inner edge of the hindwing's marginal band in **b**. However, the next to the last eyespot on the hindwing has (1) shifted distally, (2) enlarged considerably, and (3) protruded noticeably from the edge. This feature may act as a decoy (false head?) along with the nearby swallowtail (false antenna?) to deflect the attacks of predators away from the actual head [1683,2258,2454]. How does the wing "tell" one spot to act differently from all the others? Presumably, it does so via genetically designated area codes (Fig. 3.4d) – the cellular equivalent of a GPS (global positioning system) device. As usual for nymphalids, there are two submarginal stripes (E_2 and E_3 in Fig. 3.3j). On the hindwing the inner stripe is lighter (brown) and wider than the outer one, but on the forewing they are less dissimilar.

d. *Asterocampa leilia*, dorsal surface. The inner (E_3) and outer (E_2) submarginal stripes (Fig. 3.3j) are alike in width and hue, but E_3 zigzags on the hindwing. Its deepest kinks occur where spots have shifted the furthest basally, suggesting that eyespots obey the same coordinate system as E_3, as expected by the Recursion Model (Fig. 3.4), but the failure of E_2 to track E_3 is enigmatic.

e. *Polygonia faunus*, ventral surface. The ability of this butterfly to camouflage itself on tree bark is due to its variegated (brown) color and its scalloped outline. That outline is presumably carved by the "cookie cutter" cell death trick discovered recently (Fig. 3.7).

f. *Physcaeneura leda*, ventral surface. The outer stripes (E_1, E_2, and E_3) are well defined, and the medial margin of all four wings has tiger-like striations. The latter "ripple" pattern is the only motif in lepidopteran wings that displays left–right asymmetry (cf. Fig. 5.3) [123,1633]. The ripple motif is thought to be created by a reaction–diffusion mechanism (see Chapter 5) that is separate from the Cartesian coordinate system which lays out all of the other pattern elements in nearly perfect symmetry on the two wings (**a–e**).

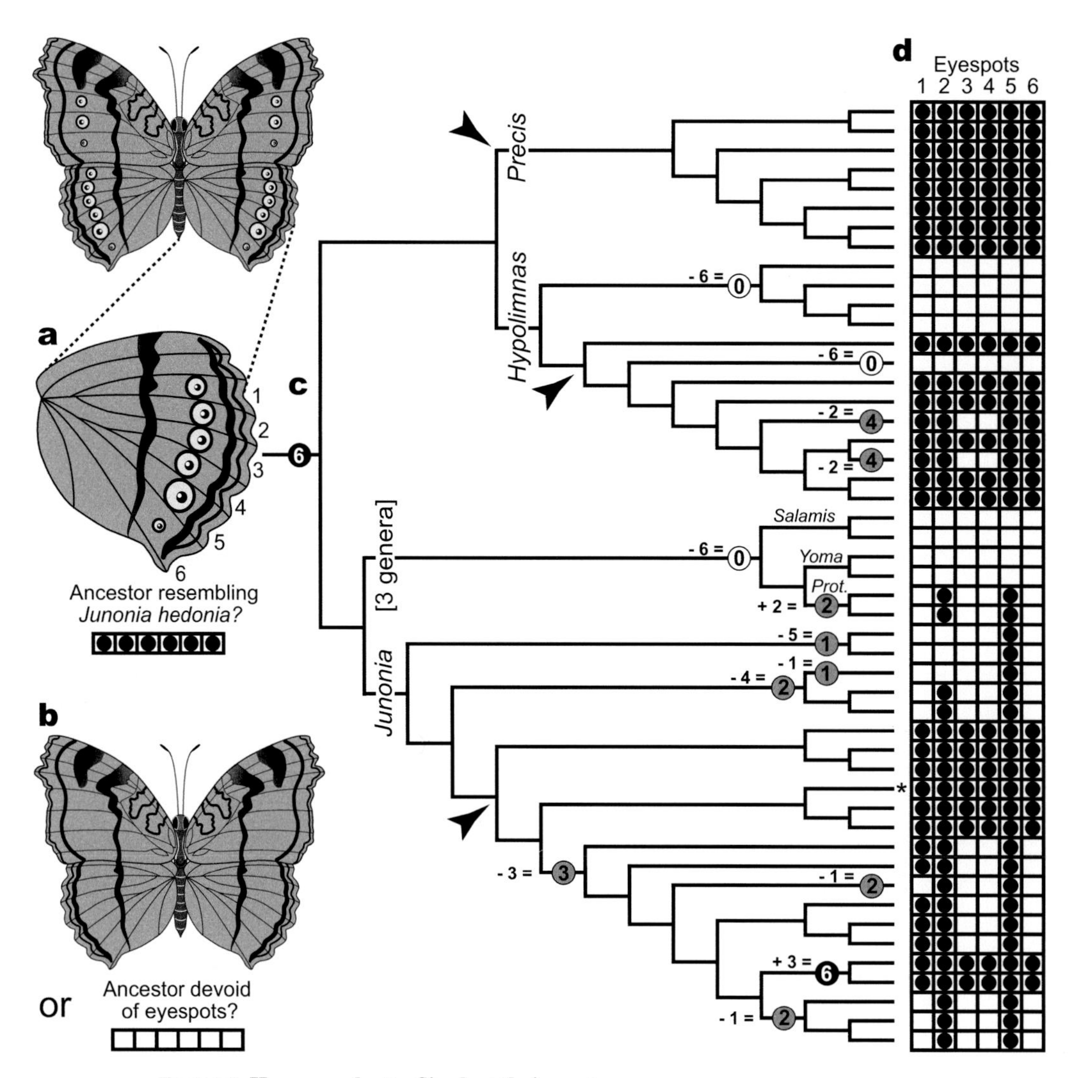

Figure 3.6 How some butterflies lost their spots.

a. The brown pansy butterfly *Junonia hedonia* (dorsal surface) is used here as a proxy for the ancestor of the Junoniini tribe because that ancestor may have also had a contiguous array of eyespots. However, this conjecture is merely one of several possibilities (cf. **b**). The alignment of forewing and hindwing stripes is an example of fore–hind blending (Oudeman's Principle) that occurs throughout the Lepidoptera [1633]. By disrupting the outlines of the wings, the compound stripe presumably serves the same function as a tiger's stripes [809,2116,2119]. Camouflage is likewise enhanced by the wing's serrations [1349].

b. A *J. hedonia* butterfly that has been stripped of its eyespots. According to this alternative interpretation of the cladogram (vs. **a**), the ancestor had no eyespots.

c. Phylogeny of the Junoniini tribe of the family Nymphalidae. Italicized names are genera. *Prot.* stands for *Protogoniomorpha*. Asterisk is *J. hedonia*. Redrawn from ref. [1171] (cf. [1681]) and approved by Ullasa Kodandaramaiah. Circled numbers denote the numbers of eyespots in descendants: black (6), white (0), gray (1–5). Pluses and minuses are net changes relative to the most recent ancestor. Numbers were assigned assuming (1) the

(Modularity Model) or with a fully spotted ancestor whose descendants lost them stepwise (Recursion Model)? The cladogram in Figure 3.6 can be viewed either way [1171], but the most parsimonious interpretation of its eyespot alterations is that the first nymphalid had a full set of eyespots [1204], as envisioned by the Recursion Model.

Despite its explanatory power, the Recursion Model is by no means a cure-all for what ails the other models. It is fraught with its own flaws, some of which are listed below [971]:

1. *The gap disparity.* The gap between eyespot rings is admittedly comparable to the space between submarginal stripes E_2 and E_3, but E_2 and E_3 arise near the top of the wing's Wg gradient. For eyespot rings to acquire a similar spacing, each eyespot's radial Wg gradient would have to span the whole wing, which seems implausible at that late stage.
2. *The recursion riddle.* Once the genome enters an infinite loop, there is no easy way out, except for time to expire and cuticle secretion to begin. What would another round of Wg secretion, *Dll* activation, and contour evocation look like? Because each eyespot has an invisible *B* ring beyond its pigment rings, that ring should intersect the *A* line at two points, but no butterflies have ever been described with two tiny (fractal) eyespots flanking a normal one [1913]. Nevertheless, there do exist rare cases where a single extra eyespot resides beside a normal one (A. Monteiro, personal communication).
3. *The impossible inversion.* Nymphalid species exist where stripe E_3 has moved so far toward the base of the wing that it is actually more proximal than the eyespot archipelago (Fig. 9 in [1986]). This "positional inversion" is a forbidden geometry according to the Recursion Model, because gene *B* should always be turned ON at a lower level of Wg than E_3.
4. *The color conundrum.* How do eyespots acquire colors never seen near the margin if they are just iterations of that algorithm [267,2434]? One way may be for them to **co-opt** color genes from elsewhere – for example, the eye [1529,1833,1834].

More challenges await any of these four models that emerges victorious [42,154]. How do eyespots enlarge or shrink [42,1441], move proximally or distally [228,1627], add or delete rings [153,1530], change shape [1531,1532,1632], fuse with adjacent spots [1533,1920], or disappear entirely [1171]? If eyespots are individualized, then why do they tend to

Figure 3.6 (*cont.*)

ancestor had six eyespots (**a**) and (2) lineages are more prone to lose eyespots than gain them. Only two gains were needed to fit the data (**d**), and each of these gains follows closely on the heels of a recent loss, suggesting that the gains entail character restorations (**atavisms**) rather than re-inventions. If, instead, the ancestor is assumed to have been spotless (**b**), then there are at least three nodes (arrowheads) where all six spots must have been acquired simultaneously (in defiance of the Modularity Model).

d. Patterns of eyespots. Eyespots 3 and 4 are always either both present or both absent, a link that also holds when nymphalids are mutagenized by x-rays (Fig. 3.2j–m) [1536], so this link may reflect a genetic **constraint**.

change size coordinately (**morphological integration**) under conditions of intraspecific variation [41] or when the size of a single spot is subjected to artificial selection [151,1538,1730]? What roles do the Hedgehog [1143] and Notch [706,1835] pathways play in these dramas? Is there no limit to the types of patterns that could have evolved [152,155,1633]? Or, if there are **constraints**, then what are the genetic bases for them [231,1920,1984]?

The foregoing saga illustrates a few general trends in evo-devo that are worth stressing:

Take-home lessons

1. Novelties such as eyespots can arise suddenly [1204,1681] with few, if any, pleiotropic side effects because they are **modularized** genetically.
2. Traits that seem geometrically distinct (e.g., eyespot rings vs. margin stripes) may actually be outputs of the same subroutine operating upon different inputs (e.g., a point vs. line source) [1824].
3. Heterogeneous elements (e.g., large vs. small eyespots) may have been identical before they acquired separate area codes (**individuation**) and diverged overtly during evolution (**dissociation**).

How the swallowtail got its tail

In 2010 a paper was published in the journal *Evolution & Development* with the intriguing title "Butterfly wings shaped by a molecular cookie cutter" [1349]. Its subtitle hinted at deeper insights: "Evolutionary radiation of lepidopteran wing shapes associated with a derived Cut/*wingless* wing margin boundary system." Indeed, the wing-shaping trick revealed in this report is profound.

Unlike other insect orders, where the wings have a smoothly convex outline [508,1700], lepidopterans display a wide variety of weird wing shapes. For instance, the hindwing of *B. anynana* has wavy serrations (Fig. 3.2f), and both the fore- and hindwings of *J. hedonia* are even more strongly crenulated (Fig. 3.6a). Still other species look as if the Mad Hatter had been cutting tiny paper dolls with his eyes closed (Fig. 3.5e). One of the fanciest frills of butterfly wings is the swallowtail, as exemplified by the hindwing of *Battus philenor* (Fig. 3.7j). It was this shape that these authors chose to analyze.

Until the fifth instar of the larval period (instars are stages punctuated by molting), the wing disc of *B. philenor* does nothing noticeably noteworthy. It transcribes *wg* and *cut* at its perimeter (Fig. 3.7g), just as flies do (Fig. 3.7b). Then something strange happens. The thin zone of *wg* and *cut* expression expands proximally like a wave coming onto shore, and the advancing front assumes a new shape that is recognizable as the final shape of the adult wing (Fig. 3.7h). The cells in this *wg*-ON, *cut*-ON zone then die (Fig. 3.7i), leaving the proximal edge of the border zone as the new distal edge of the wing. The process is indeed reminiscent of a cookie being stamped by a cookie cutter from a slab of dough.

The expansion of the *wg*-ON, *cut*-ON zone is theoretically attributable to a recursive mutation like the one that may have created eyespots. The mutation would have set up a feedback loop where *wg* activates itself above a certain threshold (T_1) of Wg concentration (Fig. 3.7f). Two other new links would have been required to finish the job: (1) activation of *cut* by *wg* and (2) induction of cell death by *cut*.

From an engineering standpoint, any kind of programmed cell death or "apoptosis" [1455] is inefficient [718] because:

1. It is wasteful to kill cells that an embryo has expended energy to create.
2. It should be simpler to produce a shape directly (e.g., by targeted growth) than to overshoot its ideal dimensions and then have to prune it back.

Nevertheless, in the case of butterfly wings, apoptosis may have offered a retrofit that was easier to accomplish genetically than a *de novo* overhaul of the wing-shaping subroutine. Another structure that evolution retroactively reshaped by apoptosis was the beetle horn [1148]. Overall, cell death may be adaptive (despite its inefficiency) because it gives embryos an additional, versatile tool in their toolkit: the cookie cutter lets them create odd shapes by carving instead of molding [1324,2524].

The cutter device cannot merely be shifting the stripe of *wg* transcription uniformly, because in that case the profile of the edge would stay the same as it moves proximally [1349]. Rather, the mechanism must draw a different contour to guide apoptosis in cutting "along the T_1 dotted line!" Perhaps Wg diffusion is *impeded* near the future tail and elsewhere to leave protruding peninsulas, and *accelerated* at other locations to carve new inlets?

Wing veins might be mediating the putative impedance, because (1) the tails of swallowtail butterflies typically contain a vein and (2) the wavelengths of wing serrations tend to match vein spacing intervals. However, the veins per se are probably not involved directly, because the new edge (marked by a lacuna) is already established before the veins are formed [1349].

Conceivably, the very same genes that are thought to provide area codes for eyespots (Fig. 3.4d) might be supplying a similar kind of **individuation** for districts that slow down or speed up Wg diffusion. Of course, it is equally plausible that area codes are not affecting diffusion at all, but instead are altering the threshold concentration at which cells *respond* to the Wg signal.

Despite these ambiguities, the curious tale of the swallowtail's cookie cutter does leave a few lasting impressions:

Take-home lessons

1. A small genetic change (e.g., shifting the wing's edge "inland") can offer a species access to vast regions of virgin **morphospace**, in the same way that a rabbit-hole afforded Alice entry to a hitherto hidden Wonderland.
2. Evolution often opts for expediency over efficiency (e.g., relying on retrofits), because it operates more like a tinkerer than an engineer [1077].

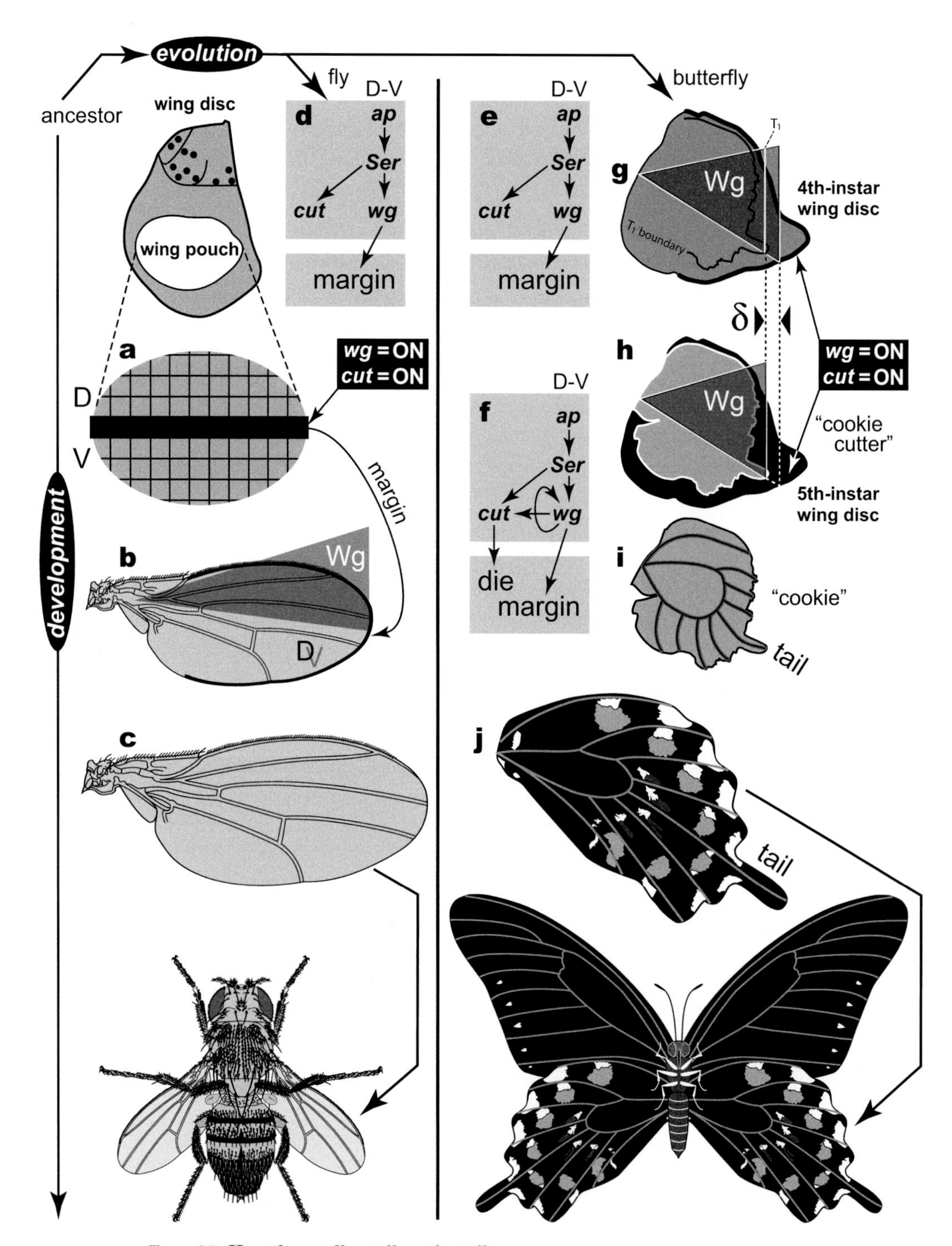

Figure 3.7 How the swallowtail got its tail.

a. Wing disc from a mature, third-instar fruit fly larva, whose pouch is schematized as an oval with square cells (grossly enlarged). Dots in the disc are sites of future macrochaetes (big

How the caterpillar got its hindlegs

It is a truism that insects have six legs, yet caterpillars flagrantly break this rule. Caterpillars typically have ten additional "prolegs" on their abdomen – four pairs in the middle and one pair at the rear (Fig. 3.8e).

Unlike thoracic legs, prolegs lack joints and do not taper to a tip [2083]. Indeed, they are so stocky that they make the caterpillar look as if two and a half elephants had been yoked in tandem behind a six-legged horse. The front half of the caterpillar can rear up on the back half because the prolegs have tiny hooks on the soles of their feet that can firmly grip a leaf or twig [1748]. Caterpillars often adopt this pose, swaying like overweight cobras in search of tender leaves to munch [2083].

Figure 3.7 (*cont.*)

bristles). The bar across the pouch is the D–V (dorsal–ventral) boundary, where *wg* and *cut* are transcribed. Wg diffuses away from this bar in both directions to establish two gradients, one of which is shown in **b**.

b. Wg gradient on the dorsal side of the wing. The gradient on the ventral side (lightly shaded "V") is omitted. The D–V axis of the disc (**a**) has become the proximal–distal axis of the adult wing due to apposition of D and V surfaces during eversion [149]. Cells in the blade assess their positions along this axis by measuring the amount of Wg at their locations (cf. Fig. 2.2). The thick line along the margin corresponds to the bar in **a**. D and V: dorsal (visible) and ventral (hidden) surfaces.

c. Right wing. Anterior is at the top, posterior at the bottom.

d. Circuitry controlling the D–V axis (abridged); see Fig. 2.2p or ref. [965] for details. Features of the margin, including bristles, are dictated by *wg*, with *cut* quashing bristles by inhibiting *sc* (not shown).

e. Circuitry governing the D–V axis of the butterfly wing disc during the fourth instar (diagrammed in **g**). Wiring appears to be the same as in the fly (**d**) [343].

f. Circuitry governing the D–V axis of the butterfly wing disc during the fifth instar (diagrammed in **h** and **i**). Wiring is the same as in the fly (**d**), except for three new links (evidence in parentheses): (1) *wg* activates *cut* (because their expression domains are congruent), (2) *wg* activates itself above a certain threshold (T_1 in **g**) of Wg concentration (because the *wg*-ON border widens), and (3) *cut* causes cells to die (directly observed). Ironically, *cut* actually blocks cell death in the fly larva [2504].

g–i. Development of the left hindwing disc in *Battus philenor*.

g. Fourth-instar hindwing disc. The Wg gradient specifies a line (T_1 boundary) where a border lacuna (not shown) will appear in the fifth instar. That lacuna will become the inner edge of the *wg*-ON, *cut*-ON zone (**h**) and will eventually assume full responsibility for secreting Wg when the outer margin dies. Redrawn from ref. [1349].

h. Early fifth-instar hindwing disc. The *wg*-ON, *cut*-ON zone has expanded basally, thus moving the peak of the gradient proximally. The gap (δ) varies along the perimeter, possibly due to regional modulation of the Wg diffusion rate (see text). The *cut* gene then induces cell death wherever it is transcribed, thus pruning the wing to a smaller size and a new shape.

i. Mature fifth-instar hindwing disc. Peripheral cells (black border in **h**) have died due to *cut*, leaving a remnant "cookie" that then inflates until it assumes the final shape shown in **j**. Black lines are incipient veins.

j. Adult hindwing of the pipevine swallowtail butterfly *B. philenor* (ventral surface). Gray lines are veins. This species is not a nymphalid, and its spots do not meet the criteria for eyespots [1681].

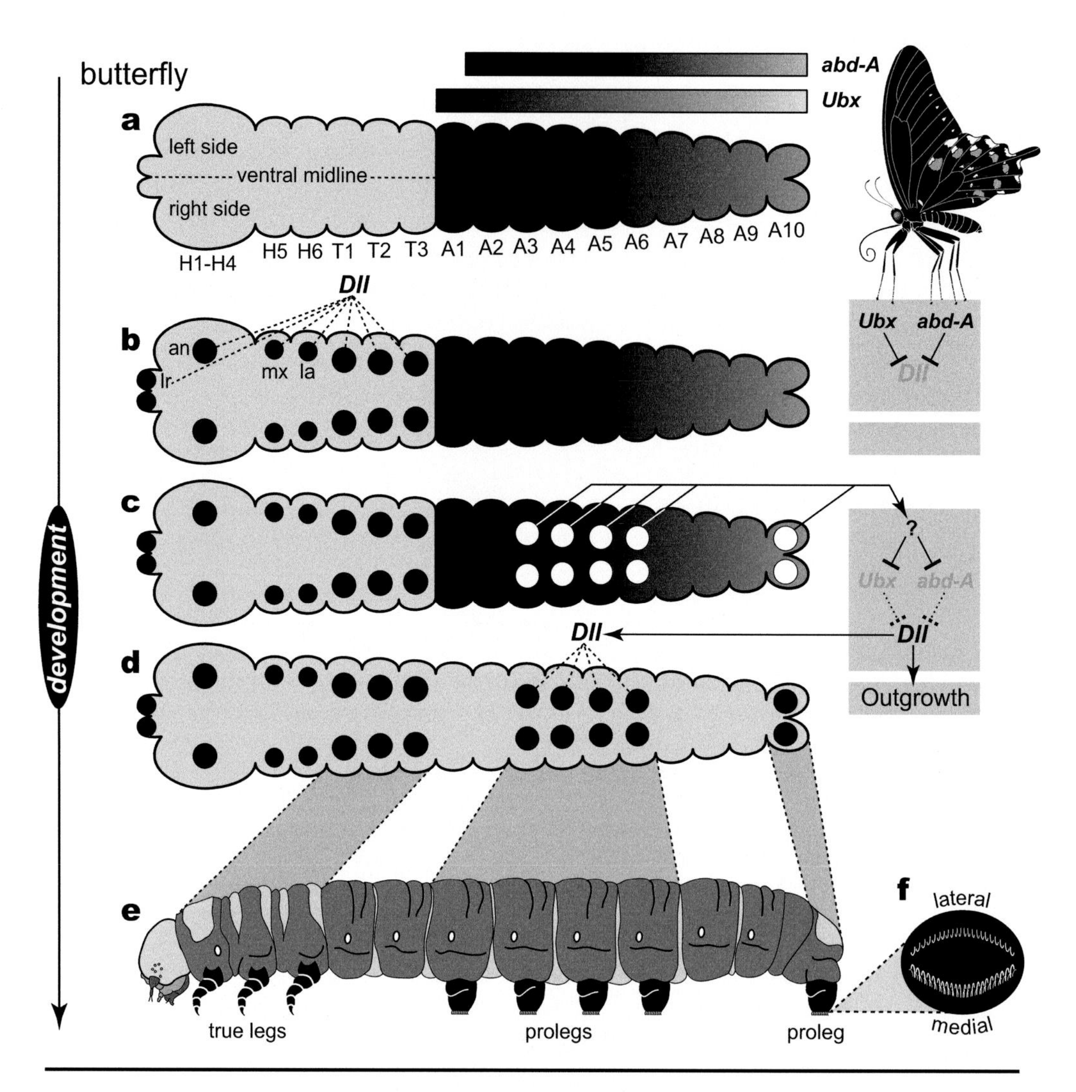

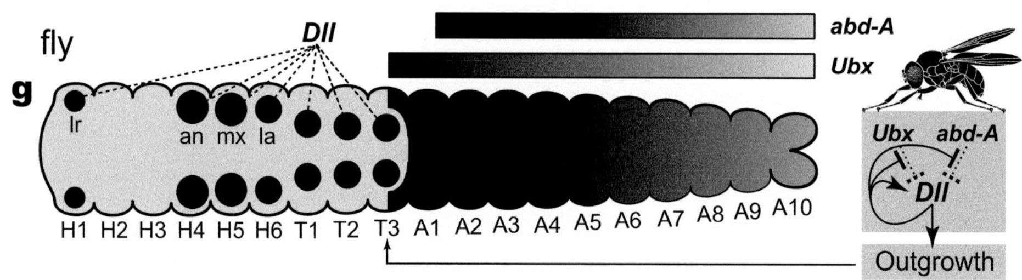

Figure 3.8 How the caterpillar got its hindlegs.

a–e. Embryonic development of the butterfly *Junonia coenia* (formerly *Precis coenia*). Segments are labeled by region: head (H1–H6), thorax (T1–T3), and abdomen (A1–A10). Embryos are oriented as indicated by the butterfly (anterior to the left). Adult butterflies look like winged, neckless giraffes with thin, coiled elephant trunks. After refs. [1714,2339, 2340].

In contrast to the lumbering plodding of corpulent caterpillars, butterflies strut like anorexic fashion models. They flit from flower to flower and tiptoe daintily on petals shopping for nectar to sip. They sail the wind on gossamer wings, flaunting their beauty for all to see. Indeed, caterpillars and butterflies act so differently that it is hard to believe they are different life stages of the same individual. In fact, caterpillars do not really walk at all (boldface added):

> The common caterpillar does not walk... **it progresses by movements of its body, not by means of its legs.** It has therefore evolved a type of motor mechanism that propels it forward while the body is close to the support, and **its manner of moving is not that of a worm or a snake**... When the resting caterpillar is about to move, the thoracic legs may first become active and somewhat stretch the anterior part of the body, but they do not bring up the heavy abdomen. Forward movement is initiated by lifting the posterior end of the body, curving it downward and forward, so shifting the anal prolegs anteriorly to a new grasp on the support. Immediately then **the deflected segments contract, straighten, and produce a hump on the back**, while the anal legs maintain their hold, though reversed in position. A wave of successive forward contraction and expansion of the segments now runs rapidly forward through the length of the body. Each segment contracts from the rear and is then expanded anteriorly by the

Figure 3.8 (*cont.*)

a. Transcription of *abd-A* and *Ubx* (bars) at ~20% of embryogenesis. (Lateral expression of *Ubx* RNA in T2 and T3 omitted.) Embryo shading mainly reflects *abd-A* expression.

b. Transcription of *abd-A*, *Ubx*, and *Dll* at 15–20% of embryogenesis. *Ubx* and *abd-A* repress *Dll* in the abdomen. (Dots of *Dll* RNA in A1 and A10 are omitted.) Abbreviations for head segments: lr (labral), an (antennal), mx (maxillary), la (labial). Gray boxes at the right depict circuitry, with arrows denoting activation, T-bars inhibition, and solid versus faded letters signifying active versus suppressed genes (upper box) or events (lower box), respectively.

c. Expression of *abd-A*, *Ubx*, and *Dll* at ~25% of embryogenesis. Holes appear in *abd-A* and *Ubx* expression where prolegs will later develop due to de-repression of *Dll*.

d. Transcription of *Dll* at ~40% of embryogenesis. Abdominal spots of *Dll* RNA appear due to relief from *abd-A* and *Ubx* repression (gray boxes at right), leading to proleg outgrowth.

e. Caterpillar anatomy (generic). Most lepidopterans have ten prolegs, but some have four (e.g., inchworms), and others have none (e.g., yucca moths) [1748]. Prolegs are also found in some dipterans (e.g., watersnipe flies) and hymenopterans (e.g., sawflies) [880,1748]. Prolegs disappear during metamorphosis, but thoracic legs form adult legs [2061,2172,2173]. The six dots on the face are simple eyes. White ovals on T1 and A1–A8 are spiracles. Redrawn from Snodgrass [2083] (his Fig. 5).

f. Rows of hooks (J-shaped white lines) on the sole of each proleg's foot act like Velcro to grip leaves or twigs or silk. Each family of butterflies or moths has its own distinctive pattern [1748]. In the one depicted (family Drepanidae) the lateral row has hooks (crochets) of uniform length, but the medial row has alternating short and long hooks.

g. Transcription of *abd-A*, *Ubx*, and *Dll* in a stage-11 *D. melanogaster* embryo, drawn with the germ band unrolled and stylized to conform to the butterfly diagrams (**a–d**) [791]. At stage 14, an autoregulatory enhancer at the *Dll* locus is activated [635,726,1436]. This enhancer makes *Dll* immune to inhibition by *Ubx* or *abd-A*, allowing growth to proceed even when the T3 leg disc later expresses *Ubx* (gray boxes at right) [355]. Abbreviations as in **b**. After refs. [354,355,1713,1714]. Head segmentation follows convention [278,937,1974], except that the antennal rudiment, whose *Dll* spot abuts the maxillary spot [1713], is canonically in H2, not H4 [1049,1152,1771].

> following contraction of the segment in front. Thus the segments successively extend forward, and finally the head is protruded a short distance. At the same time **the dorsal hump runs forward like the crest of a wave along the back**, lifting successively the prolegs of segments VI–III and carrying them forward to a new position; on reaching the thorax the thoracic legs are carried forward in the same manner. The anal legs then again are brought up for a new hold, and the whole series of events is repeated every few seconds. [2083]

Divergent anatomies such as these result from selection for distinct diets (leaves vs. nectar) in discrete parts of a life cycle [1588,2058] – essentially an ecological tug-of-war on development. What has baffled researchers, however, is how evolution managed to insert a pupal stage between them [880,2232,2477].

Digestive systems cannot be the answer, since they should be relatively easy to revamp in a single molt via new chambers and new enzymes [2189]. Muscle systems are another matter entirely. At some point the larva's locomotory apparatus must have strayed so far from the adult's that a period of immobility (bracketed by two molts) became mandatory [2001] (boldface added):

> The essence of holometabolism is the muscle transformation . . . A more reasonable theory concerning the nature of the pupa . . . holds that the pupa is a preliminary imaginal stage that has been separated from the final adult by an extra moult in order to furnish a new cuticle for the attachment of muscles reconstructed or newly formed in the pupa . . . **The pupal moult is the solution on the part of the insect to the problem of attaching new or reconstructed muscles.** The only evidence against this interpretation of the pupa that might arise would be the discovery in some insect with a pupal stage that no new muscle attachments are formed. [2081]

In this selective milieu, mutations that gave the body enough time to be reconfigured would have been retained from generation to generation. Faced with such powerful opposing forces tearing its life cycle apart, the distant common ancestor of flies and butterflies (and beetles as well) must have stumbled on the trick of hardening its larval skin to make a temporary shelter wherein it could revamp its anatomy in peace and quiet [880].

This ancestor's descendants are now called "holometabolans" in recognition of their total (holo-) metamorphosis. Inside the pupal case the body is reshaped as radically as it was during the embryonic period [1491]. After the task is completed, the ceremony concludes with the winged pharaoh emerging triumphantly from his cuticular sarcophagus. The caterpillar–butterfly duality is a dramatic testament to the amazing pliability of animal genomes [965]. Indeed, there is no greater feat in the world of evo-devo wizardry. Even the eminent morphologist Robert Snodgrass was moved to abandon his famously dry writing style to wax poetic about this phenomenon:

> So well known, in fact, is the apparent transformation of the wormlike caterpillar into the splendid winged butterfly, and so marvelous does it seem, that it has been taken as a symbol of human resurrection. [2083]

> The insect pupa is one of the most remarkable things in animate nature; within it are intimately mingled the processes of both life and death. [2081] (p. 82).

> It would indeed be a wise butterfly that knows its own child, since probably it has no memory of its own youthful life as a caterpillar. Equally certain is it that the caterpillar has no idea that it will ever be a butterfly. [2083]

Part of this reshaping process is the dissolution of the stumpy prolegs, which play no adult role. The evo-devo riddle posed by the prolegs takes us back to their inception. The dilemma concerns *Dll*, the **master gene** for appendage outgrowth throughout the insects [1714,1715]. Insects evolved from a multi-legged ancestor [1049,1283,2082] via *Hox*-mediated suppression of all but six thoracic legs (Fig. 3.9) [721,877,1889]. In modern insects, this suppression is implemented by the *Hox* genes *Ubx*, *abd-A*, and *Abd-B* [756,2255], which inhibit *Dll* in overlapping zones of the abdomen [1401,1934,2467]. Given the antiquity of the six-leg **constraint** (~400 MY), how did caterpillars manage to violate it?

Part of the answer is known. Butterfly embryos turn OFF *Ubx* and *abd-A* (and presumably *Abd-B* [1934,2219]) at all spots where prolegs develop (Fig. 3.8c) [2340]. The removal of these inhibiting factors relinquishes control of *Dll* to the default ground-state circuitry [1211,1588], which automatically turns *Dll* ON (Fig. 3.8d) [2255], whereupon *Dll* proceeds to drive appendage outgrowth.

This explanation seems tidy, but serious issues remain [1049,1588]. Which gene(s) is turning *Ubx* and *abd-A* OFF? Why isn't this gene(s) active in A1–A2 or A7–A9? Why don't the outgrowths become true (jointed) legs [1036]? How does the moth *Manduca sexta* make prolegs without turning *Ubx* or *abd-A* OFF [2508], and how do some hymenopterans (specifically, sawflies) make prolegs without turning *Dll* ON [1675,2154]? How did so many dipterans evolve prolegs independently [1588]? If it is so easy for genomes to make extra *larval* legs, why haven't any insects ever evolved extra *adult* legs [58,223]? And finally, are prolegs **atavisms** or **novelties** [337]? They may be both [1588].

A related riddle concerns the T3 leg of adult insects. Just as *Ubx* allows flies to convert their hindwing into a haltere (Fig. 2.3) [1325,2350], *Ubx* allows insects to specialize their hindleg [216,497,1905]. That is how grasshoppers and crickets got their huge T3 jumping legs [955,1358,1648]. For *Ubx* to do this job it must be expressed in the T3 leg during development, but, if so, then why doesn't *Ubx* inhibit *Dll* and thereby abort leg growth during the embryonic period?

Evolution solved this problem by banishing Ubx from the T3 leg rudiment (Fig. 3.8g) long enough to give *Dll* time to activate an autoregulatory *Dll* enhancer [635,726]. Once that enhancer is firmly engaged, *Ubx* can safely be expressed in the T3 leg, because *Dll* is thenceforth immune to *Ubx*'s interference [355]. This sort of rescheduling trick in evo-devo is called **heterochrony**.

Textbooks portray *Hox* gene expression territories as fixed, but their borders actually change dynamically during development [354]. What upstream agents modulate *Ubx* in time and space so that Ubx skirts the leg disc at just the right time and invades it thereafter? We do not yet know, but we can safely guess that they act through *Ubx*'s *cis*-enhancers. We have made some headway in dissecting the knobs and switches that control *Ubx* in *Drosophila* [1754], but that DNA dashboard may turn out to be as complex as the instrument panel of a spaceship's cockpit. After we finish mapping the *Ubx* enhancers in flies, of course, we still have to trace their evolution among holometabolans to figure out how they allowed prolegs to arise on the abdomen. These tasks rival the Labors of Hercules and will test our stamina for some time to come.

Having surveyed the proleg subfield, albeit briefly, we can draw a few conclusions about how evolution rewires gene circuitry to reshape external anatomy:

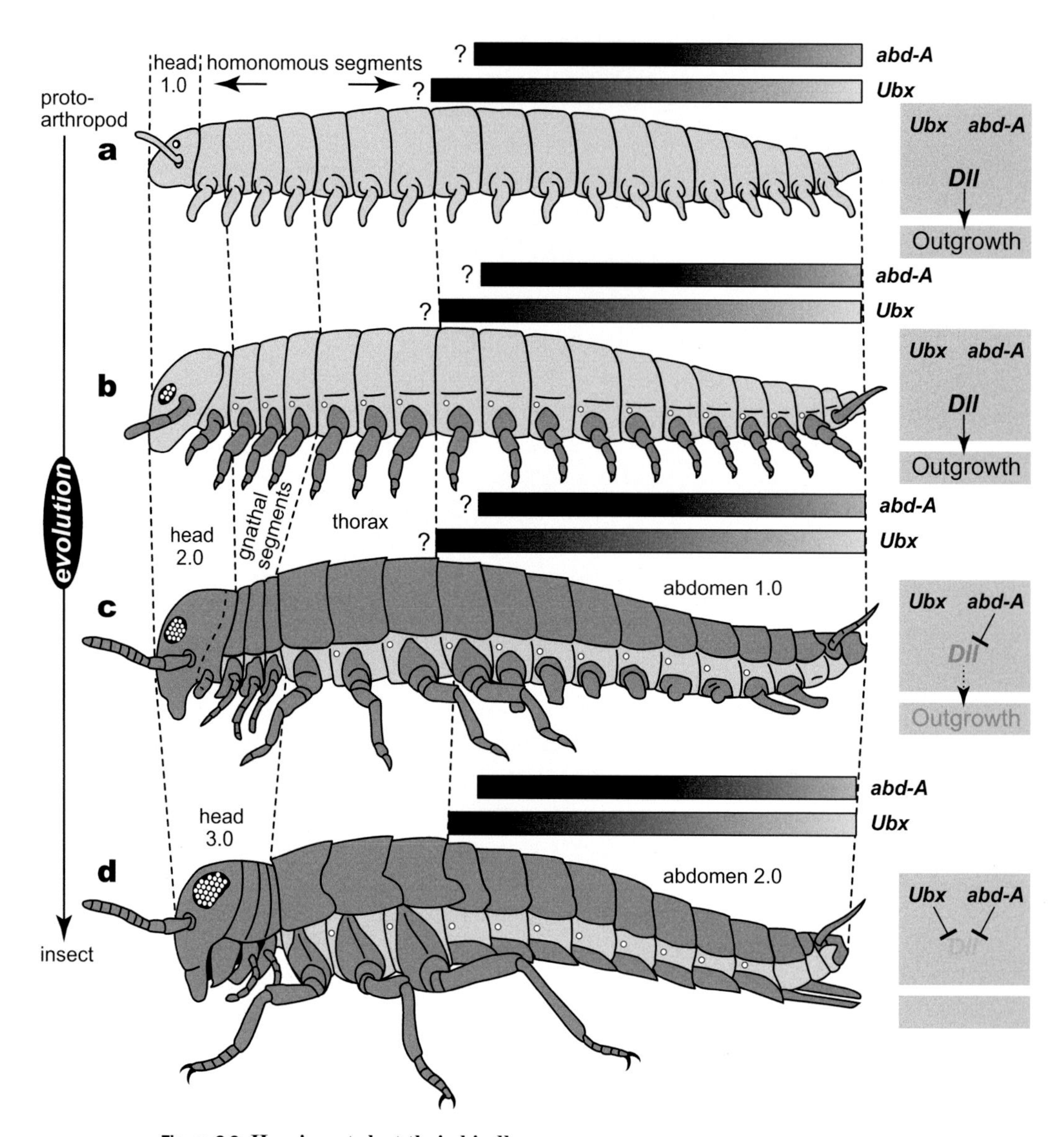

Figure 3.9 How insects lost their hindlegs.

a–d. Hypothetical stages of insect evolution after Snodgrass [2080] (his Fig. 24), but with his annelid stage omitted due to subsequent revisions in bilaterian phylogeny [28]. Successive versions of structures are numbered as if they were upgrades of an operating system (1.0, 2.0, etc.). Light gray denotes soft cuticle; dark gray is sclerotized cuticle. Tiny white circles are spiracles. Shaded bars denote transcription of *abd-A* and *Ubx* [1049]. *Abd-B* (not shown) also plays a role in suppressing legs [1934,2219]. Changes at the head end are due to other *Hox* genes (not shown) [93,1048,1771,1878]. Gray boxes depict conjectural circuitry. As illustrated here, insects may have lost their hindlegs in gradual steps. Spiders [1144] and collembolans [1594,1710] lost their abdominal legs (**convergently**) by different routes.

a. Proto-arthropods are thought to have had fleshy, unjointed legs ("lobopods" [761,2451]) like modern onychophorans [117,681,2082].

Take-home lessons

1. Although *Hox* genes usually rule as dictators, some of them have also become slaves to other genes that turn them ON or OFF in space and time.
2. Evolution sometimes resorts to inhibiting an inhibitor (e.g., turning *Ubx* OFF locally) in order to activate a pathway (e.g., proleg development).
3. Determining default states of structures can be tricky, in part because evolution has concatenated inhibitors in series to undo prior actions.

How insects lost their hindlegs

Insects belong to the phylum Arthropoda. The first arthropods are thought to have resembled caterpillar-like onychophorans [117,281,2082] – commonly called velvet worms [202,681] – insofar as every trunk segment had a pair of stubby, fleshy, unjointed legs or "lobopods" [596,2044]. In 2010 the legs of living onychophorans were shown to manifest the same regional domains of gene expression as insect legs, implying that the "area codes" along the proximal–distal axis of the leg evolved long before any joints arose to subdivide the leg into segments [1088]. This sequence of **individuation** preceding **merism** defies Williston's Law [828,1485,2418], but it may be more common than we think (e.g., heterodonty in mammals [411,1424]).

Unlike insects, where *Ubx* or *abd-A* are expressed over a broad portion of the body, onychophorans only express *Ubx* or *abd-A* at their posterior end [877]. They are "living fossils" that have essentially looked the same for ~300 MY [680,776], so we might be tempted to conclude that their *Hox* expression patterns reflect the primordial condition of the phylum. However, we can't be sure because such patterns can undergo **phenogenetic drift** (fluidity of gene circuitry) over the eons [1049,2307,2363]. Centipedes are nearly as ancient as onychophorans [880], and their segments are just as uniform, but their *Hox* pattern is actually more like that of modern insects [236,1048]. For the sake of simplicity,

Figure 3.9 (*cont.*)

b. Legs became sclerotized and jointed [2044]. Leg (and antenna [791]) segmentation may have occurred by **co-option** of the circuits for body segmentation [1500].

c. Abdominal legs were reduced but not yet lost (e.g., the fossil insects in Figs. 7 and 10 of ref. [1210]), perhaps due to partial repression of *Dll* by *abd-A* [59,791,1710], with *Ubx* merely changing the anatomy of (but not suppressing) the A1 leg [1287]. Also at this stage, the head completed its merger with the first pre-oral segment. Faded genes (*Dll*) or traits (Outgrowth) indicate reduced expression.

d. Generic anatomy of modern insects. In the dipteran–lepidopteran clade, *Ubx* became a repressor of *Dll* [1710], following in the footsteps of *abd-A* [1287,1401,2467]. With *Dll* totally suppressed, abdominal legs vanished from the adult stage [2044] (cf. **atavisms** [1947]). Tarsal segments arose via circuits separate from the leg as a whole [872,1598,1801,2162]. This schematic depicts chewing mouthparts [1974], but other insects eat differently [55,1107,2057]. For example, butterflies sip nectar via a proboscis (Fig. 3.8a) that arises from half tubes which fuse by fitting ridges into grooves (Fig. 169 in [2080]).

proto-arthropods are here assumed to have had an insect-like pattern of *Ubx* and *abd-A* (Fig. 3.9a) – not an onychophoran one.

Investigations of *Dll* expression in various insect orders suggest that the arthropod branch leading to insects lost their hindlegs in two separate stages [1710]. In the first stage (Fig. 3.9c), *abd-A* acquired the ability to partly inhibit *Dll*, thereby reducing the number of functional legs to four pairs, but leaving **vestiges** on the posterior segments. At the same time, *Ubx* insinuated itself into the circuitry of the fourth pair of legs (as it later did in dipteran hindwings; cf. Fig. 2.3) but without any appreciable alteration at this point.

In the second stage (Fig. 3.9d), *Ubx* (like *abd-A* before it) acquired the ability to suppress *Dll* [1287,2255], thereby quashing the fourth pair of legs and erasing any hint of legs on the abdomen. This inferred scenario finds dramatic support in the eight-legged flies that develop when mutations disable the *bithoraxoid* enhancer at the *Ubx* locus [1311]. In 2012 the oldest six-legged insect fossil was dated to ~370 MY ago [749,2023].

In the dipteran–lepidopteran clade, the ability of Ubx to act as a repressor has been traced to a new C-terminal domain [721,1283,1889], though other domains of the Ubx protein are also involved [756,1038,1734]. In flies, both Ubx and Abd-A have been shown to directly bind an enhancer at the *Dll* locus [355,757,2255].

Not only did *abd-A* help (together with *Ubx*) to reduce the number of legs in insects to six, it also played a comparable role in helping (together with *Scr* [1877]) to restrict the number of wings in flying insects to four. Unlike vertebrates, insects did not fashion their wings from a pair of legs (like pterosaurs, birds, and bats). Instead, they acquired them from a different source entirely. That curious tale will now be told.

How insects got their wings

The first insects, ~400 MY ago, lacked wings and so are called apterygotes. Winged insects (pterygotes) appear in the fossil record ~75 MY later [574,2449], but they left few clues as to how their wings arose [1157]. Many clever ideas have been proposed [1208,1211], but only two have been prominent [880,1106]. They are whimsically called the Flying Squirrel Hypothesis (arboreal origin) and the Flying Fish Hypothesis (aquatic origin) [483]. After 135 years of debate [464,574,2401], new evo-devo data finally settled the issue in 2010 in favor of the arboreal scenario [1647].

According to the Paranotal Lobe (Flying Squirrel) Hypothesis, wings began as flat extensions on either side (para-) of the thorax (-notum) [1674]. Paranotal lobes are still found on primitive insects called bristletails [1106]. Bristletails live in trees and can right themselves in midair after jumping, despite a lack of wings [575]. Mutations that extended the reach of such aerial escape maneuvers might have been favored by natural selection [575,597,1158], and the selection pressure could have nudged bristletail-like insects gradually from jumping to gliding [939] and then from gliding to true flying (wing flapping) [2449] – all the while upgrading their lateral lobes stepwise to fully functional wings [254].

> The palaeontological evidence, slender as it is, seems to favour the existence in the Devonian of small – c. 10 mm? – insects, whose initially protective thoracic paranota became enlarged in association with their value first as parachutes in delaying descent, next as gliding surfaces, then as steering vanes, as they developed the ability slightly to pronate, supinate, elevate and depress the pads by the action of the pleural leg muscles, and ultimately, in the case of the meso and metathoracic lobes, as flapping aerofoils, powered mainly by bifunctional leg muscles including those operating in Blattodea and perhaps supplemented by others now retained only in Odonata. [2449]

In contrast, the Gill (Flying Fish) Hypothesis has contended that wings began as leaf-like gills or gill plates that grew out from the bases of legs [2400,2401]. The larvae of mayflies, which are primitive pterygotes, exhibit such gills on their thoracic and abdominal segments [880,1106]. Because mayflies can move their gills, it is easy to see how gill muscles might have been used to move gill-derived proto-wings when insects came onto land [1210,1381,1382].

In 1997 ambiguous evo-devo evidence was adduced to support the Gill Hypothesis. The article made the cover of *Nature* with the headline "From gill flaps to insect wings" [94]. It reported that the genes *apterous* (*ap*) and *nubbin* (*nub*) are expressed in crustacean epipods (dorsal leg branches) just as they are in insect wings. Epipods are homologous to gills [485], so the implication was that gills (or gill covers) became wings.

Unfortunately, neither of these genes is an ideal marker for wing identity [1106,1474] because both *nub* and *ap* are also expressed in insect legs [1297,2243]. At the time, the authors dismissed this concern by stressing that *nub*'s ubiquitous expression in epipods matches its ubiquitous pattern in wings but not in legs, where its expression is annular. They made the same sort of argument for *ap*, although *ap*'s ubiquitous expression in epipods failed to match its pattern in wings, where it is only transcribed in the dorsal half [1106].

In 2010 any inferences that might have been drawn from the 1997 *Nature* paper were swept away by strong new evidence in favor of the Paranotal Lobe Hypothesis (Fig. 3.10) [575,1106]. Paranotal lobes of bristletails were shown to express not only *ap*, but also two other genes whose overlap is a much more reliable hallmark of insect wings than *nub* [1149,1618]: *wingless* (*wg*) and *vestigial* (*vg*) [1647]. Additional support came from further facts:

1. Unlike crustacean epipods, where *ap* is expressed uniformly, paranotal lobes only express *ap* in their dorsal half – exactly like wings.
2. The boundary of the *ap*-ON domain is sharp and coincides with the edge of the lobe where it folds back on itself – as is true for the wing margin.
3. Expression of both *wg* and *vg* appears to straddle the *ap*-ON/*ap*-OFF line, with *wg* being expressed in a subset of the *vg*-ON territory – as in wings.

The most convincing evidence of all was the startling discovery that cells along the *ap*-ON/*ap*-OFF boundary are aligned in a nearly perfect row. In *Drosophila*, such alignment is only found at the future wing margin (dorsal/ventral border) [40].

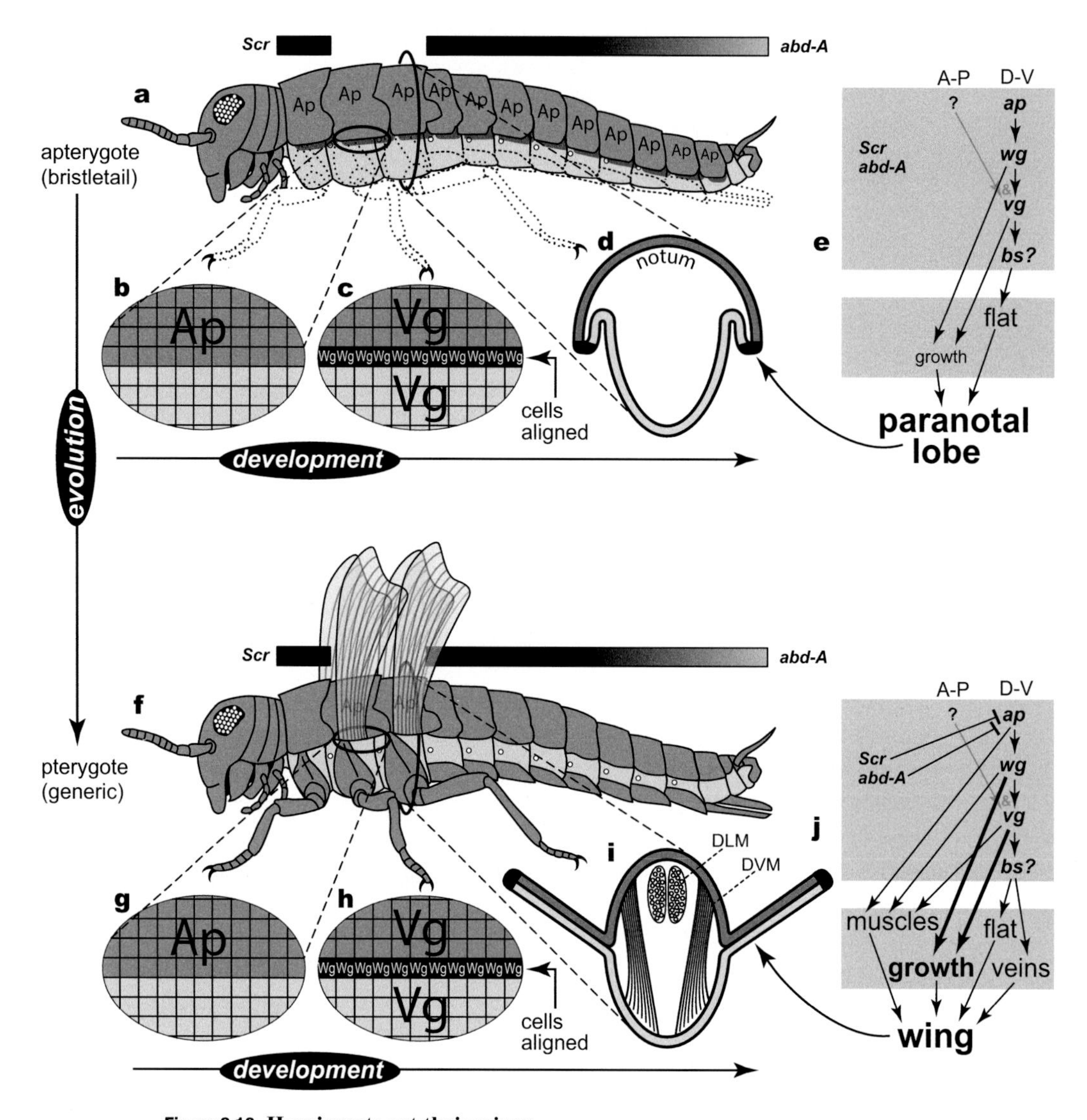

Figure 3.10 How insects got their wings.

a–e. Development of bristletails, which belong to the most primitive order of wingless (apterygote) insects [880]. Adapted from ref. [1647]. Gene abbreviations: *abd-A* (*abdominal-A*), *ap (apterous)*, *bs* (*blistered*), *Scr* (*Sex combs reduced*), *wg* (*wingless*), *vg* (*vestigial*).

- **a.** Generic insect (from Fig. 3.9d) with paranotal lobes (ventral outgrowths of dorsal sclerites with shadows shaded below to indicate an overhang) added onto thoracic and abdominal segments in accord with Carboniferous fossils (Figs. 7 and 10 in [1210]). For clarity, an adult is depicted instead of an embryo. Legs and ventral sclerites are outlined with dots. Ap is expressed in precursors of all dorsal sclerites. In leg-bearing segments the Ap area extends to a ventral line that abuts the future coxa.
- **b.** Enlarged view of the *ap*-ON/*ap*-OFF (dark/light shading) line in the second thoracic segment. Squares represent cells, but cell size and alignment are grossly exaggerated except for the alignment of one unusual row (see **c**).

In summary, bristletail lobes display an impressive suite of genetic and cellular features that are patently wing-like. These features are consistent with the notion that insect wings evolved from such lobes [575,1106].

One objection to the Paranotal Lobe Hypothesis has been that these lobes lack muscles. This absence raises the question of how proto-pterygotes moved their wings [1210,1381]. The usual rebuttal (as in the quote above) has been that flapping could have been mediated by bifunctional leg muscles that later handed off this duty to derivative wing muscles that were dedicated to this task alone [575].

Recruitment of muscles for a new (flapping) function may have been easier than we might think, because the same genes that govern paranotal lobes also elicit wing muscles in flies: direct flight muscles rely on a *wg* paralog (*D-Wnt2*) [1198], while indirect flight

Figure 3.10 (*cont.*)

c. The *wg* gene is activated narrowly at the *ap*-ON/*ap*-OFF interface, and *vg* is activated broadly around the *wg*-ON stripe. A single row of neatly aligned cells is seen at or near the *ap*-ON/*ap*-OFF boundary – assumed here to coincide with the *wg*-ON stripe (arrow). For ease of illustration, protein names (Ap, Wg, and Vg) designate areas where the genes *ap*, *wg*, and *vg* are transcribed.

d. Cross section of the third thoracic segment. Paranotal lobes are lateral outgrowths of the dorsal sclerite that call to mind the overhang of a turtle's shell. Shading conforms to expression zones in **c**. Wing development in hemimetabolous insects begins with this same cross-sectional profile [1106].

e. Probable circuitry for lobe development, based on expression data in bristletails [1647] and known circuits in flies [965]. Lobe outgrowth (albeit mild) is assumed to be driven by *wg* and *vg* [1149,1618], while lobe flatness ("flat") might be caused by *bs*, which is known to "glue" upper and lower surfaces of insect wings together [508,1873].

f–j. Development of a generic pterygote insect, showing how wings are thought to have evolved from paranotal lobes. Abbreviations as in **a–e**. Adapted from refs. [345,965].

f. Wings on the first thoracic segment are repressed by *Scr* (and can be restored by disabling *Scr*) [345,380,1035,1674,1877], while wings on the abdomen are blocked by *abd-A* [345,1674]. Archetypal vein pattern is unknown [508,1208]. The pattern here is from grasshopper nymphs [1106].

g. Same as **b**.

h. Same as **c**.

i. Schematic cross section of the third thoracic segment of a neopteran insect (e.g., butterfly), containing flight muscles that flap the wings indirectly by deforming the thorax [544,658]; flies have halteres instead of wings on T3 (cf. Fig. 2.1). Dorsoventral muscles (DVM) contract the thorax vertically to erect the wings, while dorsal longitudinal muscles (DLM) contract it horizontally to depress them; paleopteran insects such as dragonflies use direct flight muscles instead [254,574,1786]. Shading conforms to expression zones in **h**. Redrawn from refs. [1785,2080].

j. Circuitry for pterygote wing development as documented in flies (abridged from Fig. 2.2) but amended (1) to replace *dpp* with "?" because a different gene may specify the A–P axis in other insects [343,2220] and (2) to include *Scr* and *abd-A* (see **f**) [1674]. New links (not shown) connecting *wg* and *vg* to growth genes may have caused wings to grow larger (thick arrows) [104,1319]. "Flat" indicates flattening of the wing blade due to apposition and adhesion of the two halves of the wing pouch [149]. Veins probably arose by repression of *bs* at regional boundaries in the wing epithelium [508]. For explanation of links to muscles, see text.

muscles use *ap* [176,771] and *vg* [175,2141]. Moreover, the sites where muscles attach to the cuticle come from the cuticle-making cells themselves [772] and not from an outside source that might have required extra rigging. At some point flight muscles evolved novel contraction abilities via a paralog of the wing gene *spalt* [1903,1975], and neural circuits for operating these muscles were cobbled, at least in part, from existing pathways [1646,1869]. Thus, this objection to the paranotal theory is now seen as weaker than it once seemed.

If paranotal lobes were originally present on all thoracic and abdominal segments as in bristletails [1674], then how did pterygotes end up with only two pairs of wings instead of a dozen or so? As mentioned above, abdominal wings were suppressed by the same *Hox* gene (*abd-A*) that blocked abdominal legs to create insects in the first place (Fig. 3.9) [345,1674]. Wings were deleted on the first thoracic segment (T1) by *Sex combs reduced* (*Scr*) [1674,1877], which is the *Hox* gene in charge of that segment (Fig. 2.5) [380,1035].

Fossils prove that these changes did not occur all at once [345]. Conversion of paranotal lobes to wings must have preceded the denuding of the abdomen (by *abd-A*), because paleozoic pterygotes had winglets on their abdominal segments [337,1208,1209]. Also, removal of wings from the abdomen (by *abd-A*) must have preceded the deletion of wings on the first thoracic (T1) segment (by *Scr*) [1513], because some of the earliest flying insects (320 MY old) had a third pair of wings (albeit smaller) on T1 [880,1208,2452].

The notion of six-winged dragonflies would seem absurd were it not for the petrified corpses of such "impossible" insects glaring back at us from the Paleozoic. If they could talk, these fossils would mock our intuitions about what is – or is not – possible in evolution [1211]. Of course, it is equally hard for us to imagine four-legged snakes, but that is a story for the next chapter.

4 The snake

How the snake lost its legs

Vestiges are the evolutionary equivalents of fingerprints at a crime scene. They offer useful clues to the unseen past [1723]. Pythons are a case in point. They have tiny remnants of a pelvis and femur under their skin near the cloaca [1009,1827], and fossil snakes have even more complete legs (pelvis, femur, tibia, ankle, and foot [1029,2181]). These relics demonstrate that the ancestors of snakes had legs.

For other types of vestiges, it has proven possible to rescue their potential by artificially replacing the ingredients that have been lost over time [1502,1778]. However, attempts to re-create walking snakes by restoring missing **morphogens** (FGFs) to python hindlimb buds have thus far been only marginally successful [337,421,450].

Why did snakes lose their legs? The process apparently began when a lizard-like ancestor adopted a burrowing lifestyle [160,642,2395] like the one that is still pursued by primitive snakes today (Fig. 4.1) [2032,2100,2395] (boldface added):

> To nearly everybody it seems perfectly natural to assume that snakes have always looked much like the slender, tubular animals they are today, although in fact the members of this most recently evolved of the reptile groups achieved their current form only after an elaborate physical restructuring. For nearly 60 million years **their predecessors ran around as four-legged terrestrial animals** similar to contemporary monitor lizards, until, early in the Cretaceous Period, they gradually abandoned their limbs in favor of an attenuated physical form suited to a radically different way of life...
>
> In the darkness of their subterranean burrows, the [proto-snakes] came to depend mostly on smell and taste – senses combined in the specialized forked tongues they had already evolved... Since wriggling worked better than walking in narrow underground tunnels, **their limbs shrank into fetuslike stubs**, and as their bodies gradually extruded into the serpentine form of modern snakes, little by little they ceased to be lizards at all. [2187]

If this argument is correct, then how do we explain mammalian moles, which also burrow but still have their legs? The answer is that moles still use their legs for digging, whereas proto-snakes apparently did not [579]. The applicable rule here is "**Use it or lose it**" [538,1022]: structures that are not used wither away after a sufficiently long period of time – for example, the wings of kiwis [787,1028,1392] and the swim bladders of benthic fish [1429]. How long? Unused genes decay beyond repair after ~10 MY [1390], and the same may be true for region-specific *cis*-enhancers even if the genes themselves remain

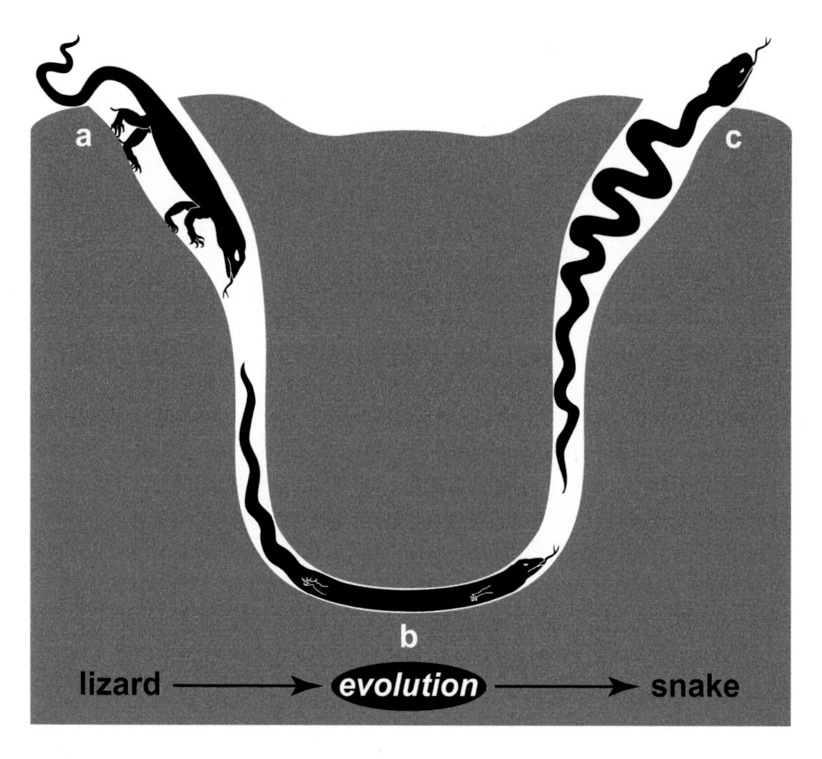

Figure 4.1 How the snake lost its legs.

a–c. In what may be evolution's greatest Procrustean feat, a lizard-like animal was squeezed (by selection pressure) into a cylindrical shape to fit a "subway system" [2187]. In this oversimplified schematic, an ancestral proto-snake enters one end of a burrow and emerges from the other end as a snake – thus compressing a drama that took millions of years into a single frame.

a. Presumptive ancestor of snakes. Snakes appear to have evolved from varanoid lizards [1264,1407,2187]. The Bornean earless monitor, *Lanthanotus borneensis*, is used here as a proxy for the progenitor, which is thought to have resembled it in various respects, including a moderately forked tongue (Kurt Schwenk, personal communication). Proto-snakes started living underground to some extent before ~125 MY ago [2187,2395]. Initially, they presumably used burrows dug by other animals [873].

b. Imaginary transitional form, based in part upon a "missing link" fossil species described in 2012 [1326]. Given the clues accrued from fossils, anatomy, and phylogeny (see text), proto-snakes are thought to have adapted to tunnels in various ways: (1) their eyelids fused, (2) their legs withered, (3) their body elongated, (4) their left lung shrank, and (5) their belly scales widened [873,1407,2100]. Essentially they became slithering stomachs – one of the most efficient eating machines on earth [1759]. The modes of locomotion best suited to tunnels are the rectilinear or concertina, but snakes also move by undulation or sidewinding [1779]. Did *all* these traits arise as fossorial adaptations? Perhaps not, but the similarity of snakes to burrowing ecomorphs suggests so [2394,2395].

c. Revamped snake returning to a life above ground ~70 MY ago [1407,2187]. From here on, natural selection would have rewarded any mutations that revived what was left of the snake's visual system (cf. Fig. 4.6). Remnants of hindlegs persist in boid snakes as **vestiges** [1009,1827], but the spurs play a role in mating, so they are not useless [2394].

intact due to their continued usage in other parts of the body. Snakes spent much of their lives underground during an exile lasting ~50 MY [2395].

A legless fate also befell other animals that burrow using only their snout [990,1431], including amphisbaenians (distant relatives of snakes) [1131,1564] and caecilians (worm-like amphibians) [1095]. Consistent with this logic, the Mexican worm lizard lost its hindlegs but kept the forelegs that it still uses for digging [89,2389].

Ironically, the best illustration of limb loss from disuse can be seen in the scariest carnivore of all time. *Tyrannosaurus rex* had ridiculously tiny arms with only two fingers on each hand [1056,1057]. Its arms presumably withered because its muzzle alone was used for capturing prey [2133].

Protruding legs on a burrowing animal would be more than a nuisance [1130,1887]. They could be a lethal liability. In a tight space, they might cause their bearer to become stuck and die. Hence, the rate at which snakes lost their legs may have been greater than expected from disuse alone. Under such conditions, natural selection would not only have failed to purge corrosive mutations from leg-making genes [686,816,817]. It would also have favored any alleles that made the limbs smaller and smaller until there was virtually nothing left of them [2130].

Amazingly, the entire spectrum of leg loss, from mild reduction to total elimination, can be seen in the gamut of lizard species living today (Fig. 4.2; see also frontispiece) [2395]. It is tempting to think of the intermediate species as "transitional" – i.e., on a path to total leglessness. However, tiny legs may serve subtle functions that forestall a **use it or lose it** demise [2394]. For example, the hindlimbs of pythons probably persist as **vestiges** because at some point they were **co-opted** for a new role as "tickling" spurs during copulation [2394].

"Legless" lizards are often mistaken for snakes not only because they lack external legs but also because their bodies are extremely elongated [89,2389]. Most of them are burrowers, but some live above ground [2032,2130,2392]. The surface dwellers have longer tails [232] and move mainly by lateral undulation [740,2398]. Because legs impede slithering through dense grass as much as moving along tunnels [2016], natural selection could have suppressed legs in that context as well [1180,2395].

The sinusoidal undulations of snakes are reminiscent of fish [889,1366] – especially eels [1454,2330]. Is this resemblance merely coincidental? Probably not. Even after fins became legs, the fish-like habit of bending the body back and forth persisted in salamanders [1887], and it is still noticeable in the slow gait of lizards [174]. Conceivably, snakes are relying on this same neural circuitry as they "swim" on dry land [291] or through the water [243]). In that case, their ancestors would have been **preadapted** to slithering along the ground without needing their legs for propulsion at all [408,1040].

Some authors have argued that snakes were swimmers from the very start [305,1265,1952], but the preponderance of evidence favors a burrowing origin more than an aquatic one [874,2492]. The fossorial theory gets most of its support from fossils [61,1981,2181] – for example, a "missing-link" from the Late Cretaceous (~70 MY ago), which is the earliest snake fossil yet found [1326]. Additional evidence comes from phylogenetics [2286], but the most convincing clues can be found in the eyes of living

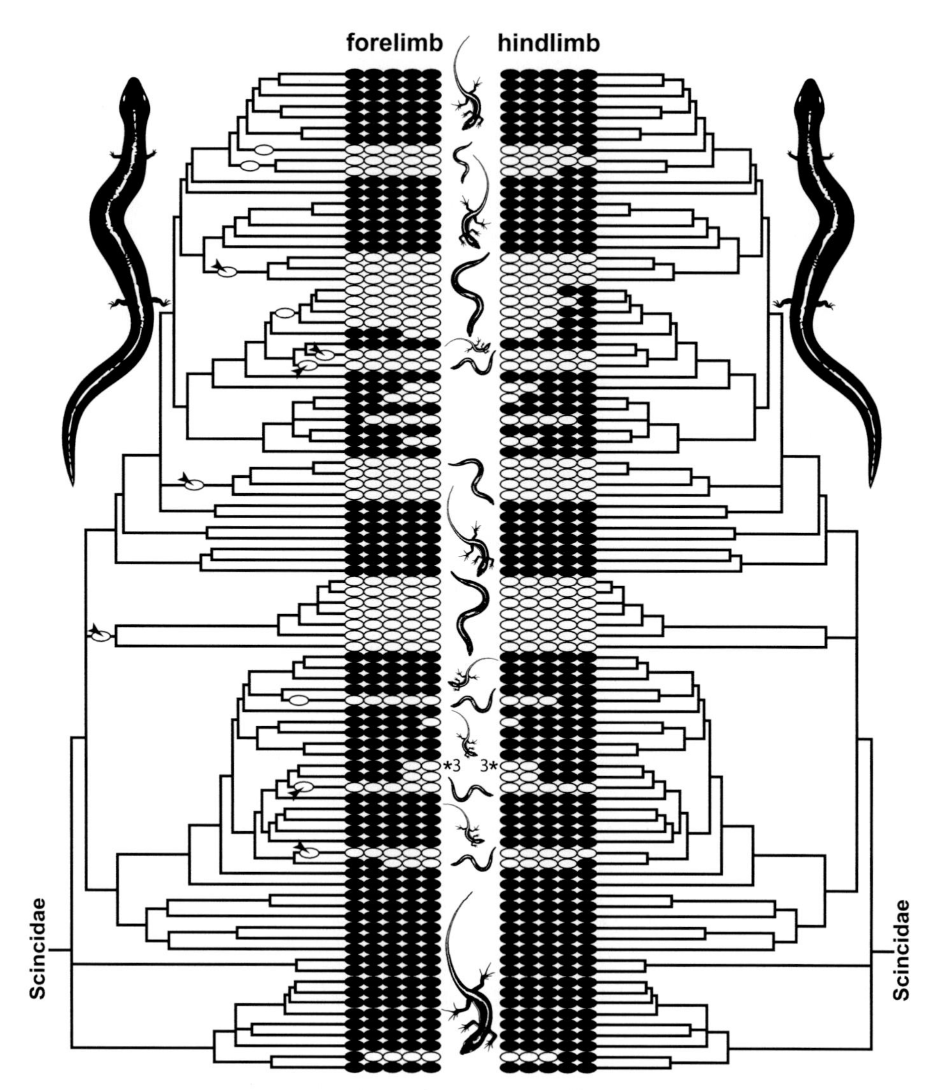

Figure 4.2 **How some lizards lost their legs**. Molecular phylogeny for a sample of 93 species in the family Scincidae (skinks). This tree, which is drawn twice as facing mirror images, is a subset of the Squamata supertree in ref. [232]. For each species (branch tip) the number of toes on the forelimb (left) or hindlimb (right) is given by the number of filled ovals (0–5), with five unfilled ovals denoting absence of an external leg. For each array, ovals are filled from the outside in, without regard to which specific digits are missing, because digit identities can suffer "frame shifts" in *Hox* gene expression [2487]. Lizards in the center column indicate blocks of species with all five toes per foot (ordinary skink) or none (legless skink). In the corners are sketches of the three-toed burrowing skink, *Saiphos equalis*, whose twig on the tree is marked by *3 (cf. the three-toed earless skink, *Hemiergis decresiensis*, in the Frontispiece, panel b). Unfilled ovals were inserted into the left tree wherever ancestors are inferred to have lost their forelegs, and arrowheads denote the subset that lost both pairs of legs. The seven instances of complete leg loss in this sample are an underestimate for the family as a whole because some limb-reduced genera were omitted [2395]. Names of the other 92 species can be found in Fig. 5 of Brandley *et al.* [232]. Matt Brandley approved this diagram prior to publication. N.B.: Lizards with vestigial legs are not necessarily destined for total leg loss, because tiny legs may have subtle roles [2394].

snakes themselves [873]. As explained below, the snake's eye harbors unmistakable signs of having decayed in a sunless habitat.

Even though "legless" lizards resemble snakes externally [2395], they retain bony vestiges of forelegs and hindlegs under their skin [2130,2398]. Snakes, in contrast, have no traces of forelegs whatsoever [160,1009,2236] except for one residual shoulder muscle [2235]. This difference between the two groups might merely reflect the longer time that snakes have had for mutations to erode their leg-making genes or the *cis*-enhancers thereof (~125 MY vs. only 4–80 MY for legless lizards, depending on lineage [2070,2395]). Indeed, at least one such gene (*Hoxa13*) does show more amino acid variation in snakes than in legless lizards [1177], suggesting a longer period of relaxed selection.

However, another possibility is suggested by the fact that no forelimb traces are found in snake fossils either: snake forelegs may have been erased by a faster process than the one that led to the legless lizards [2391,2398]. This Sudden-loss Scenario was proposed in 1999 when *Hoxc6* and *Hoxc8* were shown to be expressed all the way up to the head in pythons [421]. In other vertebrates (e.g., mouse and chicken), *Hoxc6* and *Hoxc8* are associated with rib formation, and the anterior limits of their expression zones coincide with the neck/thorax boundary [286,337,753]. The explanation for foreleg loss in snakes now seemed obvious [344,784]: mutations in proto-snakes must have shifted *Hox* expression anteriorly – putting ribs on neck vertebrae and simultaneously erasing the forelegs [450,865].

Consistent with the Sudden-loss Scenario, an episode of drastic genetic change did occur near the dawn of the squamate clade (lizards and snakes [1803,1941,2397]), when *Hox* clusters were bombarded by a massive invasion of transposons [537]. The ensuing rearrangements of regulatory elements could have been (1) chaotic enough to radically alter *Hox* gene expression zones (shifting them anteriorly) [2313] and (2) sporadic enough to have occurred much later than the invasion itself (with effects being mostly confined to the snake branch alone) [1831].

Biologists have historically been suspicious of large-effect mutations as agents of evolution – the **hopeful monster** fallacy [700,895] – so this proposal has suffered its share of understandable skepticism for that reason [1942,2398]. Of more concern are recent results that challenge the inferred trend that suggested the Sudden-loss Scenario in the first place [2444].

A 2009 study revealed that corn snakes differ from pythons insofar as they do not express *Hoxc6* and *Hoxc8* all the way to the skull [2445]. Instead, virtually all the *Hox* genes that were examined had anterior limits that matched those of lizards and mice (Fig. 4.3). Most surprising was that *Hoxc10* is transcribed in a rib-bearing part of the snake spine: *Hox10* genes stifle ribs in mice to enforce a ribless lumbar identity [325,1369,2366]. Why not here? The answer is that snakes have a mutation upstream of the rib-promoting target gene *Myf5* that prevents its *cis*-enhancer from binding Hox10 proteins [886].

In 2012 a separate team of researchers analyzed the muscles behind the snake head and found that snakes still have quite a respectable neck despite the presence of ribs there [2236]. Cervical ribs are widely assumed to be reliable indicators of *Hox*-identity,

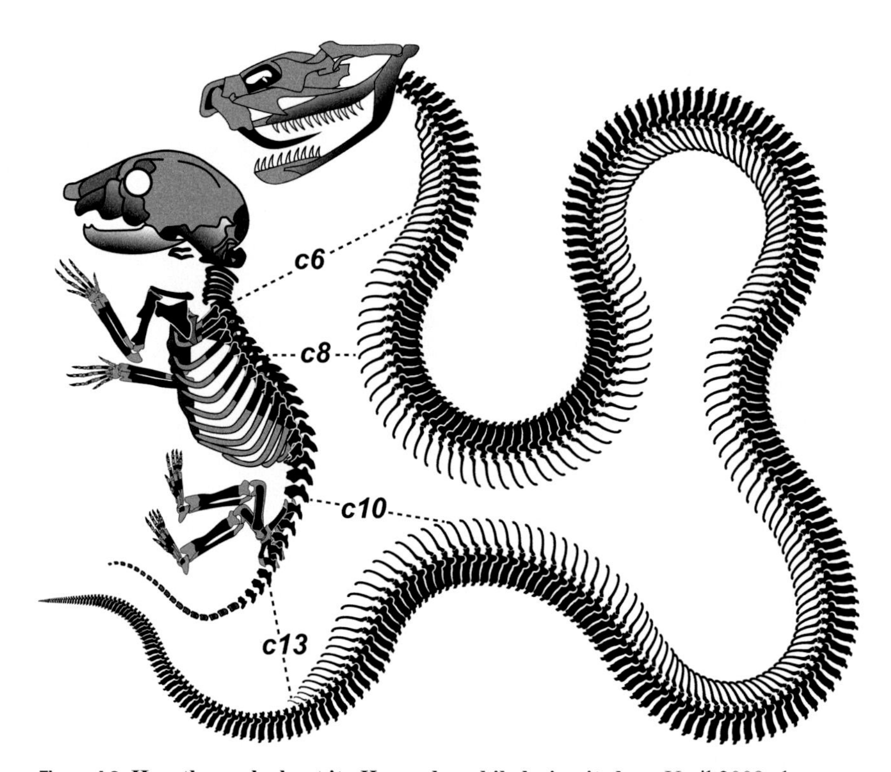

Figure 4.3 How the snake kept its *Hox* codes while losing its legs. Until 2009, the most popular theory for how snakes lost their legs was that *Hox* codes had shifted along the body column as in pythons [421]. However, in 2009 corn snakes were found to resemble mice in the seriation of their codes [2445] – i.e., no shifting. Shown here is a sample of those data. Anterior limits of transcription are marked for four *Hox* genes on the skeletons of a mouse embryo (same stage as Fig. 4.4) and a snake adult. Equivalence of *Hoxc13* boundaries is not surprising because in both species *Hox13* genes govern tail vertebrae [588]. Reasonable explanations also exist for *Hoxc6* and *Hoxc8*, but the retention of *Hoxc10* expression in snakes was startling because *Hox10* genes suppress ribs in mice [325,1369,2366]. This mystery was solved in 2013: snakes have a single base change (relative to other vertebrates) in the Hox10 binding site of a *cis*-enhancer at the *Myf5* gene locus from C<u>T</u>AATTG to C<u>C</u>AATTG, which prevents Hox10 from inhibiting ribs in the lumbar region [886]. The impotence of other *Hox* genes may also have target-gene explanations [1756], but *Hox* genes elsewhere have been shown to be blocked at a translation step [1038,1603]. Redrawn from ref. [2445], with mouse details from ref. [633]. Ribs were counted from the front (*c6*, *c8*) or rear (*c13*, *c10*) to match the original diagram [2445], but the number of intervening vertebrae is approximate, and shapes are highly stylized, as is the snake's posture. For actual shapes of snake vertebrae, see ref. [1009].

but they are so common in lizards that we should mistrust them in the absence of corroborating evidence [304,1009]. These findings force us to question whether *Hox* code homogenization was responsible for causing either cervical ribs or missing legs.

The chief difficulty in assessing the role of *Hox* genes in leg loss – or, for that matter, the role of any gene in any process – is that correlation does not prove causation. In the present situation, this means that the correlation of *Hox* codes with limb sites [286] does not prove the involvement of *Hox* codes in limb initiation [570,1942,2444]. One of the few causal links that *has* been proven involves *Hoxb5*: disabling *Hoxb5* repositions the shoulder girdle in mice [1818].

Another confounding issue is that *Hox* expression in the lateral plate (where limbs arise) is more malleable than in the paraxial mesoderm (where vertebrae arise), and corresponding boundaries can shift relative to one another during development [420,1942]. If the same slippage happens in evolution, then snakes could have lost their forelimbs in the same languid way as legless lizards [232], with *Hox* domains sliding up or down the body column at a later time for unrelated reasons [1942]. That is, the Sudden-loss Scenario may be wrong.

The best way to test evo-devo theories in general is to try to re-create an evolutionary process in the laboratory by artificially manipulating genes [1073]. The most sensational example of this approach is Jack Horner's quixotic effort to turn a chicken into a tyrannosaurus [945] by restoring its teeth [398,1502,2066], etc.

For the task at hand, we should ideally try fiddling with the genes of a lizard to turn it into a snake. Such a study was conducted in 2010, but mice were used as the starting point instead because mouse genetics is so far ahead of lizard genetics. Researchers activated the *Hoxb6* gene in the paraxial mesoderm [2291], and the resulting mice had ribs on nearly every vertebra, so their skeletons looked quite snake-like (Fig. 4.4) [1093]. Unfortunately, the transgene that was used did not force *Hoxb6* to be expressed in the lateral plate, so we don't yet know whether limbs can be erased by the kinds of *Hox* domain shifts that are envisioned by the Sudden-loss Scenario (Fig. 4.4c).

How the snake elongated its body

The number of vertebrae in snakes varies from ~150 to 400 [160], depending on the species [288,950,1769], compared with only ~60 in the mouse [1369] and only 33 in humans [1971]. How did snakes acquire such incredibly long spines?

> The elongated, snake-like skeleton, as it has **convergently** evolved in numerous reptilian and amphibian lineages, is from a developmental biologist's point of view amongst the most fascinating anatomical peculiarities in the animal kingdom. [2444]

In 2008 we learned the answer: snakes create more vertebrae mainly by using a faster oscillator in their somite-making machinery [805,2298]. Somites are the precursors of vertebrae [1142]. They arise as blocks of tissue on either side of the neural tube [535,1368]. Each pair is sliced from a slab of presomitic mesoderm (PSM: Fig. 4.5) when a traveling wavefront encounters cells in a certain phase of the oscillation cycle [1212,1220].

Somite number is dictated by the speed of the wave and the rhythm of the clock [1668]. The oscillator runs ~4 times faster in snakes than in mice [805,806], so snakes chop their mesoderm into somites like a frenetic butcher who is high on cocaine. This faster chopping rate makes the somites thinner initially, but they subsequently grow to become normal-shaped vertebrae [1009], thus lengthening the spine.

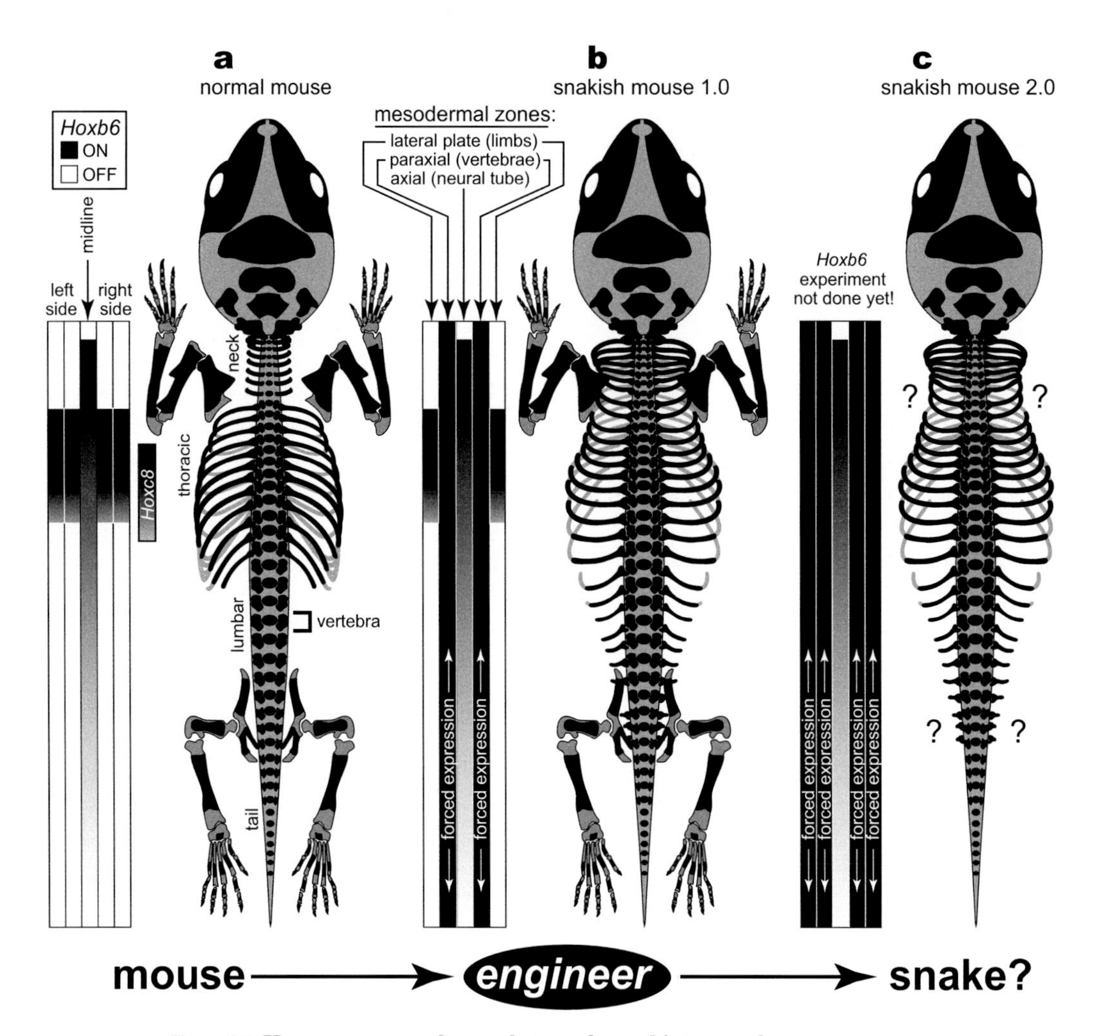

Figure 4.4 How a mouse can be partly transformed into a snake.

a. Normal (wild-type) mouse fetus [2291]. Skeleton only. Black denotes bone (clavicle omitted for clarity); gray is cartilage. Mice have ~60 vertebrae of five types [1368,1369]: 7 cervical, 13 thoracic, 6 lumbar, 4 sacral, and ~30 tail vertebrae. At this stage (E18.5) the tail is still forming. Schematic at left: neural (center) and mesodermal (flanking) areas where *Hoxb6* is assumed to be transcribed, given that its *Hoxc6* paralog is expressed in (1) somites (precursors of vertebrae) 9–17 (T2–T10) [1447] or 9–16 (T2–T9) [1093,1678], (2) the forelimb region of the lateral plate [286,570,1679], and (3) the neural tube up to the first cervical vertebra [1093,1094,2030]. For comparison, the somitic expression of *Hoxc8* is shown; it promotes rib formation like *Hoxc6* [720,1447]. Shading indicates degrees of expression. Mouse skeletons are lizard-like, with about as many vertebrae [2444], but no mammal ever evolved a snake-like spine [1431]. Why not? After all, fish, amphibians, and reptiles did so often [2444]. Put another way, why has this niche in mammalian **morphospace** stayed empty? Perhaps it is a physical **constraint**: mammals can't maintain a constant body temperature at such a high ratio of surface area to volume (M. Hutchinson, personal communication). Redrawn from ref. [2291].

The wavefront is driven by gradients of Wnt, FGF, and retinoic acid [529,1780], while the oscillator uses genes from the Wnt, FGF, and Notch pathways [1203,1668]. Similar circuits are used in different species [587,1668,1874]. We do not yet know which circuit elements were altered by the ancestors of snakes to speed up their clock [805,2298].

One candidate is the *Hes7* gene [1220], since *Hes7* transcription is rate-limiting for oscillation [181,929,1289]: removing its introns disrupts the timing of somite formation [2166] and causes spinal defects (e.g., vertebral fusions). *Hes7* encodes a Hairy-like repressor, which acts downstream of Notch [182,1220].

Additional tricks, aside from a faster oscillator, are also used by snakes to make more vertebrae, especially in their tails [2444]. Somite production normally wanes when the growing tail bud cannot replenish the PSM (at its posterior) fast enough to keep up with the diversion of tissue into new somites (at its anterior). Snakes boost the proliferative capacity of the tail bud by lifting the regulatory brakes (retinoic acid and genes in the *Hox13* paralogy group) that would otherwise retard it.

Snakes are not the only vertebrates to have "gamed the system" to generate huge numbers of vertebrae. Marine reptiles did also [1549]. Figuring out the tricks they used to do so might help us unravel developmental **constraints** in general.

Ironically, the biggest snakes that have ever lived (~13 m long; ~1000 kg) became giants without fiddling any further with timing rates [949]: their numbers of vertebrae are within the range of smaller snakes in their clades. The vertebrae themselves are simply enlarged [950,1140].

The number of vertebrae can vary widely within any given snake species [950,1203], but some species are a lot more variable than others [806,1979,2444]. We would like to know whether such fluctuations are due to an inherent imprecision in the clock–wavefront device [92] or to niche-dependent, laissez-faire oversight by natural selection. Snake genomics may soon provide the answer.

Species that evolve long thin bodies must adjust their internal organs accordingly, and snakes have mainly done so by stretching their organs into thin cylinders. The lungs went one step further: the left lung is aborted during the development of most snakes [1407]. Their distant burrowing relatives – the amphisbaenians – forfeited the right lung instead [739]. Asymmetrically missing organs are rare in nature, though visceral asymmetry per se is nearly universal among vertebrates [1707].

Figure 4.4 (*cont.*)

b. Transgenic mouse fetus (same stage as **a**) where *Hoxb6* was artificially expressed in presomitic (paraxial) mesoderm [886,2291]. Ribs or rib-like growths (cf. [1093,1435]) develop on vertebrae posterior to the thorax. This same phenotype can be created with the *Hoxb6* paralog *Hoxc6* [1093] or by disabling Gdf11 [720,1447] – a TGFβ-class ligand [52].

c. Hypothetical anatomy that might result if *Hoxb6* were expressed in both paraxial mesoderm (which makes vertebrae) and lateral plate (which makes limbs [420,1656,1916]). Forelimbs might be lost because they seem to rely on the anterior boundary of *Hox6* gene expression [286,337,753] (but see ref. [2135]), and hindlimbs might be lost because ubiquitous *Hoxb6* could overpower the *Hox* cues for hindlimb initiation [420,421,2291]. To more fully mimic a snake, we would have to also engineer more vertebrae, but we do yet not understand the oscillator well enough to do that (see Fig. 4.5).

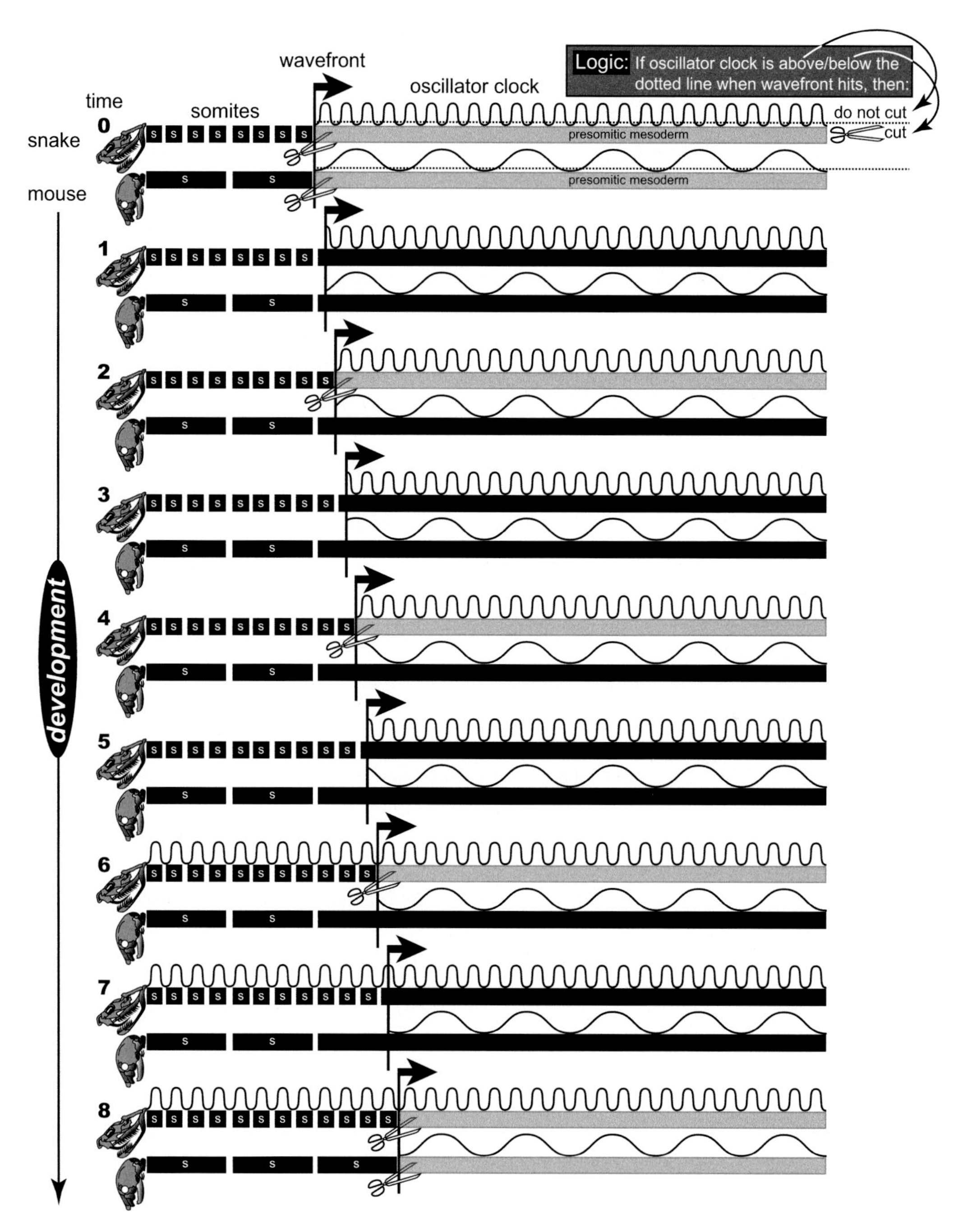

Figure 4.5 **How the snake makes extra vertebrae**. The Clock and Wavefront Model for vertebrate somite patterning, depicted here, was proposed in 1976 [443] and confirmed in 1998 [442]. In 2008 the snake's oscillator was shown to tick four times faster than that of the mouse [805,2298], yielding four times as many vertebrae. Snake and mouse embryos are compared at different times and are drawn as if they were frames in a movie (0 to 8). For convenience,

How the snake rebuilt its eyes

You are reading these words by using a spot in your retinas called the fovea. The fovea affords visual acuity because it is packed with photoreceptors. To see how dependent you are upon this tiny device, try staring at the exclamation point at the end of this sentence and, without shifting your gaze, see how far afield you can read or detect details! Not far.

Like us, lizards have a high-resolution fovea, but snakes do not [160,2321]. Snakes presumably lost their fovea for the same reasons that they lost their legs: (1) disuse and (2) liability. Their subterranean Dark Ages may have lasted ~50 MY [1239,2187,2395], so they probably suffered a **use it or lose it** penalty due to disuse [538,1022], and living in dirt-lined tunnels would have carried the added liability of eye damage and infection.

What is surprising is that snakes retained any vision whatsoever, considering that cave animals tend to go blind over a much shorter span of time [1020,1199]. Proto-snakes may have avoided this fate because they used their eyes for foraging excursions above ground at certain times of the day or night. We are too ignorant of their ecology to know for sure.

Atrophy of the snake's visual system is obvious not only in their retina but also in their optic tectum – a part of the brain that processes visual input. The superficial tectal cells are loosely scattered, and so are their fiber connections [2007]. Indeed, one tectal zone is missing entirely. Analogous impairments are found in the brains of moles [446].

Moles offer further insights into snakes. Their tiny eyes are sealed shut with skin that prevents soil from entering and causing infection [1163]. Snakes have a comparable covering – albeit one that is scaly rather than furry. Snake eyelids fuse together during the embryonic period [160] to form a horny layer called the spectacle [683].

The spectacle clearly comes in handy for the few species that still burrow [1887], but for non-burrowing snakes it appears to be merely an anachronism [2068]. Hence, the ubiquity of the spectacle among modern snakes is consistent with the hypothesis that proto-snakes were fossorial, rather than aquatic [2321], though the presence of an analogous protective spectacle in non-burrowing geckos [159,2264] undermines this argument.

Figure 4.5 *(cont.)*
oscillator waveforms are drawn as if they were static, though they actually fluctuate. The increment between frames is half of a mouse somite cycle. Somites (s) are precursors of vertebrae. They get "chopped" from a slab of presomitic mesoderm (solid rectangle) [1069,1212,1945] as a result of encounters between (1) an inductive wavefront (vertical line topped by an arrowhead) and (2) cells at a certain phase in the oscillator cycle (nadir of the sine wave) [1668], though overt boundary differentiation is delayed relative to covert boundary determination [110]. Nadirs are denoted by gray shading of the presomitic slab, and arrival of the wavefront at a nadir is marked by scissors (chopping site). N.B.: Snake somites are relatively shorter, as shown here (though the depicted shapes are distorted), but then they grow to form normal-shaped vertebrae [1009], thus elongating the spine greatly. Oscillations occur in time, though many genes (e.g., *Lunatic fringe* in snakes) also show banded expression in space [805,1668]. The slab grows posteriorly (not shown) until the process ends. Regionalization (assignment of area codes) occurs separately from somite creation and relies on *Hox* genes (see Fig. 1.1) [1979]. Resegmentation is omitted [60,1545]. Credit for this paradigm must also be given to William Bateson, who toyed with the idea of oscillator-based patterning of meristic arrays as early as 1891 [1987,2358,2361]. Adapted from ref. [2298].

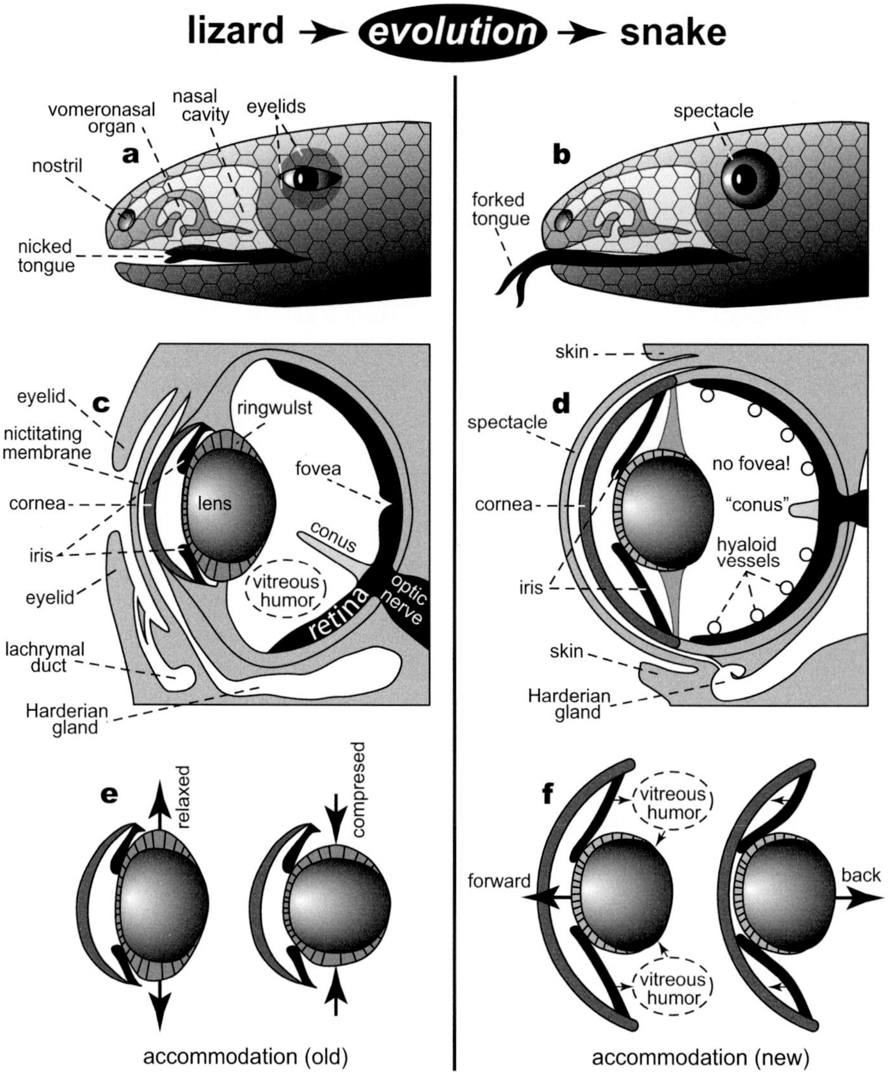

Figure 4.6 How the snake rebuilt its eye.

a, b. Anatomy of a lizard head and a snake head, showing sinuses (vomeronasal and nasal) above the roof of the mouth [884]. Teeth are omitted. In snakes (**b**) the tongue is more deeply indented (forked), and eyelids are fused to form a transparent covering called the spectacle. Geckos also have a spectacle [2264], but it is not **homologous** [159]. Most snakes have a rounded pupil (black with white dot) [1407], though vipers have slit-shaped pupils [242,1236,1431]. Redrawn from ref. [160], with scales schematized to denote their hexagonal lattice. Reptile scales are arranged so much like fish scales [961] that it is tempting to imagine a persistence of the lattice-making recipe in genomes from piscine ancestors through the amphibian interregnum to reptiles and thence to birds [591,1886], despite obvious histological differences [39,366,1268,1523,2459].

c, d. Cross section of a lizard eye and a snake eye. Snakes evolved from burrowing lizard-like animals whose vision deteriorated through disuse underground [1779]. Upon resurfacing, snakes overcame their disabilities by rebuilding their eyes, almost from scratch. Among the

The original spectacles were undoubtedly made of ordinary skin whose scales were opaque, or at best translucent [2187] – the type of scaly covering seen today in typhlopid blindsnakes [36]. When snakes eventually re-emerged into the sunlight ~70 MY ago [1407,2187], this blindfold had to be removed first, before any other optical flaws could be corrected. There were two obvious remedies, either of which would have worked quite nicely:

1. Undo the eyelid fusion that shrouded the eyes in the first place.
2. Keep the spectacle covering, but somehow make it transparent.

The latter path was taken: all modern snakes have see-through spectacles that resemble the contact lenses worn by humans [683]. Why was Option #2 pursued? Was it easier for the genome to accomplish because the old genes for making eyelids were too corroded to be salvaged (Option #1)? Or was it just the route that felicitous mutations hit upon? To resolve this riddle we will need to better understand the pathways that govern both eyelid fusion [663] and skin opacity [1072], and we will have to compare the relevant genes of snakes versus lizards to see what happened.

Once the opacity problem was solved, the next challenge may have been to enable the eye to see objects at different distances once again. The ability of the eye to adjust its focal distance is called accommodation. Lizards accommodate by using certain muscles to deform the lens and change its focal length [2250], and we humans do the same [662]. Somewhere along the way, however, snakes lost these muscles. Two alternative cures were possible:

1. Re-evolve the old muscular apparatus for deforming the lens.
2. Contrive a new method for modulating the eye's focal length.

Again, snakes blazed a new trail (Option #2) instead of retracing an old one (Option #1). Modern snakes change their plane of focus by sliding the lens forward or backward (toward or away from the cornea) rather than changing the shape of the lens itself (Fig. 4.6). Fish and amphibians use this same trick ($\approx$ a pirate's telescope) [563,976,1237], so might snakes have merely "rebooted" a dormant (fish–frog) "app" as an **atavism** to solve this problem? Probably not, because too much time must have elapsed for those decayed old circuits to be resurrected successfully [1390].

Figure 4.6 (*cont.*)

changes (**d**) are (1) a rounder eyeball (due to loss of cartilage from scleral girdle), (2) loss of nictitating membrane and fovea, (3) loss of conus in some snakes and reinvention of a novel "conus" in others at new site (shown here), (4) displacement of optic nerve, and (5) branching of hyaloid vessels (hollow circles in section) over the retina [160,2250]. Endocrine cells of the Harderian gland are omitted. Modified from ref. [160].

e, f. How the eye focuses (accommodates) in lizards and snakes. In lizards (**e**), contraction of ciliary muscles (not shown) squeezes the lens, increasing its curvature and reducing its focal length [2250]. In snakes (**f**), contraction of iris muscles (small arrows) raises fluid pressure in the eyeball, pushing the lens forward (large arrow) [160], which allows the eye to focus on nearby objects – a mechanism like the one used by fish [563,976,1237]. Based on ref. [324].

The retina is energetically expensive to maintain because of its electrical demands, and lizards nourish it by means of a vascularized organ called the conus [2489]. The conus grows out from the optic nerve and secretes metabolic fuel into the vitreous humor. That fuel then diffuses to the retina [2250]. Snakes lost the conus as their retina hibernated, so to speak. To enlarge their eyes to their former size they had to compensate for the loss of the conus somehow. As with the previous problems, two choices were available:

1. Re-evolve the old conus structure.
2. Craft a new way to feed the retina.

As before, snakes were inventive. They evolved a plexus of hyaloid blood vessels to deliver nutrients directly to the retina without any need for a conus at all. This solution sounds quite clever until we realize that those vessels were routed *in front* of the retina where they severely degraded its acuity. Some snakes reduced this blurring by supplementing the plexus with what looks like the same old conus in a different location [2489]. However, this nutritive contraption is not a true conus because it develops from mesoderm instead of ectoderm [2250]. It is really a new organ entirely.

Collectively, these crazy **novelties** (and others as well) suggest that the pathways for eyelid separation, lens deformation, conus construction, etc., were in such a sorry state of disrepair by the end of the serpentine Dark Ages that it was easier for the genome to essentially start over and cobble together a whole new eye [160]. To put it figuratively, the genome was groping desperately. It grabbed the first bandaids that mutations managed to offer it in a crisis situation.

> The resulting eye – as we see it today – presents substitutes for all the losses, remedies for all the defects, of the vestigial organ of the original snakes. And these losses and defects were so numerous that the snakes had almost to invent the vertebrate eye all over again. Nothing like this tremendous feat has occurred in any other vertebrate group, so far as we can tell... We can perhaps understand now why a legless lizard is not a snake simply because it is legless. The snake-shaped lizards such as *Ophisaurus* and *Pygopus* originated above-ground, and escaped the painful period of near-extinction which the snakes experienced and which they have so gloriously survived... If anything could make a snake-hater learn respect and admiration for this abused group of animals, it would be the study of their eyes. The writer speaks from personal experience! [2321]

As should by now be obvious, the snake eye was rescued in the nick of time. Further decay would have likely pushed it beyond any hope of resuscitation [848], though we have, as yet, no way to determine this limit precisely [1907]. In a similar vein, researchers are trying to restore the sight of blind cave fish by bypassing the genetic blockages that have accrued due to disuse [219,1778].

How wonderful it would be if what we are learning about snake eyes could someday enable blind humans to see again! That medical miracle would offer long-overdue redemption for an animal that was libeled long ago by a biblical parable.

These dramatic stories of leg loss and eye restoration offer important morals about how evolution operates in general:

> **Take-home lessons**
>
> 1. Quirks of anatomy reveal an animal's history [505,832,967]. Here, oddities of a snake's eye expose its dark past.
> 2. Structures do not maintain themselves. If they go unused for millions of years they atrophy and eventually may disappear entirely.
> 3. Complex organs such as the eye may be more **evolvable** than we think. Snakes reinvented an eye from incredibly sparse ingredients.
> 4. For any problem, there are many possible solutions. Just because a clade uses one fix does not mean others wouldn't have worked equally well.

How the snake got its (forked) tongue

Like dogs, snakes rely on scent more than sight when they hunt for food. Proof of this dependency comes from an island population of tiger snakes that feed on seagull chicks [91]. Mother gulls defend their nests by pecking the eyes out of any would-be poacher, and snakes that are blinded in this way should be severely disadvantaged in future foraging expeditions. But they are not:

> Remarkably, the blind animals were in no worse condition than normal snakes (intact or merely scarred)... Recapture rates over a 12-month period indicated that blind snakes survived as well as undamaged animals, grew as much, and gained as much in body mass... We know of only one other case analogous to that documented above, and there are striking similarities in the ecological circumstances involved. Wharton [2374] reported finding two congenitally blind cottonmouths (*Agkistrodon piscivorus:* Viperidae) on a small seabird-inhabited island in Florida, where the snakes feed on fish dropped by adult birds in the process of feeding their offspring. Both of the blind cottonmouths were in excellent body condition, and both were recaptured several times over a long period. [218]

How do blind snakes find their way? Their tongues. Snakes have forked tongues that they flick over the ground as they track their prey. Each time the tongue extends, its tips splay out and graze the ground. When the tongue retracts, the tips convey any chemicals they have picked up to receptors on separate sides of the midline. By comparing the strength of scents on the left versus right a snake can stay on the trail (Fig. 4.7) [1990]. This same "L/R disparity" strategy is used by hammerhead sharks, whose nostrils are as far apart as their eyes (~2–3 feet) [125], though they may rely more on electroreception [1116]. A bloodhound uses a behavioral version of the L/R disparity trick: it stays on a trail by swinging its snout left and right to compare scent intensities along the way.

The snake's tongue does not detect chemicals directly with taste buds like ours. Rather, it delivers the molecules to openings in the roof of its mouth by hydraulic compression [661,1992]. Those openings lead to "vomeronasal" organs (VNOs; Fig. 4.6a) [884], which then analyze the inputs [1991]. When the VNO on one side is artificially blocked, the snake can no longer follow a scent trail – thus proving its reliance on these disparity detectors for navigation. Aside from their role in prey sensing, VNOs are also attuned to

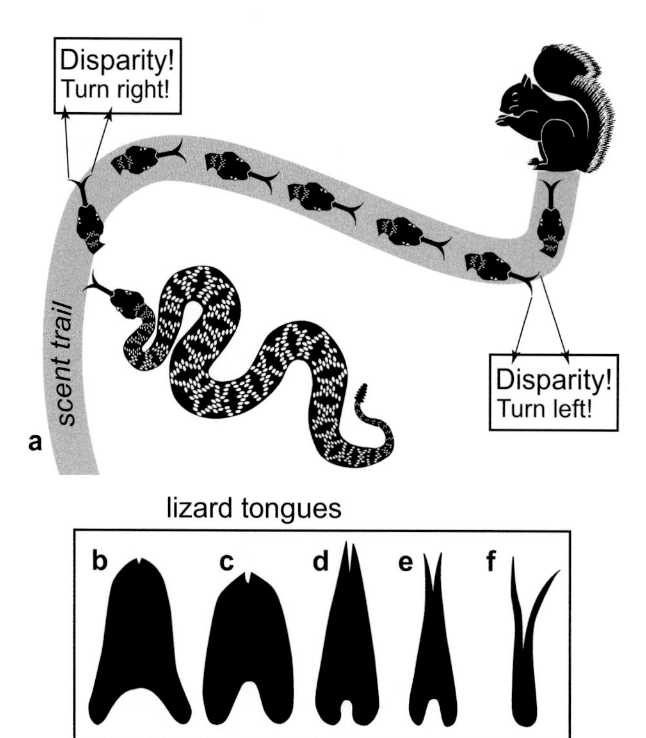

Figure 4.7 How the snake uses its forked tongue.

a. How snakes follow a scent trail [1990]. A rattlesnake is depicted, with the size of its tongue exaggerated. After coming across the scent of a squirrel, the snake tests the trail by repeatedly flicking its tongue onto the ground and pulling it back into its mouth. With each flick, the fork's outstretched tips touch separate spots that are an inch-or-so apart. When the tongue retracts (not shown), it delivers the chemicals that it has picked up to pads on the floor of the mouth and thence to detectors above the roof of the mouth. These vomeronasal organs [884] taste the left and right samples, and the brain then compares the dosages [1991]. If intensity on one side exceeds the other, then the snake turns in that direction. By such mid-course corrections the snake is able to track its prey. Diagram based on ref. [1990].

b–f. Gallery of lizard tongues, arranged in order of the depth of the dent. Silhouettes only (texture omitted). Redrawn from ref. [1991]. (**b**) The gecko *Coleonyx* (Gekkonidae, Gekkota). (**c**) The collared lizard *Crotaphytus* (Crotaphytidae, Iguania). (**d**) The wall lizard *Podarcis* (Lacertidae, Scincomorpha). (**e**) The whiptail *Cnemidophorus* (Teiidae, Scincomorpha). (**f**) The monitor lizard *Varanus* (Varanidae, Anguimorpha), which is more closely related to snakes than any other genus shown here [1407].

pheromones (attractant perfumes) emitted by female snakes [1176], which may explain why male pit vipers have longer tongues than females of equal size [2074].

Where did snakes get their forked tongues? The tongues of most lizards are notched to some extent (Fig. 4.7b–f) [1990,1992,2056], so a predisposition (**preadaptation**) for bifurcation may have existed in the founders of the squamate order. The full spectrum from a slight dent to a deep fork is spanned by extant lizard species [1981], so the transition from one extreme to the other could have been quite gradual [1991].

Finally, we must ask: how did a notch arise in the first place? Even the human tongue has a tendency, albeit slight, to split in two. At least 10 clinical syndromes manifest "bifid tongue" anomalies. Some of these defects occur in conjunction with cleft palate or cleft lip, so they may indicate a more widespread failure of fusion at the midline. One syndrome in particular offers a promising lead. It has been traced to mutations in a homeobox gene that is expressed in the facial region of mice [2185]. Whether a homologous gene causes tongue-splitting in snakes remains to be determined.

How the snake got its fangs

The return of ancestral snakes to a life above ground entailed not only challenges (e.g., the visual system) but also opportunities (e.g., dietary choices) [2067]. Chief among them was the availability of larger prey [1993,2032] – especially after the competing dinosaurs went extinct and mammals began to flourish. As snakes adapted to finding, killing, and swallowing mammals, their skull and teeth were reshaped [2032]. One drastic change was the **co-option** of a few teeth to become venomous fangs [275,1075].

A 2008 study of species along the evolutionary spectrum revealed that fangs began as ordinary teeth near the hind end of the jaw [1122,2296]. The first dental step toward "fanghood" was the acquisition of a longitudinal groove [1076,2493]. This groove allowed venom to seep into the punctured flesh of the prey from glands in the back of the upper jaw [715].

Several lineages of these rear-fanged species independently shifted their fangs to the front by a large-scale renovation [1075]. They replaced the front of the jaw with the rear portion by (1) blocking tooth initiation anteriorly and (2) increasing the rate of growth of the posterior tooth-making zone [2296]. This change would be like having your incisors and canines removed and then shifting your premolars and molars forward to fill the gap.

The next upgrade in the venom-delivery apparatus was to construct a hypodermic needle. The improvement was achieved by origami. The front fangs became hollow (Fig. 4.8) [1076] when (1) their groove sank deeper into the tooth and (2) the groove's flanking ridges fused together, leaving openings at the base and tip to receive poison from the gland and deliver it to the prey [2493]. With this new conduit the glands could now squirt venom into the victim under high pressure like a syringe.

The topology of this reshaping is reminiscent of vertebrate neurulation, where the neural plate rolls into a tube [784]. There too a groove sinks, ridges merge, and a tunnel forms [422]. Could fangs have become hollow via mutations that mistakenly cued neurulation in this odd place? Far-fetched? Yes, but precedents do exist for normal processes occurring at abnormal places (**heterotopy**) [75]. At least, this hypothesis should be easy to test by studying the expression of cytoskeletal proteins at the two sites (tooth and spine) and comparing the genetic circuits that regulate them [1529].

A viper tucks its fangs inside its mouth until it strikes [478], whereupon the fangs are erected (via rotation of the maxilla) for the plunge into the prey [2293]. After a strike, the viper re-cocks its "mouse-trap" – one fang at a time [478,536] in what seems

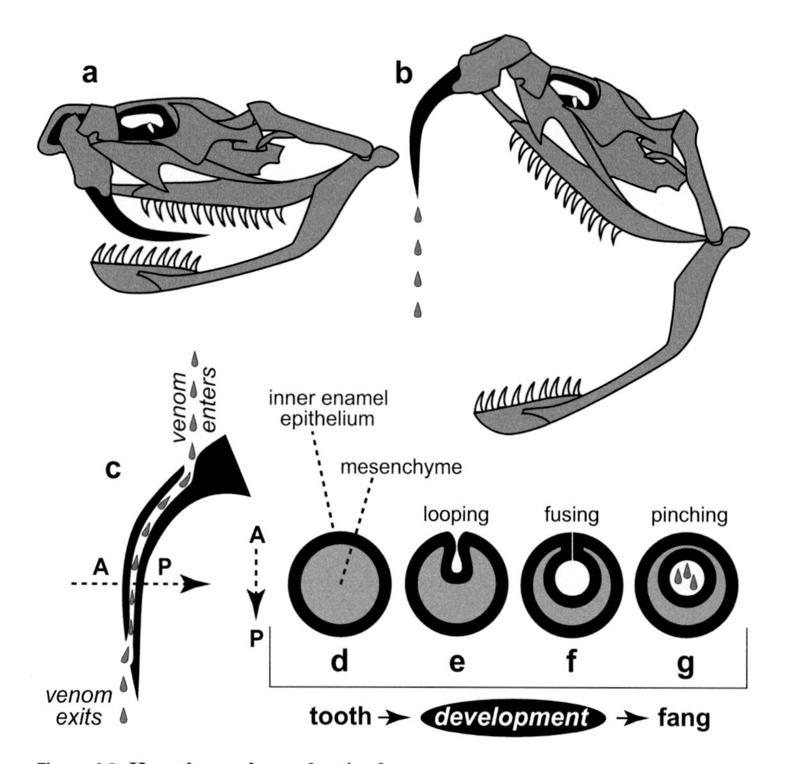

Figure 4.8 How the snake makes its fangs.

a, b. The ultimate "killer app": a viper's jaw. The skull is hinged so that the fangs swing forward automatically as the mouth opens during a strike [1162,2293], but this contraption is only one of the tools in its "Swiss Army knife" array [699,700]. Others include: (1) backward-pointing teeth that prevent escape [1162,1359,2293], (2) supple ligaments that allow the snake jaw to "walk" over its the victim by ratcheting its palate [479,1359,2100,2293], and (3) a versatile quilt of facial muscles [1075,1122]. Some of the jaw's flexibility is due to premature cessation of development before late-arising articulations appear [873,1067] – a nice example of **heterochrony** [189]. (**a**) Closed mouth, fangs retracted. (**b**) Open mouth, fangs erect. After a viper injects its poison, it separately maneuvers each fang back to the cocked position [478,536]. See ref. [1201] for how the pitviper got its pit, ref. [1456] for how the rattlesnake got its rattle, and ref. [2484] for how the cobra got its hood. In 2010 we solved the riddle of how snakes evolved their infrared detection system [863], and in 2011 we learned how a similar system arose **convergently** in bats [862]. Further mysteries posed by squamates (snakes and lizards [1803,1941,2397]) include why they so often (1) acquired venom, (2) lost their legs, (3) evolved viviparity, (4) became herbivorous, and (5) adopted a parthenogenetic mode of reproduction [2067] – an odd potpourri if ever there was one (cf. [319,320])! Redrawn from ref. [2293]; replacement fangs are omitted for clarity (cf. [1162,1407,2100]).

c. Longitudinal section of a viper fang, showing its venom canal [1162]. A and P denote anterior (front) and posterior (back), with the arrow marking the A–P axis. Venom enters via a hole near the top (from a poison gland; not shown), and exits via a hole near the bottom [1076]. This type of fang has existed in its present form for at least 23 MY [1206]. Variations on this theme are seen in the "hypodermic syringes" of other reptiles [1498,2142]. Based on ref. [2493].

like a gaping yawn. Indeed, the snake skull is a veritable Swiss Army knife of useful contraptions [1167,2145], including its back-curved teeth, which prevent escape [2293].

Moray eels resemble snakes in body form, burrowing habit, and capture strategy – all examples of **convergence**. Amazingly, they evolved an equally elegant (but *non-*convergent) device for swallowing prey: a second set of jaws [1453,1875,2146]! We cannot begin to fathom how evolution revamped the snake skull or eel jaw [697] until we learn a lot more about how vertebrate genomes dictate 3D connections among bones, muscles, and tendons in general [545,640,2215].

Questions remain about snake venoms [384,2067]. How were ordinary proteins or peptides recruited as venoms (**co-option**) in different species [95,1911], and why were the same ones often adopted independently (**convergence**) [252,714]? How do existing venom genes diversify adaptively [2297]? And why does venom contain digestive enzymes if it does not help the snake digest its prey [1426]?

Because vipers typically ambush their prey, they are critically dependent upon camouflage to give them the advantage of a surprise attack. It is therefore not surprising that many species have evolved coloration patterns that make it nearly impossible to detect them in their natural environment. The genetic basis for those pigmentation patterns (and others) will be examined in the next chapter.

Figure 4.8 *(cont.)*

d–g. How the venom canal of a viper fang arises during development [2493]. A cross section of a nascent fang (at the level of the arrow in **c**) is drawn at successive stages (A–P axis reoriented). The fang begins (**d**) as an ordinary tooth, with a sheath (black circle) of epithelial cells around a mesenchymal pulp (gray core). The anterior midline buckles inward (**e**) to form a groove (seen here as a loop) along most of the tooth. Cells at either end of the groove stop invaginating (they do not proceed to **f** or **g**), leaving holes for venom to enter and exit (see **c**). Along the rest of the groove, the bases of the loop touch and fuse to form a connecting stalk (**f**), which then disintegrates (**g**), causing the inner circle (venom canal) to pinch off from the outer one (tooth exterior). Pulp cells inside the canal later commit suicide (apoptosis) to hollow out its core. Droplets denote venom. The topological resemblance of this sequence (**d–g**) to neurulation (another ectodermal process [1972]) may not be coincidental. Mutations in the ancestors of vipers might have accidentally activated (**co-opted**) the genome's neurulation program in a novel site (the tooth) – the phenomenon of **heterotopy**. A few mammals **convergently** evolved venom-delivering fangs [688] (and cf. centipede fangs [576,577]). The platypus has a hollow spur on its ankle that secretes venom like a fang [2009], and it co-opted genes akin to those in snakes to produce venom [2341]. For reviews of how embryos make tubes in general see refs. [54,105,1012,1336]. Modified from ref. [2493].

5 The cheetah

Why animals decorate their skin

Many animals sport garish designs. Zebras, leopards, and giraffes are obvious examples [43,1480]. Some of these patterns camouflage the body to prevent detection [1788,2118], while others serve the opposite function of attracting attention [329,1010,2209]. In the latter category, the skin can act as a billboard to warn predators, attract mates, or intimidate rivals [399,1611,1692]. This signaling role is epitomized by the expressive faces of primates in general (Fig. 5.1) [1943,2107,2359] and of the mandrill in particular [1845]. Darwin was entranced by it:

> No other member of the whole class of mammals is coloured in so extraordinary a manner as the adult male mandrill (*Cynocephalus mormon*). The face at this age becomes of a fine blue, with the ridge and tip of the nose of the most brilliant red. According to some authors the face is also marked with whitish stripes, and is shaded in parts with black, but the colours appear to be variable. On the forehead there is a crest of hair, and on the chin a yellow beard . . . When the animal is excited all the naked parts become much more vividly tinted. Several authors have used the strongest expressions in describing these resplendent colours, which they compare with those of the most brilliant birds. [488] (Vol. 2, p. 292ff.)

The colors of most animals are as indelible as tattoos, but chameleons can cloak themselves to suit their background [474,2138,2139]. Chameleons may be the most famous quick-change artists [2118], but the champions of "instant messaging" are octopuses [1113,1403] and other cephalopods [925,2337,2526]. Their skin is as versatile as a television screen [142,481,1019]. *Correction:* make that *two* television screens. A male cuttlefish can display one pigment pattern on its left side to court a female while using a different one on its right side to mollify a competitor [259,1245].

How animals decorate their skin

The colors in vertebrate skin do not arise locally within the skin itself as they do in insects [1204,2434]. Rather, they come from a nomadic population of "neural crest" cells that spread laterally from the dorsal midline during neurulation (formation of the neural tube) [400,1138,2193]. Crest cells differentiate into many histological types (e.g., nerve and cartilage) [256,560,1258], including the pigment cells that tint the skin [907,1415]. Those cells are called melanocytes [445,1731,1826], and their precursors are melanoblasts

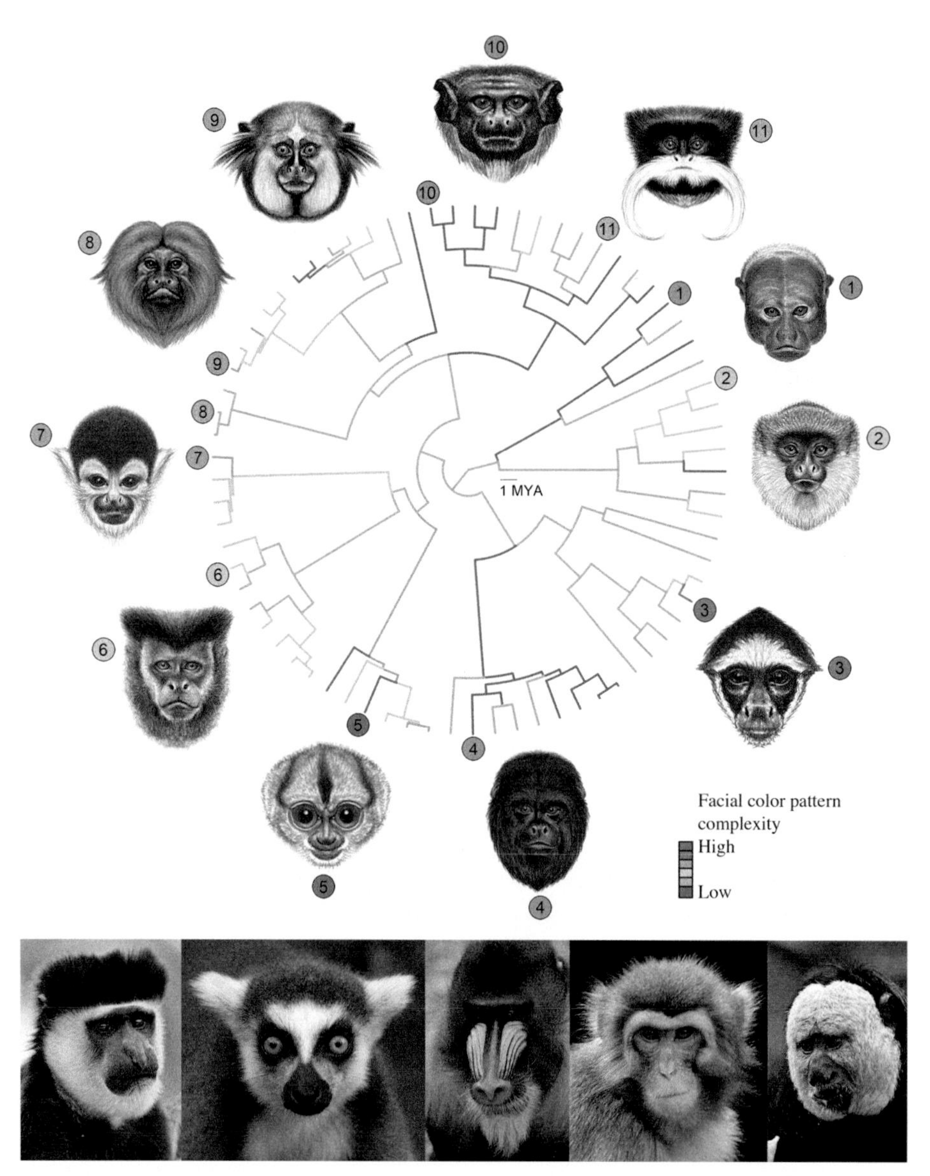

Figure 5.1 How primates decorate their faces. In the wheel diagram above, species are numbered, and shadings or colors (see color plate) denote levels of complexity (i.e., the number of facial areas that differ in hair or skin color; see key) [1943]: (1) bald uakari, *Cacajao calvus* (subspecies *rubicundus*); (2) Hoffmanns's titi, *Callicebus hoffmannsi*; (3) white-bellied spider monkey, *Ateles belzebuth*; (4) black howler, *Alouatta caraya*; (5) northern owl monkey, *Aotus trivirgatus*; (6) black-horned capuchin, *Cebus nigritus*; (7) Bolivian squirrel monkey, *Saimiri boliviensis*; (8) golden lion tamarin, *Leontopithecus rosalia*; (9) Wied's marmoset, *Callithrix kuhlii*; (10) Martins's tamarin, *Saguinus martinsi*; and (11) emperor tamarin, *Saguinus imperator*. Below (≈ the jury of the *American Idol* TV show) are more species whose faces convey signals for species recognition, sex appeal, etc. [1010,1845,2107,2359]. Left to right: mantled guereza, *Colobus guereza*; ring-tailed lemur, *Lemur catta*; mandrill, *Mandrillus sphinx*; Japanese macaque, *Macaca fuscata*; and white-faced saki, *Pithecia pithecia*. Unlike the stripes or spots of

[934,1309,1345]. The melanoblasts of mammals settle in hair follicles and thereby color the pelt [1112].

The dorsal–ventral routes of melanoblast migration are oriented parallel to the vertical stripes of tigers and zebras [942]. Might those stripes be due to these routes [1851]? No. This notion has been disproven by marking individual crest cells as they start to disperse. The resulting clones occupy zones that are far too wide [1055], and there is far too much intermixing of adjacent zones to make sharp boundaries between them [2406]. The explanation for flank stripes must lie elsewhere.

One clever hypothesis for mammalian coat patterning was proposed in 1952 by Alan Turing – the British mathematician better known for cracking Germany's Enigma Code during World War II and for pioneering the principle of digital computation [1003]. His classic paper on this topic was entitled "The chemical basis of morphogenesis" [2244]. Turing imagined that pigments are elicited by chemicals which react with one another while diffusing at different rates. His idea spawned many derivative "reaction–diffusion" (RD) models.

> It is suggested that a system of chemical substances, called **morphogens**, reacting together and diffusing through a tissue, is adequate to account for the main phenomena of morphogenesis. Such a system, although it may originally be quite homogeneous, may later develop a pattern or structure due to an instability of the homogeneous equilibrium, which is triggered off by random disturbances. Such reaction–diffusion systems are considered in some detail in the case of an isolated ring of cells, a mathematically convenient, though biologically unusual system... Another reaction system in two dimensions gives rise to patterns reminiscent of dappling.

According to Turing's equations, one chemical comes to dominate the other in discrete areas of an epithelium. The final layout of the areas depends on subtle differences in the concentrations of the chemicals, which, in turn, are subject to random fluctuations of Brownian motion. Starting with equal levels of the two morphogens, any slight disparities that arise spontaneously will be amplified to yield a heterogeneous distribution by the time an equilibrium state is reached [133]. In short, the system's noise grows to become its signal [227,413]. At the same time, the signal adopts a periodic wave profile, reminiscent of how a guitar string vibrates at an innate frequency regardless of where along its length it is plucked [2358].

Figure 5.1 *(cont.)*
other mammals, which can vary relative to body domains (e.g., Figs. 5.3 and 5.7), the facial markings of primates tend to accentuate anatomical features – for example, our own eyebrows (Fig. 5.8d). This linkage implies a process of "painting by numbers" where the genome uses *cis*-enhancers to turn ON pigment genes in existing area codes. Such a rewiring of existing circuits seems to color the fly's sex comb (cf. Fig. 2.5) and to put spots on its wings [70]. Drawings were supplied by Stephen Nash (IUCN-SSC Primate Specialist Group), and the cladogram was plotted by Sharlene Santana [1943]; both are used with permission. Photos are from foter.com: Tambako (guereza), nachop (lemur), Paolo Camera (mandrill), Lord TriLink (macaque), and Sheba Also (saki). *For color plate see color section.*

One analogy for Turing patterns is the panoply of cloud motifs that can fill the sky at any given time. Depending on weather conditions, we might look up to see a leopard-like array of puffy cottonballs [1770] or a zebra-like array of mackerel ripples [1341,1576, 1739], but no "spot" or "stripe" pattern is exactly the same from day to day because the process of cloud nucleation involves so many unpredictably stochastic factors [2355]. Like cloud formations, Turing patterns are imbued with this property of idiosyncratic variability [115].

We need not look up to see this sort of frivolity. We can just look down at our own fingertips and inspect their rippling ridges. Fingerprints are used in forensics [477] because their minute details vary so whimsically from person to person [296,1560,1615]. They even differ on the corresponding fingers of our two hands, implying that some of their variability is beyond the reach of our genome [607]. Indeed, researchers have confirmed that fingerprints must be partly epigenetic because they differ between identical twins [1081,1306] and among identical quintuplets [1347].

Given that capricious inconsistency is a telltale indicator of Turing devices [1577], we can use mismatches between the left and right sides of the body to search for patterns that might be created in this way. Some snakes have a striking asymmetry along their back [145,679] in the form of a zigzag stripe made of offset triangles (Fig. 5.2) [251,1643,2257]. Some mammals show similar slippages of left and right patterns along the midline (Fig. 5.3). Thus, zebra stripes are partly asymmetric. So too are tiger stripes. Hence, we should probably amend William Blake's iconic quatrain from his poem "The Tyger" (in *Songs of Innocence and of Experience*, 1794) to admit this flaw (boldface added):

> Tyger! Tyger! burning bright
> in the forests of the night,
> what immortal hand or eye
> could frame thy fearful **quasi**-symmetry?

Asserting that the left and right sides of an animal create their pigment patterns independently does not imply that the midline must act as a wall between them. Indeed, giraffes refute this notion. The tiled polygons of a reticulated giraffe cross the midline with a callous disregard for its very existence (Fig. 5.3). The same appears to be true for the spots of leopards, cheetahs, and jaguars (Figs. 5.3, 5.7, and 5.9) [430]. It is as if their bodies were casually draped with a factory fabric printed independently of their gross anatomy. One exception to this rule is the tail, which seems sterically **constrained** to make rings even when the rest of the body is covered with spots. Such rings are expected based on Turing's parameters [1572–1574]. For zebras and tigers, the midline *does* seem to matter, because their flank stripes often slip out of register along this boundary [296] (Fig. 5.3).

In contrast to the designs in mammal and reptile skin, the decorations on butterfly wings tend to be strikingly symmetric (Fig. 5.3). Why? Like flies, butterflies use morphogen gradients to establish a Cartesian coordinate system in each wing (see Chapter 3) [70,343,1794]. Hence, their patterns are plotted by fixed genetic formulae that assign specific colors to particular points in their planar epithelia. Axial gradients are typically generated by a *single* morphogen – not by two interacting ones (à la Turing)

Figure 5.2 How snakes use asymmetry along their backs. Some snakes have a chain of roughly symmetric diamonds down the back (e.g., **a**), while others predominantly have a zigzag stripe (e.g., **e**), and still others exhibit both motifs in successive sections (e.g., **b**–**d**) [145,251,951]. These variations imply that half-diamond pigment triangles from the left and right slide past one another so as to align front-to-front and back-to-back (diamond motif) or front-to-back (zigzag motif), but less commonly in between. Such phase shifts [815] could be due to chemotaxis [1580] or cell adhesion [296]. Most stripes are solid, but the snake in **d** has an archipelago caudally. The zigzag motif of the Eurasian genus *Vipera* is a signal for warning avian predators, but the diamond motif is not [679,1643,2257]. At first glance, these patterns seem to be mosaics of pure-color scales as in butterflies [335,1636] and some rattlers [145,1233] (not shown), but the pigment stripe actually meanders through the scale lattice as if it weren't even there. The gravid female in **a** is the Greek alpine meadow viper, *Vipera ursinii graeca*, which lives above 1,600 m [549,1639]. The long-nosed vipers in **b** (juvenile male), **c** and **d** (adult males), and **e** (juvenile female) are dwarf forms of *V. ammodytes* from the Greek Cycladic Islands. These dwarfs are classified as *V. a. meridionalis*, but they may constitute a separate subspecies based on their long isolation and genetic distance from mainland forms (S. Roussos, personal communication). Photos were taken by Stephanos Roussos and are used with permission.

[1378]. The idea that morphogen gradients can create constant patterns via coordinate systems was proposed by Lewis Wolpert in 1969 (see Fig. 1.3), and it has reigned as the paradigm of developmental biology ever since [2443]. Exceptions to the mirror rule can be found in insects (e.g., elytral spots) [1136], and even butterflies have one asymmetric type of pattern (the ripple motif: Fig. 3.5f) that probably relies on a local RD subroutine.

Figure 5.3 How mammals and butterflies differ in the symmetry of skin patterning. Left–right asymmetries can be seen in the coats of giraffes, zebras, tigers, and jaguars (upper row), as well as in cheetahs (Fig. 5.7a), but not in comparable patterns of butterflies (lower row). These asymmetries imply that the patterning processes in the two groups differ. Butterflies use **morphogen** gradients (see Chapters 2 and 3), whereas mammals may be relying on a Turing mechanism of some sort. Only the "ripple" motifs of butterflies (Fig. 3.5f) show asymmetry [1633,1775]. This giraffe belongs to the subspecies *Giraffa camelopardalis reticulata*, which has polygons instead of splotches [260,1572,1574,2041] – an example of planar tiling [883,1303,1323] (cf. dermal scales [1476,2285]). Butterflies (left to right): *Rhaphicera dumicola* (ventral), *Callicore atacama* (ventral), *Marpesia eleuchea* (dorsal), *Timelaea albescens* (ventral). All photos are from public-domain websites: Heather Bradley Photography/foter.com (giraffe), John Storr/Wikimedia Commons (zebra), ucumari/foter.com (tiger), Tambako/foter.com (jaguar); butterflies are from http://www.lepdata.org (©2012 by W. H. Piel, L. F. Gall, J. O. Oliver, A. Monteiro, and the Yale Peabody Museum). *For color plate see color section.*

How the angelfish got its stripes

Another distinguishing feature of Turing mechanisms vis-à-vis Wolpert gradients – aside from asymmetry – is scalability [1241,2218,2443]. Turing's equations enforce a fixed "chemical wavelength" between elements (spots or stripes) that does not change when the size of the field expands [120,439]. For this reason, the number of elements in patterns created by a Turing mechanism should vary with field size [438,440,441]. In contrast, the number of elements in patterns created by Wolpert gradients should stay constant, since the coordinate system should stretch like an accordion [167]. Thus,

Wolpert patterns are expected to be scalable (see ref. [361] for exceptions), but Turing patterns are not.

Angelfish stripes meet the Turing expectation. These fish gain new stripes as they grow so as to maintain a fixed inter-stripe distance (Fig. 5.4a–e) [1184,1458]. The constancy of this interval connotes a chemical wavelength. If the new stripes had been added at one end or the other, then we could have salvaged the gradient model with an ad hoc amendment that the peak rises to add coordinates as the fish grows. In fact, however, the new stripes arise between existing ones, so we can safely rule out a gradient explanation.

Figure 5.4 shows how angelfish stripes can be created by the kind of recipe that Turing concocted. Instead of using Turing's equations per se, however, the simulation relies on a derivative devised by Alfred Gierer and Hans Meinhardt [781,1461,2038]. Their Activator–Inhibitor Model postulates (1) an activator "**A**" that spreads slowly and catalyzes its own production and (2) an inhibitor "**I**" that spreads quickly, inhibits **A**, and depends on **A** for its production. This formulation is the most widely applied of all RD models [1186,1378,1457], and it has been documented in at least one real system – the mammalian palate. It appears to make regular ridges of the palate via the morphogens FGF (**A**) and Shh (**I**) [589].

Given an initially homogeneous distribution of **A** and **I**, blips of **A** will occasionally rise above the background level of **I**, whence they will grow into peaks via autocatalysis. As each peak rises, **A** produces both **I** and more **A**, but **I** diffuses away so quickly that it forms a mound that straddles the **A** peak. These mounds prevent any new peaks of **A** from arising within a certain range (= Turing's wavelength) of existing peaks [120]. Thus, the pattern acquires a crude periodicity where inter-peak intervals fall somewhere between the minimum limit (λ) and twice that distance (2λ) [1187]. As the fish grows, adjacent peaks eventually exceed the 2λ threshold, whereupon a new blip of **A** will arise between them, leading to a new peak of **A** and ultimately a new stripe [1184].

Just because a model can mimic a process does not mean that the process occurs that way, since a host of other mechanisms might be equally effective (Table 5.1) [961,1687,1699]. Indeed, there are lots of variations on the activator–inhibitor theme that can produce intercalary striping by mechanical (vs. chemical) means [933,1183,1923]. They implement short-range activation and long-range inhibition [1462,1693] in different ways that should be distinguishable experimentally [1050,1186,2254].

Instead of delving into those alternatives, however, suffice it to consider just one limitation that afflicts gradient and RD models alike. Both types of model envision cells primarily as mere spectators. The hard work of pattern formation is supposed to be done by extracellular molecules [1186]. Once the morphogens finish creating a gradient or a prepattern, then and only then do cells record the spatial information and act accordingly by adopting different internal states. Such passivity might be plausible for a static cell sheet, but we are dealing here with mobile melanocytes [328], which have a rich behavioral repertoire [1213,2214]. A recent (2012) study of a different fish shows how active (and *reactive*) these cells can be [1061]. The results of that study will now be considered.

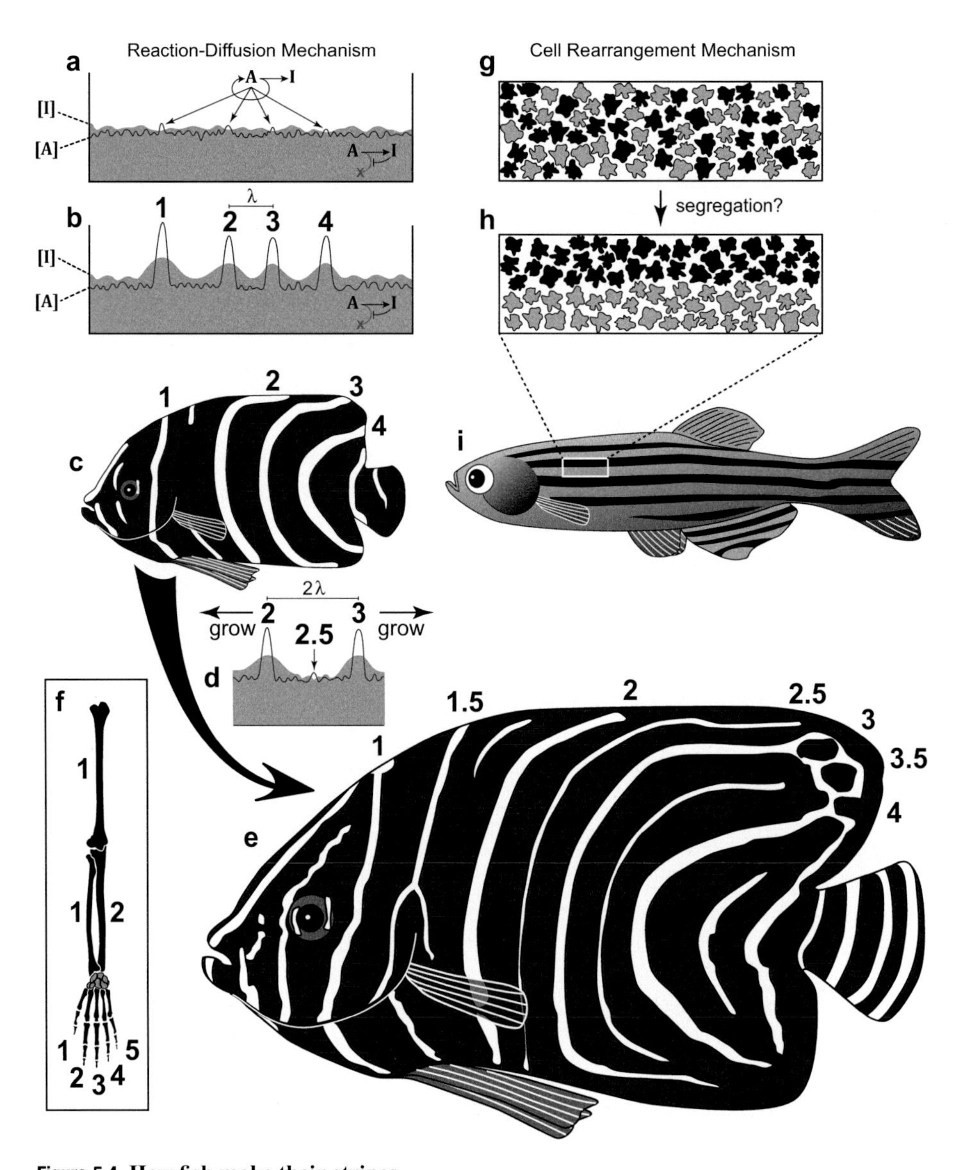

Figure 5.4 How fish make their stripes.

a–e. Angelfish have stripes that develop in accordance with a reaction–diffusion mechanism [1008,1184,1458], though the scale seems too large for diffusion to work within the available time window [1242,1566,1575]. According to the Gierer–Meinhardt Model [781,1461,2038], a slow-diffusing molecule (activator "**A**") catalyzes its own production via a feedback loop (curved arrow), while a fast-diffusing molecule (inhibitor "**I**") blocks this loop (T-bar at point **X**) [1462]. Initially (**a**) **A** and **I** have uniform levels (concentrations on *y* axis) along the fish body (*x* axis). As long as **A** stays below the "sea level" of **I** (gray fill), it will be inhibited and will not "catch fire" since its autocatalytic loop is broken (submerged circuit). However, Brownian motion inevitably causes fluctuations (wavy lines) that force **A** to locally exceed **I**, whereupon a peak of **A** emerges because the **I** made by **A** (horizontal arrow in **a**) diffuses away so quickly that it cannot extinguish the "fire." As the first peak of

How the zebrafish got its stripes

Zebrafish have parallel stripes like angelfish, but their stripes are horizontal, not vertical [1805,1826]. Black stripes alternate with yellow ones, and each color is formed by a separate population of pigment cells (Fig. 5.4g–i) [996,1354,1806]. When these cells are cultured in petri dishes, they crawl around and bump into one another. If cells of the same color happen to touch, nothing much happens, but if cells of opposite colors make contact, then the yellow cell undergoes a transient membrane depolarization and backs away from the black cell (6/10 cases) [1061].

Of course, *neurons* are more famous for "tickling" one another electrically in this way [1141,1219], but we should not be too surprised that pigment cells can do so too, because melanoblasts are intimately related to neurons within the neural crest lineage [560,1645,1731]. Indeed, melanoblasts can apparently even regenerate from peripheral neurons [282,1137,2473].

The aversive reaction of the yellow cells suggests how stripes might be forming in vivo: black and yellow cells could be sorting themselves into separate groups by repelling one another [1186,1353,2468]. A similar guess was made in the 1940s by researchers studying pigment stripes in frog tadpoles [2246–2249]. Presumably, the pigment cells would have to manifest homophilic affinity as well heterophobic incompatibility in order to make stripes that are more than one cell wide [623]. These expectations are borne out by mutations that affect zebrafish patterning [1185]:

Figure 5.4 (*cont.*)

A rises at a random site, the mound of **I** around it also rises, thus stifling any other incipient peaks within a certain range (minimum wavelength λ) [120]. The result (**b**) is an array of peaks (1, 2, 3, and 4) at roughly uniform intervals [1187] but at unpredictable locations. The angelfish *Pomacanthus semicirculatus* obeys this logic, because it acquires stripes as it grows (cf. catfish [79,1182]). A single individual (redrawn from ref. [1184]) is depicted at ~6 (**c**) and ~12 months (**e**) of age (body lengths are ~4 and ~8 cm, respectively). Major stripes are marked 1–4 at the early stage, with later stripes numbered accordingly. As old stripes grow apart (**d**), thin new ones intercalate (1.5, 2.5, and 3.5).

f. Human arm, showing how a similar process of local activation and lateral inhibition [1696,1697] might be adding bones as the limb grows distally [813,1614,2046,2403] (but cf. rebuttal [419]). Indeed, startling new evidence in 2012 supports the notion that tetrapod digits are created via a Turing mechanism [2026,2294] – albeit with a Wolpert gradient superimposing **individuation** on the otherwise identical digits [137,1867,2487], as predicted long ago [438].

g–i. Zebrafish have longitudinal stripes (**i**) that arise from black or yellow (gray here) chromatophores (amoeboid shapes in **g** and **h**). Black cells repel yellow ones in vitro, so the two types of cells may be segregating in vivo (**g**, **h**) to sharpen boundaries in the adult zebrafish (*Danio rerio*, **i**) [1061]. N.B.: Models in **a–e** versus **g–i** are not mutually exclusive [206,1507] since zebrafish show a scale-dependence of stripe number similar to angelfish [1423]. The genetic basis for vertical (**e**) versus horizontal (**i**) striping may be subtle, because closely related species can differ in stripe orientation [1721,2039], and the same conclusion applies to the basis for straight versus labyrinthine striping [1508,2037,2346].

Table 5.1 Theories for the periodic patterning of stripes or spots[a]

Model	Subtype	Principal advocates	Case studies
Positional information	Gradients	Charles Manning Child [198,383] and Lewis Wolpert [2442].	Butterfly eyespot annuli [123,1627]. Fly embryo body stripes [372,1080,1699,1773].
	Polar coordinates	Peter Bryant, Susan Bryant, and Vernon French [272,707].	Fly leg proneural stripes (see Fig. 2.4) [965].
Prepattern dynamics	Reaction–diffusion	Jonathan Bard [122], Alfred Gierer [781], Brian Goodwin [813], Shigeru Kondo [1186], Philip Maini [366,1362,1579], Hans Meinhardt [820], James Murray [1577], Barry Nagorcka [1586,1587], and finally, the polymath who founded this entire school of thought, Alan Turing [2244].	Angelfish stripes [1184]. Cheetah spots [43,1573]. Fingers [1378]. Fingerprints [1586]. Fish stripes [132,1187]. Genet coat pattern [1573]. Giraffe polygons [122,1572–1574]. Jaguar spots [43,1187]. Leopard spots [43,1187,1574,1687,2368]. Nautilus shell stripes [1586]. Palatal ruggae [589,1378]. Reptile scales [366]. Tetrapod limb bones [794]. Tiger stripes [1573]. Zebra stripes [122,307,1572,1574].
	Mechanical or mechano-chemical	Cheng-Ming Chuong [1050], Albert Harris [163,933], James Murray [1575,1695], Stuart Newman [1612,1613] George Oster [1694,1695], and Isaac Salazar-Ciudad [1930].	Brain grooves [992,1259,2017,2223]. Crocodile scales [1476]. Fingerprints [1575]. Newt stripes [296]. Snake patterns [296]. Tooth cusps [1926]. Tetrapod limb bones [2254]. Tiling patterns [795]. Zebra stripes [1948].
Determination wave	Clock and wavefront	Jonathan Cooke [443], Olivier Pourquié [1780], and Chris Zeeman [2499].	Vertebrate somite segmentation [111,1780].
	Chemical, mechanical, or both	Boris Belousov [1356,2429], Lev Beloussov [164], Richard Gordon [818,819], and Arthur Winfree [2428,2430].	Butterfly eyespots [818]. Fish stripes [132]. Fly leg proneural stripes (see Fig. 2.4) [818]. Tabby cat stripes [818]. Teats on milk lines [818].
Darwinian selection	Pruning (cell death)	Gerald Edelman [592,594].	Brain striations and neurobarrels [358]? Fish pigment patterns [1959]. Zebra stripes [1472].
	State change (cellular automata)	Cheng-Ming Chuong [1764,1906] and Stephen Wolfram [2437].	Mammal hair renewal waves [913,1764]. Newt stripes [1687].

Table 5.1 *(cont.)*

Model	Subtype	Principal advocates	Case studies
Cell rearrangement	Cell sorting	Gerald Edelman [593,595], Shigeru Kondo [1185], and Malcolm Steinberg [2108] (cf. [1363]).	Bacterial designs [1316]. Frog tadpole stripes [2246–2249]. Neural laminae [1141,1425,2468]. Newt stripes [2239,2240]. Zebrafish stripes [1061].
	Chemotaxis	Cheng-Ming Chuong [1305] and James Murray [1580].	Alligator stripes [1578]. Bacterial designs [166,283]. Feather placodes [111,1305]. Snake patterns [1580].

[a] Categories are based on the taxonomic system of ref. [961], with cell lineage models omitted because they do not seem to be involved. These devices can be used singly, but more often they are used in combination or in sequence [114,412]. For instance, neural tube cells were recently found to sort based on identities that they get from a noisy morphogen gradient [2468]. This inventory is illustrative, not encyclopedic. Richard Goldschmidt (not listed) also advocated determination wave models [800]. For other schemes, studies, and perspectives see the excellent reviews by Kondo [1186], Lander [1241], Müller [1558], Murray [1575], Othmer [1699], Salazar-Ciudad [1923,1930], Urdy [2254], and Widelitz [2386]. These models have deep roots in biology [1540] – for example, the atomism versus holism debate [811,1174,2301] and the paradox of indeterminacy at the molecular scale versus constancy at the anatomical scale [1378,1417,2365].

1. Mutations in *jaguar* prevent black cells from repelling yellow ones, and mutant fish fail to form normal stripes, perhaps for this reason [1061]. The *jaguar* gene encodes a potassium channel that controls the transmembrane potential [1071].
2. Mutations in *leopard* transform the horizontal racing stripes into leopard-like spots [79,1354]. The *leopard* gene encodes a gap junction protein [2345] that enables intercellular communication [2347], which could, in turn, foster either stripes or spots [1354]. Indeed, cellular crosstalk might mediate a mechanism that relies on *both* RD and repulsion [1185,1591,2346].
3. Mutations in *seurat* turn stripes into irregular spots. The *seurat* gene encodes an adhesion protein that may mediate the homophilic binding needed for cell sorting [623] (but see ref. [2163]). It also promotes melanoblast migration and survival.

The zebrafish parable imparts a moral that we did not learn from angelfish: cells often behave like robots [961,2065]! They are programmed with a few conditional rules [2441], such as "IF this happens, THEN move, divide, die, emit a different signal, or ELSE take some other action" [482,1251]. Cellular interactions provide cues that elicit reflexive responses. In this way cells steer one another to certain sites, states, and future actions. After many such episodes of "social networking," the final pattern **emerges** [1923,1930].

This communal dynamic defines a class of models called "cellular automata" [892,946,2437]. These models inhabit a different world from either "Planet Turing" or "Planet Wolpert" [798,1448,2386]. They require that we think of pigment patterning in terms of "the games that melanocytes play" (cf. Conway's game of *Life* [108,747,748]), and they force us to delve into the sociology of melanocyte behavior [65,363].

Such investigations have only just begun, and this quest will occupy the next generation of evo-devotees for some time to come. The depth of our ignorance about pigment

Figure 5.5 How the zebra makes its stripes. There are three extant species of zebra [169,898,931]: (**a**) the mountain zebra, *Equus zebra*; (**b**) Grevy's zebra, *E. grevyi*; and (**c**) the plains zebra, *E. burchelli* [835,836]. The quagga (**d**) was a subspecies of *E. burchelli* [988,1279,1332] until the breed went extinct in 1883 [931]; the one here was photographed at the London Zoo in 1870. Note the various striping idiosyncrasies [835,1775]: (1) the pin stripes of Grevy's zebra (**b**) imply a later onset (**heterochrony**) of the patterning mechanism [121]; (2) the shadow (incipient?) stripes of the plains zebra (**c**) [122] suggest a patterning period long enough for more than one round of stripe initiation; and (3) the sergeant's chevron or "triradius" atop all four legs in Grevy's zebra (**b**) [307,1572,1573] and atop the forelegs of mountain and plains zebras (**a**, **c**) [122] indicates clashing tissue polarities. One riddle that has so far defied explanation concerns horizontal stripes on the rump [1575]. These stripes extend far forward in *E. burchelli* (**c**), less so in *E. zebra* (**a**), not at all in *E. grevyi* (**b**), and they vanish in the quagga (**d**) [1775]. Photos are from foter.com: Arthur Chapman (a), Dirk-z20 (b), FurLined (c), and Recuerdos de Pandora (d).

patterning in general [1010,1043,1452] and regarding mammals in particular [1007,1687] is made painfully clear by the case studies presented below: the zebra, the cheetah, the mouse, and the clouded leopard.

How the zebra got its stripes

The new stripes that arise between the old ones as angelfish grow are thin at first (Fig. 5.4e) [1184]. This disparity is expected based on Turing's model. It is therefore of interest that similarly faint intercalary stripes are also seen in Burchell's (plains) zebra (Fig. 5.5c). Evidently, the time window during which patterning occurs in *Equus burchelli* is a long enough span to permit a second – albeit partial – round of stripe initiation, whereas the available time is so short in other zebras that only a single round can occur [121,122].

A comparable argument, based on timing differences (**heterochrony**), may explain why the stripes of Grevy's zebra (Fig. 5.5b) are especially narrow [121]. Assuming a fixed distance between stripe-inducing morphogen peaks throughout the genus (Fig. 5.4b), such pin stripes imply that patterning starts later in *E. grevyi* than in the other species. The Grevy's embryo would thus acquire more stripes due to its larger size at that stage, and they would widen less because there is less time available for growth before birth [121].

The idea of a *spotted zebra* seems absurd, but a few such individuals have in fact been seen in *E. burchelli* herds (Fig. 5.6a) [122]. Moreover, there must be some underlying tendency to make spots in the genus *Equus*, because certain horse breeds [1427] are as spotted as a dalmatian (Fig. 5.6b) [161,400,1798]. It is equally hard for us to try to imagine a *striped giraffe*, but there too we are deluded because the okapi – the nearest living relative of the giraffe [1790,2424] – has stripes that are remarkably zebra-like on its hindlegs (Fig. 5.6c) [936].

Evidently, the mammalian genome does not make as strong a distinction between stripes and spots as we do in our own minds. Indeed, some mammals are *both* striped *and* spotted [1053,1655,2423], as if they just could not decide which pattern to adopt [1573,1575,1775] (cf. whiptail lizards [2177]). These "strotted" mammals include genets and civets in the family Viverridae [87,1155,2252], baby tapirs [2319], badger-colored bats [1836], spotted skunks [2266], and 13-lined ground squirrels [2202] (Fig. 5.6d).

With reaction–diffusion models, it is relatively easy to shift from stripes to spots or vice versa by tweaking the parameters of the morphogens or the dimensions of the fields that they inhabit [133,1575,1587]. For this reason, these models are our best guess at present for how mammals make stripes or spots or both [1457]. However, none of the proposed models has yet been subjected to definitive experimental tests, so we can't yet be certain of what is actually going on.

How the cheetah got its spots

Analogous to the "shadow" stripes of *E. burchelli*, cheetahs display what appear to be incipient spots in the spaces between their larger prominent spots (Fig. 5.7) – suggesting that they too are using a Turing mechanism to paint their pelage. If so, then we might expect to encounter a *striped cheetah* every now and then, just as we've seen spotted zebras. Sure enough, there is a rare variety of *Acinonyx jubatus* called the "king" cheetah that has unmistakable stripes along its back and enlarged, irregular spots on its flanks [1053,2423]. In 2012 the gene that toggles between these outputs – normal versus king cheetah – was identified as *Transmembrane Aminopeptidase Q* (*Taqpep*) [1115,2077]. Mutations in this same gene cause tabby cats that are striped to instead become blotched.

Figure 5.6 Spotted zebras, striped giraffids, and "strotted" squirrels. It is hard to imagine a spotted zebra, but such animals have been seen – albeit rarely – in herds of *Equus burchelli* (**a**),

Again we are forced to admit that the difference between stripes and spots may be trivial from a genetic perspective [603,1587], but at last we may have found the key to unlock this distinction. *Taqpep* encodes a protein belonging to a family of enzymes that regulate peptide hormones [1115,2237]. Conceivably, we have been fooling ourselves all along by envisioning Turing's agents as chemicals that interact *directly*. They could instead be inert peptides that obey Turing's rules by interacting *indirectly* via feedback loops between signaling pathways [220,1865]. Indeed, a few odd peptides have been shown to be instrumental in animal patterning [727,1893,1947,2224].

Peptides that are only ~10 amino acids long could obviously diffuse a lot faster than the cardinal protein **morphogens** (cf. Table 1.1) [1566], which are an order of magnitude bigger than peptides [965]. If this notion is correct, then genetic screens for patterning mutations could soon lead us to the cell-surface receptors that (1) bind peptides as inputs, (2) convey these signals to downstream transcription factors, and (3) turn pigment genes ON or OFF. The conservation of the *Taqpep* gadgetry (whatever it may turn out to be) in cheetahs and domestic cats gives us hope that we have stumbled upon not just a quirky feline pathway, but a deeply mammalian mechanism.

How the mouse makes waves

The analyses of angelfish and zebrafish that were presented above come mainly from the laboratory of Shigeru Kondo at Nagoya University in Japan. This next story also comes from his lab, but it is stranger than the other two by far.

In 2003 the Kondo lab described a mouse mutant that has hair follicles but no hair because of a splicing defect in the *Foxn1* gene. This defect causes hairs to fall out as soon as they start to grow [2152]. Any changes in pigment associated with hair follicle renewal can thus be seen directly in real time [1361].

Figure 5.6 *(cont.)*
and some horse breeds are as spotted as a dalmatian (Knabstrupper; **b**) [1798]. It is equally hard to imagine a striped giraffe, but the nearest relatives of giraffes are okapis [1790,2424], and they have stripes on their rump and legs (**c**). Finally, there is the riddle of the thirteen-lined ground squirrel. *Ictidomys tridecemlineatus* (**d**, top, male) has six stripes and seven rows of spots [2205] and hence could be called "strotted" (a blend of striped and spotted). When it interbreeds with the Rio Grande ground squirrel, *I. parvidens* (**d**, bottom, female), which has more of a checkerboard pattern, the hybrid offspring display stripes that break up into spots (**d**, middle, female) [2202]. The overall lesson from these examples is that stripes and spots may be created by a common process (reaction–diffusion?) that can easily be switched from one mode to the other (see text). We can also infer, based on the background color of spotted zebras, that normal zebras must be black animals with white stripes – not the other way around [835]. The zebra photo was taken in Botswana in 1967 and was provided courtesy of Jonathan Bard, the horse photo was furnished by Monika Reissmann, the okapi photo is from Dave DeHetre/foter.com, and the squirrel photos are from Frederick Stangl. All pictures are used with permission. *For color plate see color section.*

Figure 5.7 **How the cheetah got its spots.** The king cheetah (**a**, right; **b**, below) differs from the spotted cheetah (**a**, left; **b**, above) insofar as (1) its spots are larger, (2) the spots appear to have fused into blotches on the flank, and (3) there are five stripes down the back (cf. the dorsal "stripes" of the clouded leopard: Fig. 5.9) [1115]. Similar conversions from spots to stripes or vice versa are seen in tabby cats [1115], zebrafish mutants [2345,2346], and catfish species [1187].

All mammals renew their hair periodically, but typically they do so sporadically over the body surface so that bare areas are too tiny to be noticeable [717,1263,1269]. Mice are unusual insofar as their hairs cycle in waves across their skin [109,1764,1781].

In the *Foxn1*^{tw} mutant these waves are distinctly sharp because the renewal cycle is drastically accelerated [397,1687,1763]. Hence, the trajectories of the waves can be charted precisely. The waves were found to start in the armpit and spread out in all directions over the body – like dropping a pebble in a pond and watching ripples emanate from that point concentrically [1765,2152].

Figure 5.8m shows how hair-replacement waves sweep across the face of a *Foxn1*^{tw} mouse. The pigment bands change shape in weird ways from week to week. At particular stages, the markings bear a striking resemblance to the pelage patterns of certain mammals. Could the latter patterns be due to a similar wave that gets frozen in place during development [397,2386]? Perhaps. Further research will be necessary to decide this matter [1687].

How the clouded leopard makes clouds

Given that mice have an "app" for launching pigment waves from their armpits, might other mammals be using this same app to make other patterns from different foci? The clouded leopard seems like a good candidate (Fig. 5.9). *Neofelis nebulosa* has polygonal "clouds" on its flanks that look as if they were made by (1) launching ripples from sparsely spaced leopard-spot foci, (2) allowing these ripples to spread until they encounter ripples from neighboring centers, (3) drawing a straight line along the interface, and (4) painting a shadow inside each resultant polygon. The last step may rely on the ancient bilaterian PCP (planar cell polarity) pathway to target the shading to a specific end of the polygon [373,812,871].

Another animal where melanocytes might be playing this sort of game is the reticulated giraffe (Fig. 5.3). *Giraffa camelopardalis reticulata* looks as if the splotches of a Masai giraffe somehow expanded to abut their neighboring splotches at linear boundaries, culminating in an irregular lacework of abutting polygons.

All of these musings are, of course, just idle speculation, but therein lies the fun of strolling leisurely along the evo-devo frontier. We can gaze into the unknown and daydream endlessly about the treasures that are surely just beyond our reach.

Figure 5.7 *(cont.)*
Most pigment patterns in animals consist of spots or stripes [1135,1139,1187,1480,1635], and some felids display odd mixtures of both [43], including ocelots [332], servals [122,760], fishing cats [122], and domestic cats [603]. Why the cheetah has spots is unclear, since it lives on open plains where they should be useless for camouflage [43]. Note the tiny (incipient?) spots between the larger spots. This phenomenon of new pattern elements arising in the biggest spaces between older ones during growth was first documented in 1940 for insect bristles [2399], and it is theoretically explicable in terms of a Turing mechanism (see text). Both photos were taken at the Ann van Dyk cheetah preserve and were provided courtesy of Greg Barsh. *For color plate see color section.*

Figure 5.8 How the mouse makes waves. Facial pigmentation of a single *Foxn1*^{tw} mutant individual (**m**) [2152], showing its traveling wave (dark zone) of hair formation at different times over a 12-week period (numbers = weeks). These eight frames (≈ the rock group *Kiss* [1866]) are enlarged above as **l**, **i**, **g**, **e**, **j**, **f**, **k**, and **h**. Some of the transient wave forms resemble static pigment patterns in dogs (**a** ≈ **e**; **b** ≈ **f**), raccoons (**c** ≈ **g**), or humans (**d** ≈ **h**). The path of the wave may be constrained by the polarity of the epidermal cells [373]. The black mask of the raccoon (**c**) is thought to be an aposematic warning signal [1611]. Scottish actor Sir Sean Connery (**d**) is included because of his prominent eyebrows, not for having a unibrow per se. This photo was taken by Stuart Crawford at the 2008 Edinburgh International Film Festival. (*Aside*: I was afraid that Stuart would be appalled by my request to put Sir Sean's distinguished picture next to those of a mouse and a raccoon, but he was tickled and graciously consented. Whether the venerable Mr. Connery is similarly amused remains to be seen. My witty copy-editor, Hugh Brazier, is guessing that the unflappable Mr. Bond will be shaken but not stirred.) Panels **a**–**d** are from public domain websites. Panels **e**–**l** (= **m**) along with their temporal data were furnished by Shigeru Kondo and are used with permission. It was Shigeru's idea to compare these zones to dogs and humans. If he is right that some animal patterns evolved by freezing these waves in transit (**co-option**), then they could be considered **spandrels**. *For color plate see color section.*

Figure 5.9 How the clouded leopard makes clouds. The clouded leopard, *Neofelis nebulosa*, which is seen here from various angles (cubs at upper right and lower left), is a walking enigma. No one knows how it makes the rounded, polygonal "clouds" on its flanks. Most puzzling is the shading at the back of each cloud, which creates the illusion of concavity. The riddle of the clouded leopard is one of many Gordian knots awaiting the next Alexander along the evo-devo frontier. See http://kittysites.com/breeds/bengal/breeder.html for domestic cats of the Bengal breed that look eerily like clouded leopards. They may be our best hope for discerning the genes beneath the clouds. (An equally murky mystery is the curious case of the painted wolf, *Lycaon pictus*, which looks like a hyena that was tortured by a kindergarten art class.) Two species of clouded leopard are recognized [1053,2423], based partly on pelage differences: *N. nebulosa* and *N. diardi* (not shown) [276,393,1161,2425]. Incidentally, as for why big cats have round pupils while domestic cats have slit pupils, theories do exist [1370]. Pictures are from foter.com (clockwise from upper left): Sibtigre2, Smithsonian's National Zoo, cliff1066, San Diego Shooter, and MrGuilt. *For color plate see color section.*

Take-home lessons

1. Having multiple models allows us to devise incisive experiments that distinguish among contrasting predictions [1762].
2. Melanocytes behave like robots, and we are only now starting to gauge the breadth of their behavioral repertoire.
3. Names can dull our sense of wonder [1011,1231]. Just because we call something a "zebra" doesn't mean we understand it, but the act of naming lulls us into thinking we do.
4. As Darwin showed us, much can be learned by asking simple questions, noticing subtle details, and studying quirky traits.

These first five chapters have delved into the meatier aspects of evo-devo, and stalwart readers who have reached this point have chewed many gristly facts and digested many tough lessons along the way. The next chapter has tastier treats.

6 An evo-devo bestiary

This chapter recounts how sundry animals got their quirky traits. The explanations are all provisional, and any one of them could be overturned by new data at any time. But isn't that the nature of science in general?

The topics were chosen mostly from evo-devo literature of the past decade. The animals are listed from A to Z, with some cross-referencing where necessary. Some of the more familiar animals may rekindle the curiosity we felt as children when we first saw them at the zoo or the circus or, for the smaller ones, in our own back yards. Others will be new to many readers, and some are so exotic that they might delight even seasoned zoologists. Overall, the menagerie shows that earth's animals are at least as wondrous as any alien creatures ever imagined by science-fiction writers [468,553,2383].

Like the parade of pilgrims in Richard Dawkins's *The Ancestor's Tale* [504], each of the animals discussed here has its own unique story to tell. Sadly, some of those tales will be their last gasps as their habitats vanish [915,1539], and we will never fully learn how their features arose. Among the species most endangered are the gentle manatee [2376], the nimble cheetah [430], and the fearsome polar bear [1052]. Because the vignettes in this chapter are meant to be more informal, no lessons will be listed, but key concepts will still be set in boldface.

One aim of this book has been to trace the edge of evo-devo's Terra Incognita for future explorers who might wish to seek its buried treasures. Nowhere is the allure of those sparkling gems more apparent than in the 50 cameos of this chapter. With animal genomes being sequenced at such a dizzying rate, we are on the verge of answering the kinds of questions that tormented Aristotle ~2300 years ago [127]. To live in such a Golden Age is a blessing. To play any role in this grand quest is a privilege. To forego the great feast of insights would be a crime. Below are just a few of the tastier confections.

A

How the ant lost its wings

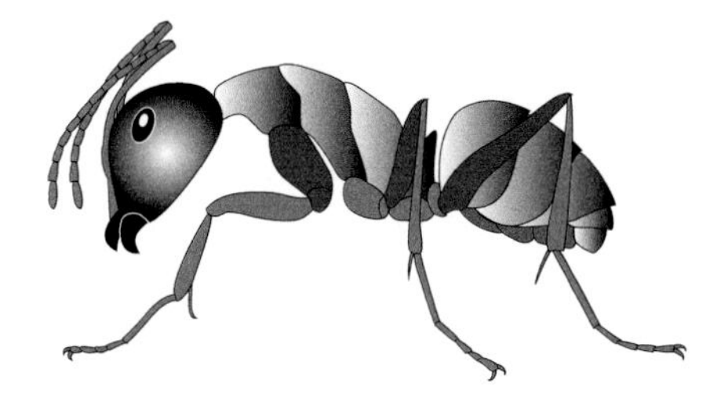

Ants evolved from wasps (which they still resemble) but, like fallen angels, lost their wings ~100 MY ago [1248,2334]. Ants probably lost their wings for the same reason that snakes lost their legs: these appendages became a hindrance in the tunnels which had recently become their homes [1017,1876,2304]. Also like snakes, living underground took its toll on their eyes, which have become mere **vestiges** compared to those of wasps. Presumably, as their vision waned, their chemical senses assumed a greater role to the point where modern ants are as adept as snakes in following scent trails. Wing discs still develop to some extent in larval workers [6], so why don't these discs go on to make wings? Until recently the evidence seemed to implicate malfunctions at various nodes of the wing gene network [1589,2020], but in 2010 we learned the primary reason: worker wing discs undergo programmed cell death before adulthood [2021]. We still need to figure out (1) how the cell death pathway gets triggered in workers and (2) how other castes (queens and males) manage to avoid this doom.

How the ape lost its tail

In the science-fiction movie *Avatar*, the humanoid aliens have long tails like monkeys. If our own monkey ancestors had not lost their tails when they became apes, then we might still have a tail like the Na'vi on Pandora [2150]. Our forebears lost their tails when they came down from the trees [1244,2245] and abandoned their habit of swinging in the jungle canopy [538,1912]. That loss was consistent with the **use it or lose it** scenario. We actually do have a tail when we are embryos, but those vertebrae stop growing, and we are left with only a **vestige** called the coccyx [967]. Rarely, babies are born with what looks like a tail, but these growths typically lack vertebrae, so they are probably not **atavisms** [2283]. It would be surprising if we still have any neural circuits for wagging our immobile "tail bone" considering that it has been ~20 MY [1445,2331] since such talents would have been of use. This question is not as silly as it sounds, because we can still wiggle our ears despite the futility of this twitching for swiveling our ears in a particular direction so as to better hear an incoming sound [1244]. Other vertebrates that forfeited their tails at some point in their history include frogs [44], birds [134,752], and certain breeds of cats [520] and dogs [1062]. In all of these cases, the evo-devo riddle is: how was the tail curtailed genetically [922]? Studies of lizards and snakes show that the tail is molded by a separate module in the genome [2444]. One aspect of that module was revealed in 2003 when *Hoxb13* mutations were found to lengthen the mouse tail [588]. Then in 2009 a new analysis revealed the kernel of this circuitry [2488]: Wnt drives tail elongation (goaded by trunk *Hox* genes and *Hox*-related *Cdx* genes) until a brake is applied by *Hox* genes in the 13th paralogy group – *Hoxa13*, *b13*, *c13*, and *d13*. Given this recent insight, we should now have enough clues to know where to search in our own genome for telltale traces of how our own tail got trimmed (cf. [1177]).

Frontispiece Legless or partially legless lizards, some of which resemble snakes to a remarkable extent. Numbers next to legs in each panel are the number of digits per foreleg or hindleg respectively if present.

Figure 3.5 **How butterflies decorate and sculpt their wings.**

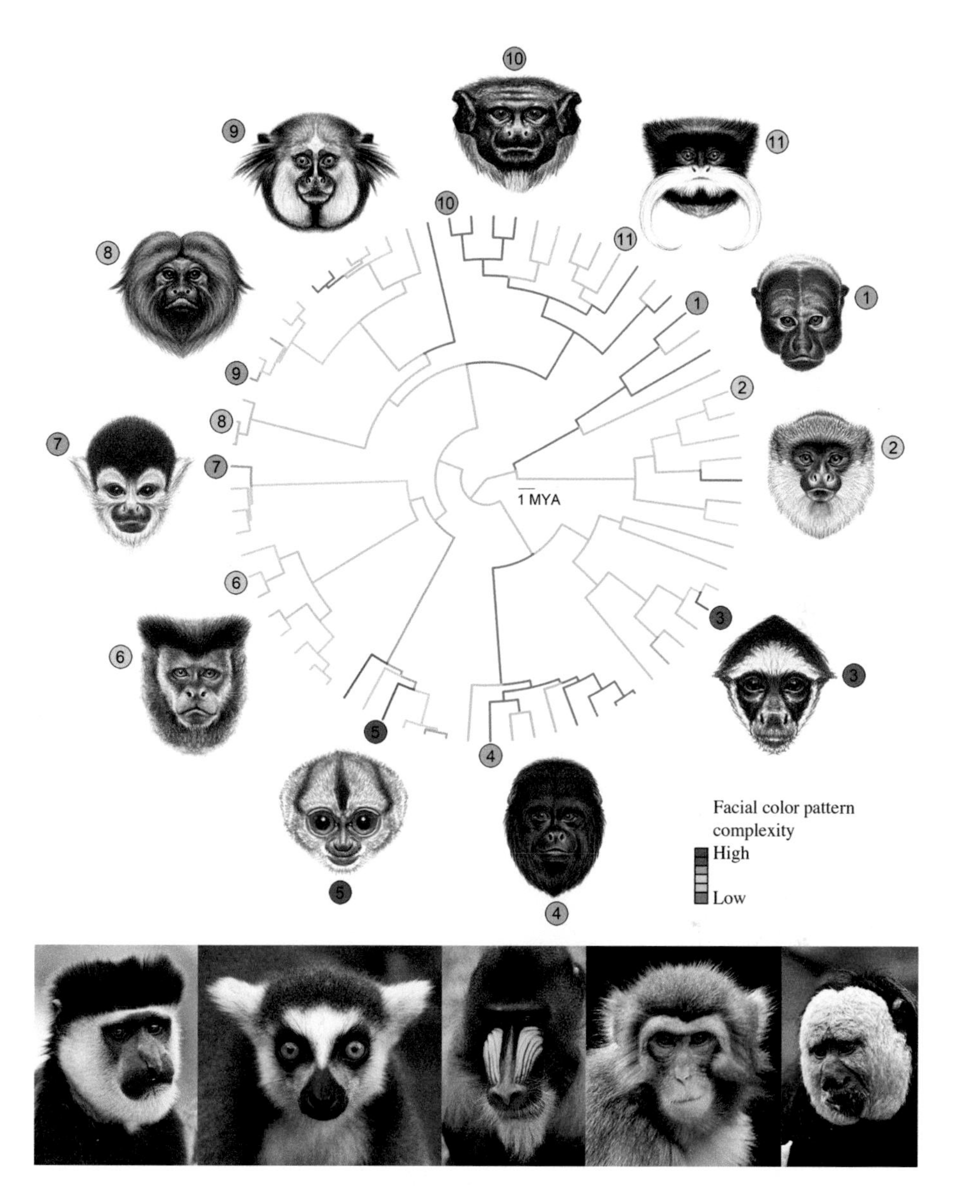

Figure 5.1 **How primates decorate their faces.**

Figure 5.3 How mammals and butterflies differ in the symmetry of skin patterning.

Figure 5.6 Spotted zebras, striped giraffids, and "strotted" squirrels.

Figure 5.7 How the cheetah got its spots.

Figure 5.8 How the mouse makes waves.

Figure 5.9 How the clouded leopard makes clouds.

B

How the bat got its wings

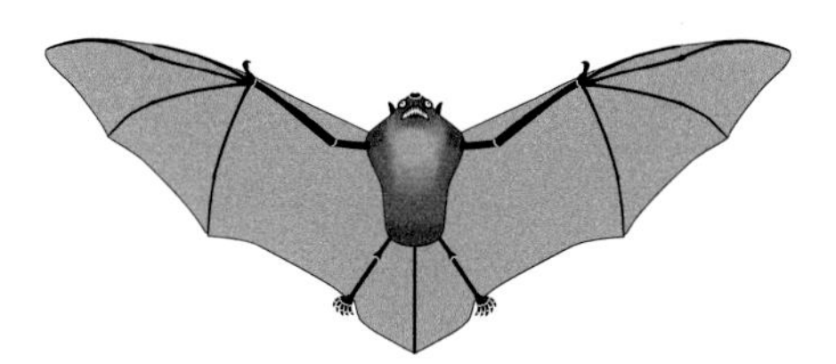

Bats are the only flying mammals [15,656,2052]. However, similar airfoils evolved **convergently** in a few gliding mammals such as "flying squirrels" [2206,2207] and "flying lemurs" [1085,1232]. Hence, bats may have acquired wing membranes as an aid to gliding before they mastered flying ~50 MY ago [196,295,2053] (cf. four-winged birds [118,2507]). Those membranes resemble the webbing of a duck's foot [737,1054,1364]. Curiously, all vertebrate limbs start to make interdigital webbing in the embryo, but this webbing self-destructs whenever it receives a BMP signal [1706,2524]. Such pruning occurs in humans and other vertebrates whose toes are not webbed [2042]. Duck feet emit BMP but prevent it from triggering cell death by also secreting the BMP antagonist Gremlin [1469]. In 2006 Gremlin was shown to be used in a similar way in bat wings to preserve the webbing there [2348]. A mitogenic FGF8 signal was also discovered that could be stoking further membrane growth. We do not yet know how the membrane comes to span the thumb/shoulder, hand/leg, or leg/tail junctures [2348]. Nor do we know how a membrane manages to sprout from the dorsal midline in two bizarre "naked-back" bat species [14]. As for why digits 3–5 of the bat hand grow so big [2052], four genes have been implicated: *BMP2*, *FGF8*, *Prx1*, and *Shh*. The findings are as follows. (1) In 2006 *BMP2* was found to be more highly expressed in bat forelimbs than hindlimbs, and BMP2 was shown to make bat fingers grow longer [1998]. (2) In 2007 *FGF8* was found to be transcribed in an area of the apical ectodermal ridge (the engine driving limb outgrowth [450]) that is three times wider in bat forelimbs than mouse forelimbs [465] – implying that FGF8 lengthens bat fingers [1997] aside from its role in the webbing [2348]. (3) A 2008 report examined the homeobox gene *Prx1*, which causes shorter toes when mutated [449]: mouse toes can be made to grow longer by putting a bat's *Prx1 cis*-enhancer into a mouse [466]. (4) In 2008 a second wave of *Shh* transcription was found in bat wings [1001]. This "rebooting" of the feedback loop at the rear of the forelimb could be what is elongating digits 3–5. The bat wing harbors many more secrets that are worth investigating [467,1002,1147,1649,2329].

How the bee got its pollen basket

Bees have been pollinating plants for ~125 MY [323]. Worker bees collect pollen in a naked area on the hindleg tibia called the basket. Queen bees, in contrast, do not collect pollen, and this part of the queen's hindleg has a normal bristle density. In 2012 it was discovered that this spot expresses the *Hox* gene *Ubx* at a level ~25 times higher in workers than in queens [216]. This Ubx intensity suppresses bristles, probably by blocking the proneural genes *ac* and *sc*, though this inference needs to be confirmed. A similar **co-option** of *Ubx* for the role of making naked patches (devoid of trichomes rather than bristles) was previously documented in the fly midleg [2111], where the extent of bare cuticle is likewise proportional to Ubx dosage. Another *Hox* gene (*Scr*) is used in an analogous, dose-dependent way to establish sexually dimorphic patterns at certain sites on the fly foreleg (see Chapter 2). (*Footnote:* To answer my sister's nagging question once and for all as to whether honey is really "bee spit or bee poop," it is *spit*, and it is made from *nectar*, not pollen [1620]!)

How the beetle got its horns

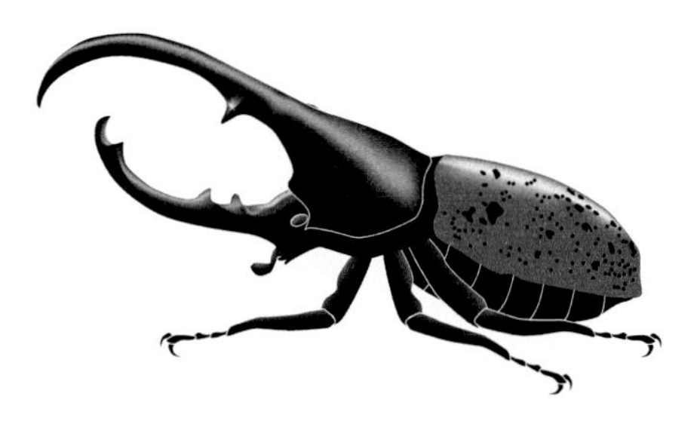

Many beetles have outgrowths that look like rhino horns or deer antlers [614]. As in their mammalian counterparts, these structures can serve as weapons or ornaments or both. The Hercules beetle depicted here wields a powerful pincer [262]. Its two halves develop from the head below and the prothorax above (cf. the humeral discs that make the fly's shoulders: Fig. 2.1). The eye is above the antenna, which hangs down like a tiny golf club below the mouth in this sketch. Raising the head closes the bottom half of the pincer. In this way rival males can be grasped, wrestled, and flipped during jousting contests [88,584,610]. In 2007 the genetics of prothoracic horns was revealed [613]. They apparently evolved when *Scr* (the *Hox* gene for the first thoracic segment) accidentally **co-opted** genes for leg outgrowth [1517,2045,2344], yielding a leg-like appendage at an ectopic site (**heterotopy**). A similar co-option created "helmets" in hemipterans (see the treehopper story). **Allometric** increases in horn length [617] may stem from an enhanced sensitivity of horn-making cells to insulin-like growth factors [618,619,2033] (cf. horned lizards [173,2485]). Bigger horns come at the expense of smaller eyes, wings, or antennae, depending on their body location [610,612,1637], but surprisingly, larger horns do not appear to hinder flying [1428]. Horn shapes are sculpted by cell death [1148]. Research on beetle horns – especially in dung beetles [610] – is one of the most fertile subfields of evo-devo [1190,1511,1516].

How the bird got its beak and lost its teeth

The ancestors of modern birds lost their teeth ~100 MY ago [499,1027,1664]. Despite this huge span of time and the consequent erosion of enamel genes [2066], the oral epithelium retains a residual ability to make teeth [1100]. Evidence for this ability was first adduced in 1980 when chick oral epithelium was supposedly shown to make tooth-like structures when recombined with mouse molar mesenchyme [1181]. However, the possibility of tissue contamination has haunted this claim ever since [1277], and the issue remained unresolved until 2006. In that year the ability of chick tissue to make dental structures was clearly demonstrated by injecting mouse neural crest cells into chick embryos [1502]. The resulting "teeth" abort development prematurely [499], so there must still be blockages to tooth culmination lurking somewhere in the bird genome [1329]. **Atavistic** mutations that reboot tooth initiation in chickens suggest that birds lost their teeth due to a physical gap between the inductive and responsive oral tissues [935]. So much for the *internal* situation. What *external* factors might have fostered tooth loss? Six other clades of extant vertebrates also lost their teeth, and in every case the *loss* was apparently preceded by the *gain* of a new means of food processing. Initially, each new device coexisted with the teeth, and this redundancy would have allowed natural selection to refine the new mechanism gradually [2417]. Eventually, the improved device would have obviated the need for teeth at all and would have led to their demise by the **use it or lose it** rule [499]. What kind of auxiliary device are we talking about here? For anteaters the **novelty** was a long, sticky tongue [37,1593]. For baleen whales it was frilly outgrowths of their gums [1939]. For birds it was their beak plus their gizzard [712,1329,2512,2513], though the gizzard may have already existed as a **preadaptation** [1329,2277,2431]. Some pterosaurs lost their teeth **convergently**, perhaps for similar reasons [1329]. Bird beaks may have begun as opposing scales in the snout of a reptile ancestor [2213] since the beak's horny exterior is made of the same keratinous material as scales [39,1329].

C

How the camel got its hump

The camel's hump stores reserve fat, not water as is commonly believed. Together with other physiological adaptations [1963], this "larder" allows long treks across vast deserts [1790]. The genetics of fat allocation has been studied extensively in our own species for obvious health reasons [685,919,2287]. Humans store lipids at different places in our bodies, depending on our sex [377]. Women store excess fat in their hips [684], while men pack it around their waist as a "spare tire" [1729]. Our first expedition into the fatty subroutines of the camel genome was made in 2005 with the cloning of the genes for leptin hormone [2490] and its receptor [1951]. Both genes were found to be expressed in the hump as well as in the liver and mammary gland, but we don't yet know whether leptin acts in a paracrine way to regulate fat at these sites per se. Interestingly, women's breasts are shaped by adipose tissue more than those of any other primate [1244], so some of what we learn about fat sequestration in humans may shed light on how camel humps evolved. Why did the camel grow a hump (or two) on its back, rather than a breast-like organ (or two) on its chest or belly? No one knows. However, there may be a predisposition for humps in ungulates in general because dorsal humps evolved not only in camels but also in cattle (see the zebu story). One last riddle: What is the genetic "area code" for the hump region, and how did it **co-opt** the fat deposition machinery? Our best guess at this point is that the insulin pathway helped to target this site [619,2033].

Why the centipede has odd segments

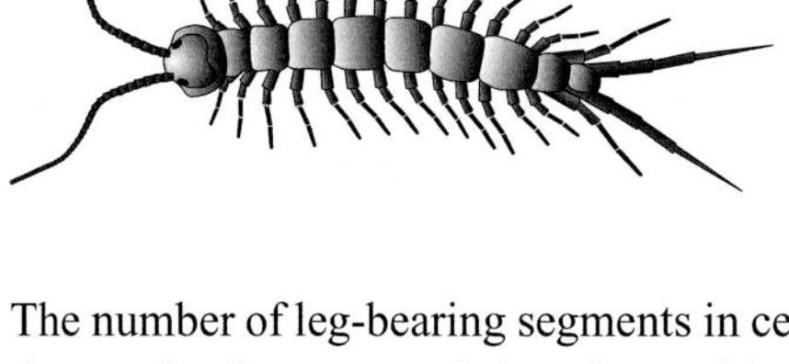

The number of leg-bearing segments in centipedes ranges from 15 to 191, depending on the species [78,217,1490], but the total is *never* even [75,484]. Why? The riddle of this odd **constraint** was solved in 2004 [387]. It turns out that centipedes undergo two phases of segment formation during development [235]. First, a certain number of segments is made in the head, and this number is invariant and odd. Then additional segments are added two at a time: a double-segment "prepattern" unit is made, and then subdivided into two segments. The number of these double-segment units varies among species, but the final number of segments added is always even, and so the sum remains odd. This "double addition" method of segmentation differs from that used by other arthropods, including millipedes, which also have lots of segments but suffer no such constraint [387,484,2022]. How do centipedes and millipedes escape the standard arthropod limits to make a virtually unlimited number of segments [2417] bearing up to 750 legs [1383]? The answer may be the same as for how snakes defy the rules for vertebra number [2298]. Centipedes, like snakes, use a Notch-based oscillator clock [386] to make their segments from a posterior growth zone [604,1079,1487], so their clock may simply tick faster [719,1491]. It is baffling that certain centipede species can make a fixed number of segments when that number seems far too large for an imperfect oscillator clock to ensure [77,217]. Indeed, the enigma of how embryos do math in general has eluded our bravest Beowulfs for 100 years [139,438,1412,1413,1485]. Is there no hero out there who can slay this dragon for us?

How the crocodile got its (facial) scales

In 2013 a craggy crocodile face glared out from the cover of *Science* [1476]. The authors of the cover article reported that the array of scales on the crocodile head is asymmetric, unlike the symmetric lattice of the snake head. Moreover, the patterns vary a lot from one individual to the next in crocodiles but not in snakes. As discussed in Chapter 5, asymmetry and variability imply reaction–diffusion (RD) mechanisms, which appear to be commonplace. So why all the fuss? What was newsworthy here was that RD models could be ruled out based on other idiosyncrasies of the patterns. Instead, the patterns matched expectations that are based on how surfaces crack when they are subjected to tensile forces. The notion that aspects of anatomy can be explained by physical forces (like expansion cracking) was advocated ~100 years earlier in D'Arcy Thompson's 1917 *On Growth and Form* [2203] and in Theodore Cook's 1914 book *The Curves of Life* [437]. Over the intervening century, various traits have been proposed to arise *mechanically* rather than *genetically* [677,1612,1925,1930,2254]: brain convolutions [992,1132,2101,2265], cartilage condensations [124,1696], flower corrugations [2017], tooth cusps [1098,1927], and fish otoliths [2457]. To this kooky list we can now add the crooked smile of the crocodile, or at least the cracked skin that surrounds it.

D

How the deer regrows his antlers

Deer antlers are the only appendages in mammals that are regularly shed and replaced. For that reason they have been used as a model for regeneration in general (see also the newt story). Their utility as a model spread to the realm of stem cell research in 2008 when the process was found to rely on stem cells at the antler base [1882]. Antlers that are in the process of regrowing retrace their own embryonic path not only in terms of cartilage and bone formation but also with regard to hormones and morphogens [649,1551]. These parallels fostered the notion that antler regeneration is merely a "rebooting" of antler development. However, this hypothesis was disproven in 2010 by a series of experiments involving transplants of presumptive stem cells [1295]: stem cells from the base of *regenerating* antlers could not induce an ectopic antler when transplanted to a non-antler site, whereas induction *was* possible when stem cells from *developing* antlers were used. Further work will be needed to ascertain (1) why these cell types differ in their potency and (2) whether the less potent cell type can be converted into the more potent one experimentally [379].

How the duck got its bill

The duck bill is essentially a flattened chicken beak. A study of duck-versus-chick embryos in 2004 uncovered the basis for this difference [2461]. In both species the frontonasal tissue mass (FNM) has two lateral growth zones initially [1039,1315]. In chick embryos these zones converge and fuse medially to make a pointed beak, but in duck embryos they endure longer, resulting in a wider bill. In both cases, the distal outgrowth is attributable to the **morphogens** Shh and FGF8, but the beak rudiments of ducks express BMP4 at a higher level. Acting on this clue, the researchers tried adding extra BMP4 to the FNM and other facial bulges [103] of a chick embryo and found that they could reshape the beak into a facsimile of the duck's bill. This ability to mimic a genetic trait by non-genetic means is called "phenocopy" [802,899], and it is a good way to test evo-devo hypotheses in general. Transplants between duck and quail embryos to create "quck" or "duail" chimeras [583,2225] show that beak shape is dictated by immigrant cells from the neural crest [1100] – *not* by resident facial tissue [1965,1967] – and the same is true for jaw muscles [2215]. BMP4 controls beak shape in other birds [2460] such as Darwin's finches [10,1367], and it regulates jaw shape among vertebrates more generally [867,1724]. Beak *length* (as opposed to shape per se) is affected by the calmodulin pathway [9,1724,1966] in addition to the morphogens listed above. Duck-like bills evolved **convergently** in the hadrosaur (a dinosaur) [625,1431] and in the platypus (a mammal) [879]. It is the peculiar platypus that we consider next.

How the duck-billed platypus got its bill

The platypus is a Frankenstein-like mammal [842,2194] that looks as if a duck's head had been grafted onto a beaver's body [87,879,2009]. However, the platypus bill has a "killer app" that the duck bill lacks: it bears rows of sensors along its length that can detect electric fields and thereby find prey [669,879,1787]. A similar prey-finding "antenna" evolved **convergently** in the sawfish, and its sensors reside along its "saw" [2463]. We know very little about how the platypus bill develops [1372] and even less about how it evolved. When the platypus genome was sequenced in 2008, enamel genes were identified [2341], but this discovery was not surprising since the juvenile has teeth [499], and so do the fossils of potentially ancestral species [2134]. Here we see a limitation of genomics in general: we are a long way from being able to plug a genome into an informatics algorithm so as to chart either its embryonic trajectory or its evolutionary history. Indeed, we may *never* be able to do so if, as some authors argue, the computer metaphor for development is inherently flawed [365,655,1931,2408].

E

How the echinoderm got its pentamery

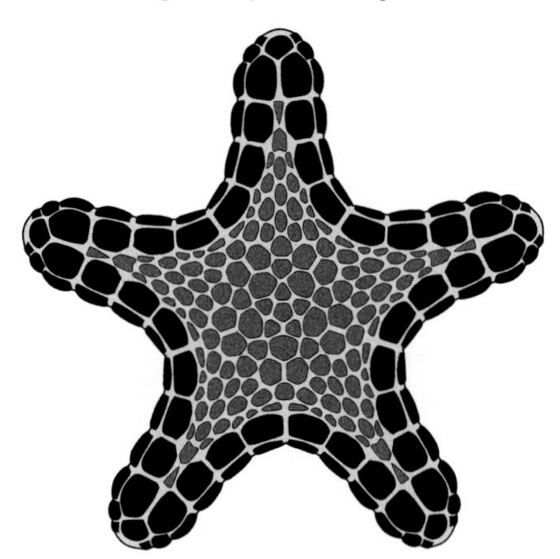

The first echinoderms had mirror symmetry like other bilaterians [45,2496], but the only ones alive today (e.g., this biscuit star) have *radial* symmetry as adults [1101]. In 2008 a theory was proposed for this transformation [2073]. According to that theory, the chief benefit of two-sidedness (fast locomotion [926]) was forfeited when a sessile lifestyle was adopted. The selection pressure for two-fold symmetry was thereby relaxed, and the stage was set for its complete loss at a later time. Then the larva, for unknown reasons, attached itself to the substrate by its front end so that the mouth did not have enough room for efficient feeding. This cramping was remedied by an asymmetric torsion of the internal organs to relocate the mouth (cf. torsion in snails [1698]). Modern echinoderms abandoned the torsion but kept the asymmetry. Incredibly, they fashion an adult rudiment (a miniature version of their adult selves) from tissue on only one side of the larval body [183]. (*Disclaimer:* Aspects of this argument seem flawed to me, but readers can consult the paper [2073] and judge for themselves.) The geometry of sea urchin metamorphosis [491,1494] is as strange as noticing a pimple on your left cheek that grows larger every day until it bursts to unfurl a five-pronged umbrella. The genetic basis for this asymmetry [983] was revealed in 2005 [571]: *two* adults develop per larva (one twin on each side) when the gene *nodal* is disabled. This gene is normally expressed on the right side, where it suppresses that adult rudiment. This discovery also helped solve the deeper mystery of D–V inversion (see Chapter 1) [568,984]. In 2012 BMP was identified as the signal that induces the adult rudiment [1337,2336]. Most extant echinoderms are pentameric, but we still don't know why five-fold – as opposed to six-, seven-, or eight-fold – symmetry evolved [1546], nor how it develops [63,513,1495] (cf. sea anemones [711]). In any event, the recipe can't be perfect, since six-armed sea stars are occasionally found that survive with no obvious adverse side effects. This same counting riddle applies to flowers that have characteristic numbers of petals (cf. four-leaf clovers) [519,1524]. Unlike the eight arms of an octopus (see the octopus story), the five arms of a sea star do not appear to be **individuated** by unique *Hox* codes [2073], though certain arms are preferentially used for specific types of motions [1101].

How the elephant got six toes

Ever since 1710, when elephant feet were first dissected in detail [199], the cognoscenti of the scientific community have known that elephants have a sixth toe, but the rest of us have remained ignorant of this quirk. Until recently, the extra toe was presumed to be a bona fide digit, like the extra toes in Hemingway's cats, whose preaxial polydactyly has been traced to an aberrant enhancer at the *Shh* locus [1281]. In 2011, however, an evo-devo analysis [1058] showed that the sixth toe is more like the ersatz thumb of the giant panda [495,621,832]. Like analogous "thumbs" of the mole [191,1168,1499], the pterosaur [1709], the sea turtle [731], and the hairy-legged vampire bat [1982], the sixth toe of the elephant is merely an enlarged sesamoid (wrist or ankle) bone [2284] that was **co-opted** to play the role of a toe [1058,2522]. Because this pseudo-toe bears so much of the body's weight inside the foot cushion, it is able to prop the heel up to an angle between plantigrade (e.g., human) and unguligrade (e.g., horse). Despite their girth, therefore, elephants have been strutting around in high heels all these years! Could that be why they kept their **vestigial** toenails – as part of an overall fashion statement? (See ref. [1475] for how the elephant got its trunk.)

F

How the firefly got its flashlight

Until 2003 the flashlight of fireflies was presumed to have evolved for its current role in courtship [1291]. In that year a detailed phylogenetic tree was charted for fireflies and their kin [233]. The tree revealed that bioluminescence arose with a different duty entirely. It served as a signal to warn predators of the larva's toxicity, analogous to the stripe of a skunk [1611,2324]. Only later was this *larval* flashlight **co-opted** for *adult* usage in sexual display. **Character displacement** [986,1751] may have then led sympatric species to emit light from different abdominal regions and to diverge in their flashing rhythms. From an evo-devo standpoint, we would like to know how bioluminescence came to be confined to the ventral compartments of particular body segments [775]. The posterior *Hox* genes must have been involved, but how?

How the flatfish got its (funny) face

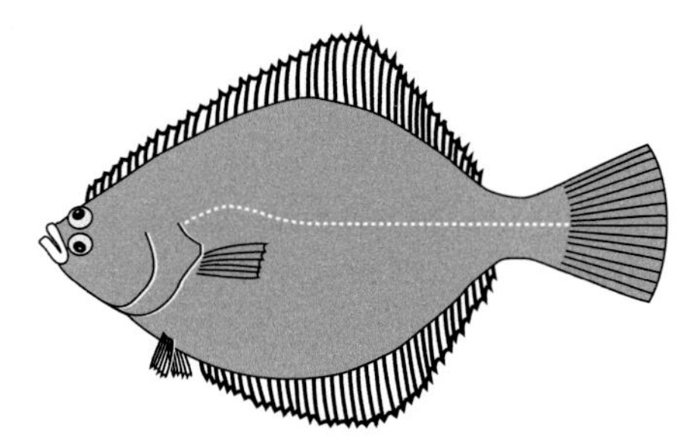

Juvenile flounders look like ordinary fish, but their anatomy changes radically when they undergo metamorphosis [1978]. One eye migrates across the midline to create the piscine equivalent of a Picasso painting [1918]. Adults nestle on the ocean floor [1814] with their eyes peeking out above the sand to scan for predators or prey [1767]. The upper side is camouflaged, but the lower one is not [2343]. When disturbed, flounders swim by undulating in this same "magic carpet" posture [864,1977]. Until 2008 flatfish were thought to have arisen by a sudden mutation [2520], but in that year fossils were discovered that stunned the scientific community. These fossils had "missing link" phenotypes where the migrating eye barely reached the midline [710,1089]. The obvious implication was that the enabling mutations initially had only mild effects, so the **hopeful**

monster idea had to be wrong [700]. In 2002 a mutation called "*reversed*" was found in the Japanese flounder *Paralichthys olivaceus*. It switches the direction of eye migration so that both eyes end up on the *right*, instead of the *left*, side of the head [940] in 20–30% of homozygous offspring. Such mirror-image fish, which lie on their left side, do not reliably have a reversal of their abdominal organs, so *reversed* can't be affecting the *nodal* pathway that creates *embryonic* asymmetry [96,1282,1592]. However, a study in 2009 showed that one key gene in the pathway (*pitx2*) is re-expressed in the *post*-embryonic brain, and that *reversed* is acting via *pitx2* [2153]. Once upon a time, the genetics of flatfish asymmetry – and of vertebrate asymmetry in general – seemed as simple as a fairy tale [1467], but now it is starting to look as convoluted as a Dickensian novel [2273]. Untangling it will take some time.

How the frog got its (jumping) legs

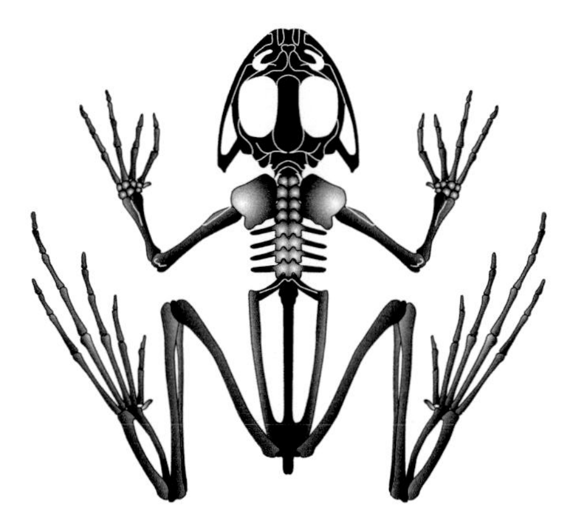

Frogs broke one of the oldest rules in the tetrapod handbook [2046,2135] by evolving what looks like an extra tibia ("tibiale") and fibula ("fibulare") in their hindlegs [924]. Having an extra link in their chain of leg bones makes sense because it gives them more leverage during liftoff [1168,1885]. This same device evolved **convergently** in two leaping mammals: the tarsier and the galago [609,758,1894]. The frog's new leg bones evolved from old ankle bones [334,2230] by a cute genetic trick. The successive sections of the tetrapod leg (femur, tibia–fibula, and ankle–foot) are encoded by different *Hox* genes from the base to the tip [2494]. In 1998 it was revealed that *Hoxa11*, which bestows long-bone identity to the tibia and fibula, is expressed for a longer time in the developing leg than in the developing arm [201]. This temporal difference (**heterochrony**) apparently causes two ankle bones to adopt the same identity (**homeosis**) as the tibia and fibula [1850]. From a functional standpoint, we are left to wonder (1) how the circuitry of the frog brain got rewired for bipedal hopping from an initial state of sinusoidal swimming (the "evo" issue), and similarly (2) how the circuitry of a tadpole's brain gets rewired for hopping during metamorphosis (the "devo" issue).

Figure 6.1 How the giraffe got its long neck. Head and neck skeletons of (left to right) an okapi, a giraffe, a duck, and a swan (not to scale). The okapi and giraffe both have seven neck vertebrae, while the duck and swan have 16 and 23 neck vertebrae respectively [728,1616]. The number of cervical vertebrae is **constrained** in mammals, but it varies freely in birds [728]. The reason for this difference has been traced to muscle cells that migrate from the neck to make the diaphragm [273] – a mammalian **novelty** that birds lack. Giraffes probably evolved a long neck due to sexual selection [2054], whereas the driving force for swan neck elongation was presumably natural selection for benthic foraging [2411].

G

How the giraffe stretched its neck

Giraffes evolved from short-necked ancestors that resembled okapis [1496,1790]. Until recently, the most popular explanation for the long neck was that taller individuals were better equipped for reaching higher leaves during longer droughts [2411]. However, the actual foraging behavior of giraffes during dry seasons refutes this notion. Instead, the driving force was more likely to have been sexual selection [857,2054]: (1) males use their necks in jousting contests to win mates, (2) they have longer necks than females, and (3) their necks exhibit the positive **allometry** expected for a sexually selected trait [614] (but cf. [1497]). Naively, we might think that longer necks demand more vertebrae, but that is not so [2318]. Giraffes have only seven cervical vertebrae (Fig. 6.1) [2086,2267] like every other mammal [1597] except sloths (7–10) [668,943] and sirenians (6–7) [274,1597,2275]. The only other animals with any obvious neck

constraint were pterosaurs [2180]. In contrast, cervical vertebrae vary freely in birds [487,730,1616] and reptiles [1565,2180]: there can be as many as 17 neck vertebrae in stegosaurs [1402], 19 in sauropods [554,953,1879,2008,2179], 25 in swans [1616,2180], and 76 in plesiosaurs [1549,1944]! Until 2011 the favored explanation for the neck constraint was that mutations causing cervical ribs in mammals (but not other groups) have harmful side effects (cancer, etc.) [728] that are selected against [729,732,2186]. A rebuttal of this paradigm was published in 2011 [82], and in 2012 a new idea was proposed. The new hypothesis stems from the fact that a critical organ depends on the neck somites that form neck vertebrae (cf. [1406]). Mammals differ from other vertebrates insofar as they breathe using a diaphragm [331]. This **novelty** arose ~200 MY ago [1746]. Oddly, the diaphragm muscle comes *not* from nearby *thoracic* somites but rather from the distant *neck* somites C3–C5, whose myoblasts migrate over the intervening distance [1244]! Because of this quirk, the mammal neck specialized into three **modules** whose **morphological integration** may be too difficult to disrupt without risking death: C1–C2 (skull articulation), C3–C5 (diaphragm musculature and innervation), and C6–C7 (forelimb musculature and innervation) [273].

H

How the horse got its hooves

Horses walk on tiptoe like ballerinas [991,1389]. The single toe per leg that bears their weight corresponds to our middle finger or toe (digit 3) [505,641], and each hoof is a thickened toenail [225]. Toes 2 and 4 arise in the embryo, but their growth is stunted, so they become **vestigial** splints in the adult [644,645,904]. Rarely, these flanking toes reappear **atavistically** in adults like the side toes of a pig [348,2103]. Pigs and other animals that seem to have "cloven" hooves [450,1999] actually have *two* hooves per foot [850,1888,2016]. These *even*-toed ungulates evolved hooves independently of their *odd*-toed relatives in a nice example of **convergence** [37,1775]. Hooves are made of the same keratinous material as the reptilian claws from which they evolved [39,2085,2459]. By 2001 the agents that govern keratin deposition had become clear: (1) BMP4 recruits mesenchymal cells to digit tips, (2) Shh induces a claw-making matrix, and (3) transcription factors Msx1 and Msx2 [170] control matrix growth, thereby regulating hoof size [917]. How the links among these players arose evolutionarily remains to be determined.

I

How the insect got its (stiff) upper lip

Insects typically have a shelf above their mouth that acts as an upper lip to retain ingested food. The origin of this "labrum" has been disputed for decades [897]. The issue was whether or not it is **serially homologous** to thoracic legs [896]. One fact that seemed impossible to reconcile with such homology was that it is not paired. How could a *medial* organ be related to *lateral* pairs of limbs? In 2006 this mystery was solved. The labrum actually develops as a *pair* of buds, but then each bud rotates 180° and fuses with its partner at the midline to form a single unpaired outgrowth [1152]. The appendicular nature of the labrum was confirmed by its gene expression territories, which match those of the legs [1772,1776]. The matter seemed settled until 2009, when investigators looked for *wingless/hedgehog* boundaries that trigger leg outgrowth. Such boundaries were nowhere to be found in the labrum! Evidently, it does not need this cue for initiation [1776]. Indeed, the area of the head where it develops lacks any hallmarks of a body segment. Overall, therefore, the labrum resembles a leg that got shoehorned into an unsegmented part of the head. Clearly, we can't infer either **homology** or **co-option** in this case (or any other for that matter) without assessing the *entire* gene network that has supposedly been recruited [1529]. At this point, we don't even know which came first – the labrum or the leg.

J

How the jerboa got its (jumping) hindlegs

Jerboas look like mice but hop like kangaroos [600,601,1129]. In all, there are 33 species of these desert rodents [447]. Hopping evolved **convergently** in kangaroo rats, presumably for the same reason as in jerboas: it is a better way of traversing blisteringly hot deserts than scampering on all fours [447]. How hard was it for evolution to rewire the jerboa's ancestor's brain for hopping? Not hard at all, perhaps. In 2007 a single-gene mutation was found that causes ordinary mice to hop! The growth cones of the mutant's motor neurons are led astray from their normal targets in the spinal cord [1070]. Behavioral changes often precede anatomical ones in evolution [2372], so the proto-jerboa may have adopted hopping before its feet were modified, rather than the other way around. In the lesser Egyptian jerboa the feet (1) elongated the metatarsals of digits 2–4, (2) fused the metatarsals into a cannon bone as in the horse, and (3) eliminated digits 1 and 5 [447]. The ability to raise jerboas in captivity [1111] has made them an ideal system for studying the evo-devo of legs [447]. The mystery of how the jerboa's metatarsals manage to grow even longer than its femur was solved in 2013 (cf. [583,1883]). It turns out that jerboas elongate the metatarsals of their legs in the same way that bats lengthen the metacarpals of their wing [648]: cartilage cells in the growth plates swell extremely rapidly after birth [448].

K

How the kiwi almost lost its eyes

The kiwi is an odd bird in many respects [288,302]. Its wings are **vestigial**, its feathers look like fur, and its egg can weigh a pound, even though the kiwi itself is no bigger than a chicken [303,2075,2161]. Kiwis hunt nocturnally, but unlike owls, whose eyes are huge [257,1032,1464] (cf. tarsiers [1464,1894]), kiwis modified their eyes in the opposite direction. Indeed, kiwi eyes are so tiny that they are virtually useless at night [454]. Instead of relying on vision, kiwis probe the soil using a slender beak whose nostrils are at its tip – not at the base as in all other birds [2161]. How the kiwi relocated its nostrils is an evo-devo riddle, akin to the migration of nostrils in whales to make a blowhole in the roof of the skull [178,1450,2197]. In 2007 dozens of sensory pits were discovered around the bill tip [1394]. The concentration of smell and touch sensors at this site may have forced a trade-off in the brain, where fewer neurons could be devoted to vision as more were diverted to the bill. Analogous trade-offs (**constraints**) exist in beetles, where eyes shrink as horns enlarge [612,1637]. At the base of its bill the kiwi has sensory bristles that look exactly like mouse whiskers, and they may serve a similar (**convergent**) function at night [480].

L

How the ladybird got its elytra

Elytra are hardened forewings. They cover the hindwings [88,694], and they can be quite gaudy, as in the ladybird beetle shown here [1136]. (The gray background is actually red.) The elytra are spread during flight to permit hindwing flapping. Coleopterans evolved from insects with flimsy forewings [1211], so how did they transform their forewings into protective shields? In 2009 we found out [2220]: cuticle-stiffening genes were recruited (**co-opted**) to the wing circuitry on three separate occasions during evolution [64,653]. Given that the hindwings rely on this same genetic circuitry, how did they resist becoming just as thick as the forewings? They were able to do so because the *Hox* gene *Ubx*, which is active in the hindwing, thwarted each of these new links (downstream from *ap*, *bs*, and *sc*; Fig. 2.2) as the links arose . . . or shortly thereafter. For that reason, disabling *Ubx* in a beetle converts its hindwings into elytra, rather than the other way around [2221]. Oddly, the elytra of some desert beetles are fused at the midline, making flight impossible, even though the wings underneath look perfectly functional [1957]. How the medial edges of the latter elytra fuse during development is a mystery, considering that they come from separate discs (cf. [1481]).

How the lion got his mane

Lions are the only cats that have manes [1704], but some male baboons do also, and male ornaments of other kinds abound – for example, goatees in goats and beards in men.

Darwin was the first to explain dimorphic traits in terms of sexual selection [488], and research has confirmed his proposal in many animal groups [51]. In the case of lions, life-size decoys show that fellow males assess prowess by both mane length and color, but lionesses seem to care about color alone [2370]. The latter result must be interpreted cautiously, however, because female preferences can change [2390]. No other felid looks close enough to a lion to be confused with it, so **character displacement** [986,1751] seems unlikely as a driving force, but we can't be sure since lion-like subspecies might have existed and then perished. So much for the "evo" side of the mane issue. On the "devo" side, what we want to know is how genes make manes, so to speak, but available approaches are limited: (1) among big cats in the *Panthera* genus, lions are most closely related to leopards [494], but leopards show no hint of a mane; (2) hybrids between lions and tigers cannot lead us to mane genes because they are sterile; and (3) no mane has yet been bred in domestic cats. Ironically, the best clue so far comes from a 2010 analysis of our own species [968]. Wnt appears to assign "area codes" to body regions much like the ***Hox* Complex** subdivides the body axis [1650]. Once the lion genome is sequenced, it should be possible to compare the regulatory motifs of Wnt pathway genes with those of the cat to seek differences. Suspected control elements can then be assayed in mice to see which ones give the mouse a mane [2040]. If we can put a mane on the mouse, then making a mouse that roars might be possible as well.

M

How the manatee lost its hindlegs

Like their distant whale cousins [148,450,2195], manatees turned their forelegs into flippers while losing all but a few **vestiges** of their hindlegs and pelvis. Manatees belong to the order Sirenia, which is closely allied with elephants [1666,1790]. Fossils of partly aquatic, four-legged (walking) sirenians date back ~50 MY [556], implying that the transition to aquatic life occurred around that time. In 2004 we found out how the hindlimbs vanished. They were suppressed in the same way as the pelvic fins of the stickleback (see the stickleback story) [2012] – namely, by disabling an enhancer of the homeobox gene *Pitx1* that dictates hindlimb identity (**individuation**) in all vertebrates [570,1063,1493]. *Pitx1* cannot be just acting as a **selector gene** here, however, because in that case its null phenotype should be **homeosis** of hindlimb into forelimb – not hindlimb *loss* per se [1321,1916]. Loss- and gain-of-function studies in mice [2157] and chicks [1322] show that *Pitx1* does indeed affect hindlimb growth as well as limb identity [523,866,1235,1377]. In 2013 the mystery of the longer rib cage in manatees was solved [1597]. The extra ribs turn out to be due to one base change (CTAATTG to CCAATTG) in the Hoxa10 binding site of a *cis*-enhancer at the *Myf5* gene locus [886], and a **convergent** mutation explains why snakes have lumbar ribs. Darwin was amused by the "rudimentary nails on the fin of the manatee" (p. 454 in [486]). Such silly **vestiges** on an otherwise respectable flipper prove a terrestrial ancestry better than any fossil ever could. These fingernails might also help explain why seafarers – including Columbus himself – often mistook manatees for mermaids [1124,1861], though it is more likely that they were just drunk and suffering from beer goggles syndrome.

N

How the newt regrows its legs

There is no evo-devo mystery in more urgent need of a solution than this one [1064,2129]. If only we could find out how newts regenerate their limbs [2127], then we might be able to mimic the process in humans [248,1777,2169] so that amputees could regrow their arms or legs with no need for prosthetics [327,1568,2126]. The prospect of a breakthrough is nearer than ever, given exciting advances that have been made recently [1308]. In 2012 the regrowth of the radius and ulna was accomplished in amputated forearms of neonatal mice by exposing the stumps to the growth factor BMP7 [1060]. The chief roadblock has always been a propensity for mammalian wounds to form scars instead of blastemas [69,1568,1571], but in 2012 this obstacle was finally overcome. Two obscure mouse species were found to be able to regrow their skin without scarring [2002] – thus offering a mammal system where blastema formation might be decipherable . . . and ultimately inducible [2168]. Moreover, axolotl skin was coaxed to heal without scarring by raising the level of tenascin and lowering the level of fibronectin [2004] – thus paving the way for similar remedies in mammals. Less progress has been made on the "evo" side of this riddle. What we would like to know is whether limb regeneration is a primitive ability that was lost in our lineage [165,733,2003] (cf. tooth regeneration [696]) or a talent that arose uniquely in newts and other salamanders [249,751,1582,2167]. Currently, the evidence is equivocal [1327], and it is likely to remain so until the "devo" side of the matter is settled once and for all.

O

How the octopus got its arms

Amazingly, the eight arms of the octopus evolved from the foot of ancestral mollusks [213,214,1202]. We do not know how this transformation occurred [119], but based on what we learned in 2003 about the ten arms of the squid [1267], each octopus arm is likely to possess a unique identity (**individuation**) that is encoded by a combination of *Hox* genes. The constancy of arm number in octopuses – always eight, never seven or nine – could therefore be due to an "inertia" of this *Hox* code after it "congealed" in the genome [1489]. Evolution is often **constrained** by development, which, in turn, can be limited by gene circuitry [582,670,1924]. This same logic may explain why tetrapods never exceed five digits per hand or foot (see the tetrapod story) [731,852,2159], except in rare cases where a wrist or ankle bone has been **co-opted** to function like a digit (see the elephant story) [2284].

P

How the pterosaur got its wings

Pterosaurs were pelican-like reptiles that went extinct along with non-avian dinosaurs ~65 MY ago when a wayward meteor collided with earth [53,370,1522]. Unlike bats, whose wing is braced (like an umbrella) by four splayed digits [1609], the pterosaur wing was supported (like a sail) by a single spar along its leading edge [1246,2251]. That spar consisted mostly of the fourth finger, which was 10 times longer than the other fingers and nearly as long as the entire body [158]! How was one digit singled out for such excessive growth [967,2124]? Similar mysteries of **mosaic evolution** also apply to the huge middle toe of the horse [644,645], the thick middle finger of the giant anteater [1168], the thin middle finger of the aye-aye [2522], the thin fourth finger of the striped opossum [1410,2263], and sundry other extreme digits in bandicoots [654,2278], seals [177,178,2465], and dinosaurs [444,2253]. All of those digits somehow managed to grow (or shrink) without adding (or losing) any phalangeal bones – implying a **constraint** on the number of phalanges per digit [843,1849], though that rule has been broken on occasion (see the whale story). The best clue we have to the riddle of the pterosaur spar comes from a 2012 study of congenital human macrodactyly [1253]. All of the patients had elevated levels (34-fold excess) of pleiotrophin in their enlarged fingers. This secreted protein is a diffusible growth factor whose mitogenic effects may be mediated by the Wnt signaling pathway. How is pleiotrophin targeted to one finger alone? The targeting must rely on digit **individuation** as enforced by *Hox*-mediated area codes [524,1527,1650], but we do not yet know how area codes get linked to growth factors [967,2025]. Fossils of pterosaur embryos reveal that the fourth finger enlarged before hatching [381,1102,2325]. There remains the puzzle of how the wing membrane arose [606,2251,2415], but we may be able to figure that out by studying the bat wing [2348].

Q

How the quagga lost its hind stripes

The quagga was a brown zebra that lacked stripes on its hindquarters (see Fig. 5.5d). Before it was hunted to extinction in 1883 it overlapped with the plains zebra, *Equus burchelli*, in South Africa [931]. When its DNA was sequenced from a pelt in 1984 – the first such recovery of DNA from an extinct animal [987] – it was thought to be a separate species (*E. quagga*), but later tests showed that quaggas were actually a *sub*species of *E. burchelli* [988,1279,1332]. Once researchers realized that quaggas must have interbred with fully striped zebras before dying out, they decided to try to resurrect quaggas by artificial selection. The idea was to distill any ghost alleles that are still floating in herds of *E. burchelli* to homozygous purity by breeding rare individuals that have quagga-like striping [1725]. Beginning with 19 founder animals in 1987, the program yielded ~25 third-generation offspring by 2009, some of which look quite quaggish [931]. Once the fourth generation forms an autonomous herd it should be easy to figure out the genetic basis for zebra versus quagga striping. Ideally, this analysis should also include zorses and zonkeys (hybrids between zebras and either horses or donkeys) [1802,2210], whose stripes differ from those of the quagga [834,835,2456]. As for *why* the quagga lost its hind stripes, we must first address the question of why zebras acquired stripes in the first place. Most theories are speculative [329,1139,1917,2120], but in 2012 zebra stripes were convincingly shown to repel biting flies [598], so quaggas may have lost the need for this defense mechanism when they abandoned the African region that is most strongly infested with these pests [2300].

How the quetzal got its crest

The resplendent quetzal (yes, that's its real name: the resplendent quetzal), *Pharomachrus mocinno* [288], is so beautiful that Aztec kings adorned their head-dresses with its iridescent tail feathers [1749], and they worshiped a quetzal-feathered serpent god named Quetzalcoatl [2432]. The male has a fuzzy tuft of hair-like feathers on his head. As luck would have it, the same type of crest occurs in mutant chickens [2328] (cf. fancy pigeons [2013,2014]), so we can use the chicken as a proxy for the quetzal, which cannot be raised in captivity. In 2012 the mutant's crest was traced to the gene *Hoxc8*, which normally specifies thoracic identity (**individuation**) in the vertebral column (see Fig. 4.4). In the mutant, *Hoxc8* is ectopically expressed in the scalp, where it apparently alters the polarity of the feathers so that they stand erect instead of lying flat. Whether the mutation acts via a *cis*-enhancer in the ***Hox* Complex** [2328] is not known. Its most likely downstream targets are the planar cell polarity (PCP) genes [373,812,871] that are part of the urbilaterian toolkit. Indeed, the cowlicks that afflicted many of us as children (see Fig. 6.2, panel 5) might turn out to be due to similar PCP flaws [97,373,1763] that are incited by misexpressed *Hox* genes (**heterotopy**). If so, then we and the quetzals should be thankful we don't have ribs growing out of our heads (the normal role of *Hoxc8*). The quetzal wouldn't look quite so resplendent then!

Why the quoll kills its babies

According to Greek mythology, Cronus ate his own children [431]. The northern quoll, a marsupial about the size of a cat, is not much better as a parent [288]. It gives birth to ~16 babies per litter [1605], but half of them starve to death because there are only eight teats. Quolls can have as many as 30 babies, in which case the carnage is even worse [87]. Shouldn't natural selection have ensured as many teats as babies? In principle, yes [141]: the rule for placental mammals (obeyed by our own species) is that teat number exceeds average baby number by a factor of two [782,1736,1737,2024]. However, this **constraint** is less relevant for marsupials because their neonates are born prematurely, making the investment per baby virtually negligible [1604]. How easy would it be to remedy this discrepancy? Teat number should be quite **evolvable** [147,1150,2335], according to a 2008 paper [1411]: by lowering BMP activity in mouse nipples researchers converted those nipples into ordinary furry skin. Based on this result, only a few genetic steps should be needed to evolve in the other direction – that is, to add more nipples to the belly of a mouse or a quoll.

R

How the remora got its sucker

The remora is a fish with a sucker on its head. The sucker allows it to hitchhike rides on a larger host [1667]. The host provides protection and, in many cases, leftover scraps from its meals. Sharks tolerate remoras well, but dolphins do not. Spinner dolphins dislodge remoras in a spectacular way: they leap several meters in the air, whirl around wildly like an olympic figure skater, and splash broadside back into the ocean with a "thwack" [2388]. The sucker looks like an oval, slightly concave grease pan that is lined with ~20 pairs of parallel ribs, and the ribs branch laterally from a central septum. In 2012 the sucker was shown to be **homologous** to the dorsal fin of an ordinary fish [247]. The septum corresponds to the fin itself, while its ribs are lateral outgrowths from the bases of the fin rays. The muscles that erect or depress the rays in other fish are apparently used by remoras to affix the sucker, though the muscle matrix has subtleties that remain enigmatic. We do not yet know how the dorsal fin got **co-opted** to make a sucker in the first place. One possible scenario is: (1) mutations caused bifid dorsal fins with the same tiny hooks on the rays' hind edges as modern remoras, (2) those hooks allowed these proto-remoras to latch onto their hosts loosely, and (3) the attachment strengthened when the outer edge of the fin acquired a sealing rim, and the muscles enforced a vacuum. (*Confession:* Before writing this vignette, the author had heard about remoras but had mistakenly assumed that they used their *mouths* as suckers.)

How the rhinoceros got its horn

Unlike cattle horns [1293] and deer antlers [1295], which are made of bone [39], the horn of an Indian rhino is made of bundled hair [226]. So are the *two* horns of two-horned rhinos in the other black, white, Javan, and Sumatran species [893,1338,1789]. In *On Growth and Form* D'Arcy Thompson argued that the shape of the rhino horn – a planar logarithmic spiral – can be explained if cells near the front make hair faster than those near the back, since such a gradient would cause the horn to curve backward [2203]. His idea makes sense, but it does not explain why the horn is conical rather than cylindrical. Might the growth zone at the horn's base start as a point and enlarge to its final diameter as the horn grows (cf. mollusk shells [1824])? Perhaps, but if so, then how do horns that are amputated regenerate a conical shape from a wide base [1807]? This mystery was solved in 2006: the core of the horn is hardened by melanization and calcification, such that the tip can be sharpened (like a pencil point) by scraping it against the ground (an obsessive behavior) or by clashing it against a rival's horns (in jousting contests) [985]. In two-horned species the rear horn is typically smaller than the nasal horn [2028]. To explain this trend Thompson argued that the growth-rate gradients spanning each horn are sections of a larger snout-spanning growth gradient that slopes in the same direction.

As yet, we have no data to confirm or refute that idea. Where did rhino horns come from in the first place? Given their keratin composition and mode of growth [226], they might be misplaced (**heterotopic**) hooves! Once the rhino genome is sequenced, we can test this far-fetched "pedoceros" theory. For a demented look at how nasal *locomotion* (i.e., using the nose as a leg) could evolve, see Stumpke's fanciful book *The Snouters* [2140].

How the rodent got its incisors

Rodents (e.g., mice, beavers, and squirrels) use curved, chisel-like, self-sharpening incisors to gnaw seeds and other hard objects [463,1129,2356]. These teeth pose tantalizing riddles. In 2003 the ability of rodent incisors to grow continuously throughout adult life [1164,2326,2327] was traced to a cocktail of signaling factors (FGF, Notch, and probably BMP) that maintains the epithelial stem cell niche in an active state [1296,2241]. In 2008 the effectors of those signals were found to include heparin-binding cytokines [1503]. Progress has also been made on the "evo" aspect of the mystery. Placental mammals originally had three incisors in each quadrant of the jaw [1666]. Rabbits, which belong to a separate order, reduced this set to two (like us), while rodents whittled it down to only one. In 2011 the signals that suppress extra incisors in rodents were found to belong to the RTK (receptor tyrosine kinase) pathway: disabling this pathway doubles the number of incisors [369] – **atavistically** revealing an ancestral condition [1747].

S

How the sabertooth cat got its canines

Smilodon fatalis was the top predator in America for millions of years [1808] until its demise at the end of the last ice age ~13,000 years ago [129,559]. Its "sabertooth" canines were shaped like blades rather than cones – suggesting that they were used to *slash* the flesh of its prey, rather than to *stab* them as a lion does [49,1449]. Similar weaponry evolved **convergently** in a contemporary marsupial [1086,1431,2268], though the teeth of *Thylacosmilus* may have been used differently [269,672]. How did evolution convert a radially symmetric cone into a bilaterally symmetric blade? This puzzle of how teeth add axes also applies to snake fangs (see Fig. 4.8), but we know very little about this process or, for that matter, about tooth shaping in general [923,1099]. This portion of the evo-devo frontier is replete with unsolved riddles [930,1505,1929,2123,2364] concerning the sundry tusks of elephants [1823], narwhals [1660], walruses [514], and warthogs [1318]. One clue to the sabertooth mystery was published in 2012 under the enticing title "The making of a monster" [394]. When fossil *Smilodon* skulls are compared with modern felid skulls at corresponding (juvenile and adult) stages, a common trajectory of **allometric** transformation is observed. The deciding factor is the size of the infant's canines before the skull enters the shaping pathway. Hence, a sabertooth cat is basically a tiger whose canines began to grow prematurely – a *dramatic* case of dental **heterochrony**.

How the seahorse got its shape

The seahorse is a fish that looks like a horse. Ironically, both animals were sculpted by grass. Yes, *grass*! Proto-horses became grazers when grasses spread across the Great Plains [1790], and seahorses diverged from their pipefish ancestors [1510,2422] when seagrasses spread across the West Pacific [2190]. The latter spreading followed a tectonic uplift that allowed sunlight to foster grass growth on the (higher) seafloor [2190]. Proto-seahorses were able to exploit the grass blades for *camouflage* by holding their body upright [2190]. Camouflage helps seahorses ambush tiny crustaceans, which they suck into their long snouts with a lightning-fast buccal expansion [1292,1890,2271]. Seahorses hold their head horizontally to capture prey [1891,1892] while keeping their torso vertical [1292,2269], thus arching their neck like a horse. Their horsiness is also due to their cheekbones, which give them a horse-like jaw, even though they have no teeth [1292]. Seahorses anchor themselves to grass with a prehensile tail that evolved in place of their caudal fin [2270]. In a 2010 paper entitled "An adaptive explanation for the horse-like shape of seahorses," the prehensile tail, arched neck, and long snout were shown to constitute a "killer app" that makes seahorses better at catching tiny shrimp than their pipefish cousins [2270]. The "ears" of a seahorse are actually its pectoral fins. Though seahorses look like horses, they mate face-to-face like humans [1895], and they raise their young like kangaroos [22,2089]: the male (yes, the male) has a pouch on his belly where he broods as many as 2000 babies until they're ready to fend for themselves, and his wife visits him regularly until then to check on the little darlings [682].

How the stalk-eyed fly got its stalks

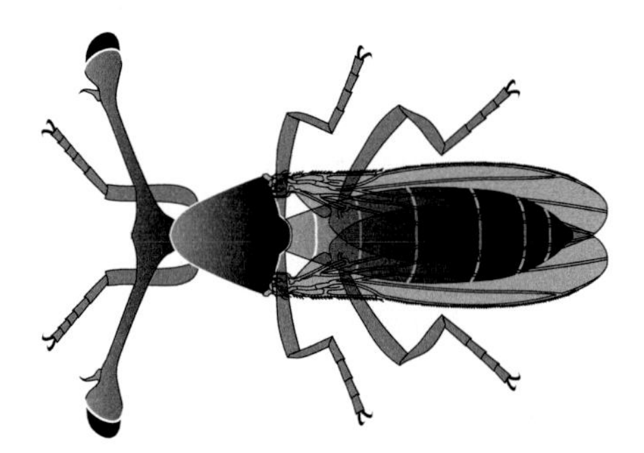

In what appears to have been a hysterical spree of **runaway selection** [2069,2413], male flies of the family Diopsidae went to ridiculous lengths to outdo their rivals in contests for females [368,617,2338]. Those females, for some reason [51,431], liked broad heads [614,2412,2520]. The outcome was the insect equivalent of a hammerhead shark [880]. The stalk-eye trait arose partly from **allometry** [112], but it was also likely abetted by hormonal stimulation [716]. Genetic analyses are ongoing [2414], but we can venture a guess based on evidence amassed in 2012 [619]: stalks were probably targeted for growth due to a higher-than-normal density of receptors for insulin-like growth factors [797,1677,2033]. Stalk eyes evolved **convergently** in snails [204], crabs [2500], and other crustaceans [2506].

How the stickleback lost its pelvic spines

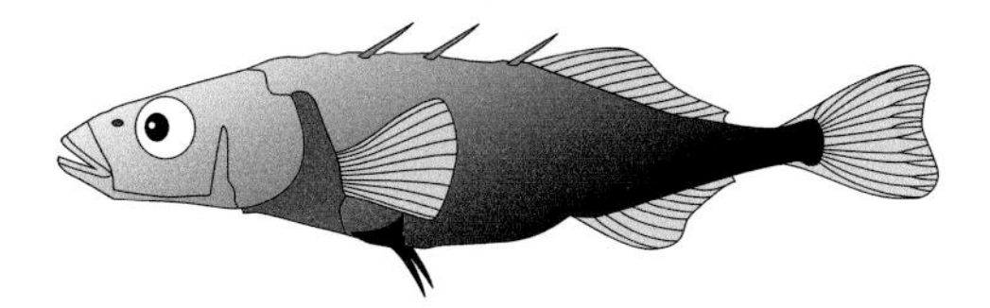

Tetrapod legs evolved from fish fins [906]: one pectoral (front) pair and one pelvic (back) pair [557]. The threespine stickleback drawn here has pectoral fins where our arms would be, but its pelvic fins are not where we would expect to find them [1775]. As in most teleosts, the pelvic fins start developing on the rear flank where our legs would be, but they soon move anteriorly and ventrally to a site below the pectoral fins [1570,2174]. There they act as steering rudders [2102], rather than as propulsive paddles [557]. In ocean-living sticklebacks, each pelvic fin was converted (**co-opted**) into a down-pointing spine (here shaded black), analogous to the erect spines on its back (for which it is named) [2475]. These spines protect the fish from marine predators in the same way (**convergence**) that a porcupine's quills protect it from land predators. After the last ice age, some sticklebacks colonized lakes and became bottom feeders [1109]. This benthic niche allowed them to escape their old enemies, but it confronted them with a new foe. Dragonfly larvae ambush sticklebacks and chew them to pieces, rather than swallowing them whole. Under these conditions the spines actually became a liability because they helped the smaller attacker grip them. For this reason, presumably, lake-dwelling sticklebacks lost their pelvic spines, though in theory they could have instead adapted **atavistically** by turning the spines back into pelvic fins. In 2004 this spine loss was traced to a disabled hindlimb enhancer of the *Pitx1* gene [2015,2047], which endows hindlimbs with a separate identity (**individuation**) from forelimbs in all vertebrates (see the manatee story) [570,1063,1321,1916,2012]. How did so many clades of sticklebacks manage to independently lose their spines so quickly? In 2010 that riddle was solved. The *Pitx1* locus in sticklebacks resides near the fragile tip of a chromosome, and deletions of the hindlimb enhancer can occur easily (**evolvability**) through non-homologous end joining after DNA breakage [362]. Such mutants are bona fide **hopeful monsters** [389]. For an *actual* monster, no animal can top the hitchhiker (see the remora story), whose dorsal fin became a hat that looks like a grease pan [247].

How the swordtail got his sword

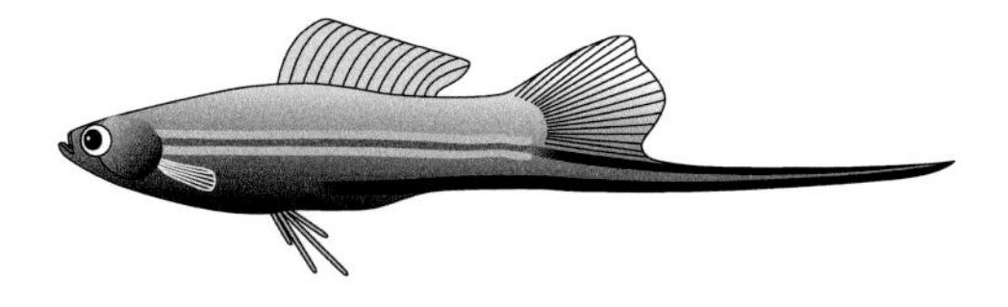

In some species of the genus *Xiphophorus* the male has a pointed extension of his caudal fin [1119]. This "sword," which is accentuated by pigment stripes, is reminiscent of the tail of a swallowtail butterfly (see Fig. 3.7), but it arises differently. Males do not grow a sword until they mature sexually, and females, who do not make a sword, can be forced to do so by exposing them to testosterone. How swords get made was revealed in 2008. When fin rays from the future sword area are transplanted to dorsal parts of a juvenile's caudal fin, they induce neighboring rays to make an extra sword at this location [599]. The inducing signal is FGF [1671], but the downstream factors under FGF control appear to be novel [1672]. Given FGF's pervasive role as a mitogenic agent in both fin development [1799] and fin regeneration [101,222], the caudal fin was likely **preadapted** for sword induction. All it took, apparently, was one focal signaling center. Hence, swords are easily **evolvable**. If females love them so much, then why don't more fish have them? Zorro's mystique notwithstanding, the sex appeal of a sword is not universal. Indeed, the allure even wore off in at least one swordtail clade. Females of *X. birchmanni* actually prefer males that *lack* a sword [2446]. Mate choice in general is a fickle factor in evolution [146,431,1840]. It accounts for a huge range of weird traits [610,619,622] – some of which have gone out of fashion [51,2164,2390], much to the chagrin of their hapless bearers (e.g., bald men [967]).

T

How tetrapods got their toes

Until 2012 the chief role for *Hox* genes in digit development in tetrapods (four-legged animals) was thought to be **individuation** [524,1527,1650] – i.e., the creation of differences among digits after they are specified by a **morphogen** gradient [137,1867,2025,2487]. It has been known for some time that denying mouse cells access to their Shh gradient causes the number of digits per paw to increase from five to about eight [378,1313] – implying that the gradient enforces the pentadactyly **constraint** that governs all living tetrapods [731,852]. (*Fossil* tetrapods had up to eight digits [409,2045]; cf. the *talpid* mutant in chickens [692].) In 2012 it was found that disabling *Hox* genes in Shh-compromised mice raises digit number even further (up to 14 per foot) [2026]. Even more surprising was that the increase is proportional to the *number* of *Hox* genes disabled – as if *Hox* genes are setting the wavelength of a Turing mechanism in a dose-dependent way (see Chapter 5) [2294]. The digits in these polydactylous feet look remarkably like the radiating rays of a fish fin [2479], suggesting that tetrapods have kept the antecedent algorithm for fin formation as a starting template for building their own (much derived) feet [498,702,1108]. A separate investigation in 2012 showed that it is amazingly easy to "beef up" an ordinary fish fin by merely overexpressing *Hoxd13* [701]. Thus, the first steps from "finhood" to "leghood" (and hence from sea to land) might have involved tweaks in the timing of *Hoxd13* activation – tweaks that are much subtler than anyone ever suspected [1964]. (*Aside:* you *must* see the walking shark video [906].)

How the tiger makes its stripes

Felids with a tiger-like body have been around for >2 MY [1421,1610], but striping does not fossilize, so we must seek the origins of tiger stripes elsewhere. No data are yet available from tiger embryos, nor have any theoretical models of tiger striping been tested directly [1575,1577]. However, a similar pattern was studied in 2012 as a handy proxy [868]. Mammals have ridges ($\approx$ stripes) on their hard palate that help grip food [292]. The number of ridges is constant within a species (four in humans vs. eight in mice), and ridges develop in a rigid sequence parallel to one another [2367]. Because they arise at fixed intervals, evolution was able to increase their number by merely extending the patterning period (**heterochrony**) [1717]. When mutations or chemicals are used to alter FGF or Shh signaling, the effects on the ridges implicate FGF as the activator and Shh as the inhibitor in a reaction–diffusion system [589], but other signals may also be involved. One odd result has yet to be explained: new ridges continue to arise between older ridges even when growth is suppressed, but inter-ridge distances should not be getting smaller if they are created by a Turing mechanism (see Chapter 5). As for how the rare white tiger lost its orange background, the answer was found in 2013. The white tiger trait is due to a single amino acid change (A477V) in a transporter protein (SLC45A2) involved in making the pigment pheomelanin [2470]. Similar mutations were found in cream horses [2470], silver chickens [2470], and one pale German lad with dark blond hair [1908]. White tigers still have black stripes due to the continued presence of eumelanin, so they are not true albinos [1670,2433].

How the treehopper got its helmet

Treehoppers are sap-sucking relatives of cicadas that look like they're wearing silly party hats. They display a zany assortment of ornate outgrowths called helmets [1513,1775]. In 2011 genetic evidence suggested that helmets develop like modified wings [1514,1796], but in 2012 stronger *counter*-evidence showed that they develop more like beetle horns [1474]. Helmets appear to have arisen when *Scr* (the *Hox* gene for the first thoracic segment) accidentally activated genes that normally function in leg development (**co-option**) in a new place (**heterotopy**) – namely the dorsal thorax [1474]. The species sketched here (*Cyphonia clavata*) looks as if an ant is riding backward on its back [1355], but this "ant" is actually an outgrowth from the treehopper's own thorax. Other species bear replicas of tree thorns, dead leaves, or even bird droppings [1355,1796], and this entire psychedelic menagerie evolved in only ~40 MY. Helmets probably started as freakish deformities that were free to "explore **morphospace**" because they were harmless [1796]. Any shapes that happened to mimic inedible or repulsive things were rewarded by selection, leading to refinement and diversification by **adaptive radiation**.

Nowhere in the biological world is there anything so much like an avant garde (3D) sculpture studio, aside from the (2D) painting studios for butterfly wings (Chapter 3) and mammal skin (Chapter 5). Indeed, no other animal group better exemplifies the Darwinian dialectic between (1) the blindly groping, endlessly inventive, and eternally playful nature of mutations and (2) the context-dependent, obsessively demanding, and compulsively judgmental character of natural selection. In short, the genome proposes, and nature disposes [967]. In the computer world, Richard Dawkins's *Blind Watchmaker* game offers a *virtual* playground of equal charm [501], as does the field of artificial life in general [1247,1690,2065]. Some treehoppers stumbled upon a separate, *non-visual* trick for deterring predators: they secrete honeydew (made from the sap that they suck) to attract ants, who return the favor by guarding (and "milking") them like a herd of tiny, six-legged cows [245]. Those ants are the insect equivalent of cowboys.

How the turtle got its shell

Turtles and tortoises are reptiles of uncertain ancestry [952] that resemble miniature Volkswagens . . . and move nearly as slowly. The turtle's shell consists of a dorsal dome called the carapace and a ventral plate called the plastron. Based on the earliest turtle fossil known (220 MY old), the plastron evolved before the carapace. The carapace arose by flattening of the ribs, followed by ossification of the dermis [1294,1843,2285]. The turtle shell has traditionally been viewed as a **novelty** [700,2419], partly because it appeared to violate a cardinal rule of vertebrate anatomy [1585,1859] – namely, that the pectoral girdle always resides *outside* the ribcage [1218]. In turtles the scapula was thought to be located *inside* (cf. the shell in octopuses [1202,1267,1698]). In 2012, however, a comparative study of amphibians, reptiles, birds, and mammals overturned this conclusion. The turtle scapula was found to actually lie in *front* of the ribcage, and its vertical orientation (typical of turtles) was shown to be the *primitive* (not derived) state for reptiles [1343]. In 2013 this same scapular angle was documented in pre-turtle "missing link" fossils [1342]. Thus, the turtle shell is not quite as novel as it once seemed. The shell prevents turtles from breathing by ribcage expansion and contraction, just as we find it hard to breathe with broken ribs. In theory, turtles could have evolved a diaphragm as mammals did [1746]. Instead, they reconfigured their abdominal muscles to fashion (1) a new contraption behind their lungs that acts somewhat like our diaphragm to draw air into the lungs and (2) a sling around their lungs that squeezes air out of the lungs when it contracts [1238].

U

How the uakari went bald

The uakari is a New World monkey that looks like a little old bald man in a thick fur coat (Fig. 6.2; cf. Fig. 5.1, face 1) [981]. In *our* species ~50% of men go bald by age 50 [1915,2059], but women rarely do [2060,2096]. In *their* species both sexes are bald. Why? Their baldness cannot be serving the same function as baldness in vultures [977,1030,1374] because uakaris don't dig their heads into carrion. They are vegetarians [128]. Our best guess is that this trait was selected via **character displacement** [986,1751] in the distant past. Primates are notorious for using facial markings to tell one species from another (see Fig. 5.1) [1943,2359], and bald areas are equally useful in this regard [1479], so the uakari may have recruited this trait for that role. How did such a **co-option** occur genetically? Here is where human baldness comes in handy. The sharp outline of our bald crown is so congruent with theirs that a shared "area code" probably exists in our genomes that delimits the pate [1479]. If so, then we can learn about their circuitry by studying ours [968]. Much progress has been made in analyzing male pattern baldness [995,1437,1663,1819], and it should soon be possible to search for *our* baldness genes in *their* genome because a shotgun sequence library of the uakari genome became available in 2012 [1082]. Another question we might be able to answer is how the *rubicundus* subspecies colored its head red (Fig. 5.1, face 1) [100]. We know that the color stems from (1) a network of capillaries just beneath the skin and (2) transparency of the skin itself [128,981], but how did the uakari's genome install these new "apps"?

How the unicorn got its tusk

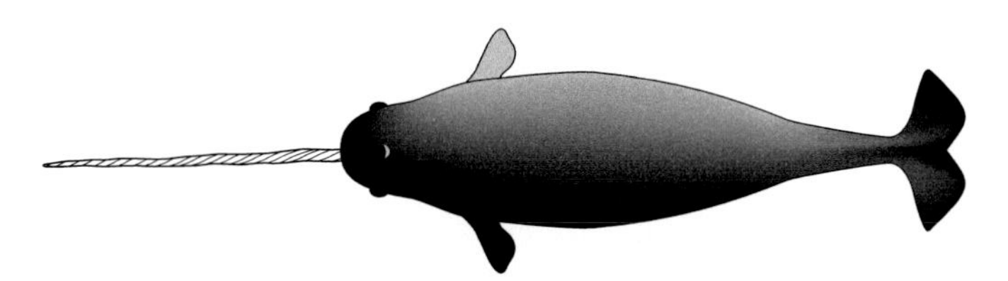

Nicknamed the "unicorn of the sea" [266] for its spear-like tusk [675,1619], the narwhal, *Monodon monoceros*, is a type of whale that lives in the Arctic [1438]. The tusk can reach 2.6 metres on a body of 5 metres [102,184]. Males typically have *one* tusk, but rarely (1.5%) they have *two*, while females usually (85%) have none [1661,1837]. The tusk is sometimes used for jousting [184,2049], but its chief role seems to be to sense external stimuli (e.g., salinity) via nerve endings in its dentine [1661] just as our own teeth are hypersensitive to cold or pressure when we get a cavity. The tusk has helical grooves that twist leftward along its axis to the tip [1156]. Paradoxically, both tusks of two-tusked males have sinistral grooves [403,1837] – suggesting that the spiral is governed by a *molecular* chirality of some sort [263]. However, no obvious chirality is detectable in the ivory's grain [234,1318], so the nature of the helix remains a mystery. The tusk is a tooth that grows out from the left side of the upper jaw, while its partner on the right aborts development [695]. For that reason every tusk is slightly offset from the midline, unlike a rhino's horn. Evo-devo has made headway in deciphering the genetics of *visceral* asymmetries [96,97,2272], but *superficial* asymmetries like the narwhal's tusk remain enigmatic (see also the flounder story) [415]. Until 2012 the tusk was thought to be an upper left *incisor*, but a detailed investigation showed

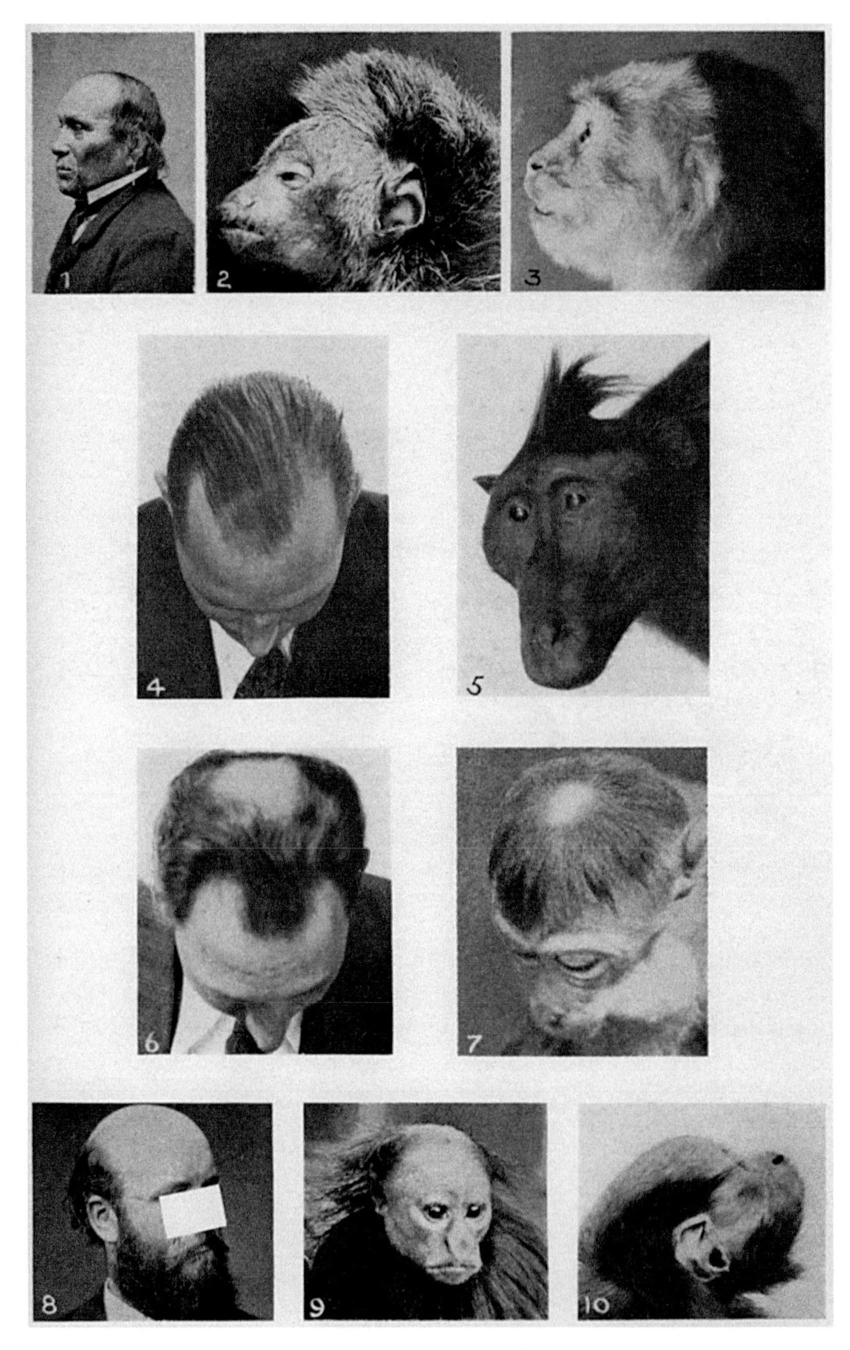

Figure 6.2 How the uakari went bald. Like many middle-aged men (e.g., photo 8), the South American uakari (*Cacajao calvus*) has a bald crown (frontal and side views in 9 and 10) [2319]. Other monkeys mimic various stages of human balding [1663,2319]. A raised hairline (1) is seen in the South American monk saki, *Pithecia monachus* (2; photo 3 is a conspecific with light hair). Receding temples (4) characterize the Celebes crested macaque, *Macaca nigra* (5). A tonsured

that it is in fact a *canine* [1660]. This finding is important because it implies (1) a loss of incisors at some point in the narwhal lineage and (2) a molariform identity for the lateral vestigial teeth in the narwhal maxilla. **Individuation** of tooth types in mammals is due to different signaling pathways [411,1505,2364] and homeobox transcription factors [2238], so these new clues might help us figure out how tooth loss and tusk elongation led to this beguiling whale, which is every bit as fabulous as a unicorn. Of course, to the unicorn, *Alice* was the "fabulous monster," not the other way around [333].

V

How the vampire bat reinvented running

Bats are adept fliers, but most are clumsy on the ground. One exception is the vampire bat, *Desmodus rotundus*. When tested on a treadmill these bats run amazingly well. An analysis of their gait, published in 2005, showed that they "bound" into midair like mice, with their forefeet and hindfeet touching the ground at different times [1862]. The study's authors argue that (1) the founders of the bat order had no need to run after they specialized as aerial insectivores, so (2) the neural circuits for running atrophied over millions of years (**use it or lose it**), but (3) vampire bats adapted to a new (blood-drinking) niche by reacquiring a running capability [1863]. Whether this talent is an **atavism** (old circuits revived) or a **novelty** (new circuits created) is unclear. Despite how *human* vampires are portrayed in movies, vampire *bats* do not suck their victim's blood. Rather, they land near their sleeping prey (e.g., a cow), crawl to a leg, pierce the skin with razor-sharp incisors (not their canines!), let the wound bleed, and then lap up the blood like a cat lapping up milk [1982]. If their host awakens and moves away, then their ability to run allows them to (1) avoid being trampled, (2) keep up with the host, and (3) resume feeding on the same victim [1982]. The evo-devo mystery here is how genes rewire neurons for different gaits. Based on a mutation that makes mice hop like rabbits (see the jerboa story) [1070], the rewiring may be easier than we might think.

Figure 6.2 *(cont.)*
vertex (6) is seen in the toque macaque, *Macaca sinica* (7). Evidently, human balding had roots in our monkey past. Once upon a time, a bald pate may have been as sexy to women as it still is in uakaris, though its appeal has clearly waned, notwithstanding Sean Connery's movie-star good looks (Fig. 5.8d). Why? Cues for mate choice are fickle [431], and they often shift (**character displacement** [986,1751]) to permit pre-mating isolation during episodes of sympatric speciation [2359]. Our *Homo* genus was once much bushier than the single twig we have today [24,203,1950], so it is not surprising that our ancestors may have used balding to set themselves apart [533,967,2048] with a signal like "Check me out! I'm not a Neanderthal!" [916]. This montage was constructed by the renowned mammalogist Gerrit S. Miller, Jr., in a 1931 paper [1479] that has sadly faded into obscurity (cited fewer than five times since it was first published). His Plate 3 is reproduced here (unaltered) with permission of the Smithsonian Institution.

This conclusion was underscored by a startling finding in 2012: one of the five gaits that Icelandic horses can perform (pacing vs. walking, tölting, trotting, and galloping) was traced to a premature stop codon in *Dmrt3* [50] – an ancient toolkit gene that normally functions in sex determination [1192]. We do not yet know how a gender gene got **co-opted** for the task of horse pacing, but this discovery at least gives us hope that the problem of bat bounding might be equally as tractable.

W

How the whale got its flippers

The evolution of whales from hippo-like ancestors [211] is well-documented in the fossil record [360,1556,2197,2447], and well-described in the scholarly literature [178,790,1791,2196,2198]. However, one lingering question concerns their flippers [148]. The fingers of whales and dolphins have extra phalanges [451] – a condition termed "hyperphalangy" [652]. Hyperphalangy may be an easy way of making a paddle, because it evolved **convergently** in plesiosaurs and ichthyosaurs – two other vertebrates (both reptiles) that also returned to the water [2194]. In 2002 a study of flipper development in the dolphin revealed the basic trick [1854]. Dolphin digits begin to grow under the control of an apical ectodermal ridge as in other mammals [450,569,1041], but then they launch a second growth phase confined to digits 2 and 3, and ultimately to digit 2 alone. By extending the growth period (**heterochrony** [1444]) dolphins – and presumably whales – were able to elongate certain digits [1852], but we still don't know how they managed to break the **constraint** on the number of phalanges along each digit that governs other vertebrates (see the pterosaur story). Evolution of the cetacean flipper must have also involved suppression of interdigital cell death, but the manner of this suppression has not yet been determined [652].

X

How the xenarthran stretched its neck

Xenarthrans are a basal order of placental mammals [1666]. The animals in this clade are a motley hodgepodge worthy of a Dr. Seuss book [195,1593,1741,2149]: armored armadillos, toothless anteaters, and sluggish, stinky sloths [525,993]. Sloths are intriguing from an evo-devo standpoint because they seem to violate an ancient **constraint** of mammal anatomy [82,728,1565,1597,2275]. Three-toed sloths in the genus *Bradypus* have 8–10 (vs. the canonical seven) cervical vertebrae (see the giraffe story) [274]. In 2010 the extra 1–3 neck vertebrae were unmasked as thoracic vertebrae in disguise [943]: they only *look* cervical because they lack ribs, but they are still thoracic based on their time of ossification. Why would these sloths go to such lengths to elongate their necks? Hanging upside-down all day poses certain problems [1662], not the least of which is the difficulty of seeing what lurks below. Having a longer neck allows *Bradypus* to turn its head through a greater angle (270°) than other mammals [274] – the same arc that owls achieve via an entirely different (but no less fascinating) trick [512].

Y

How the yeti lobster got its mittens

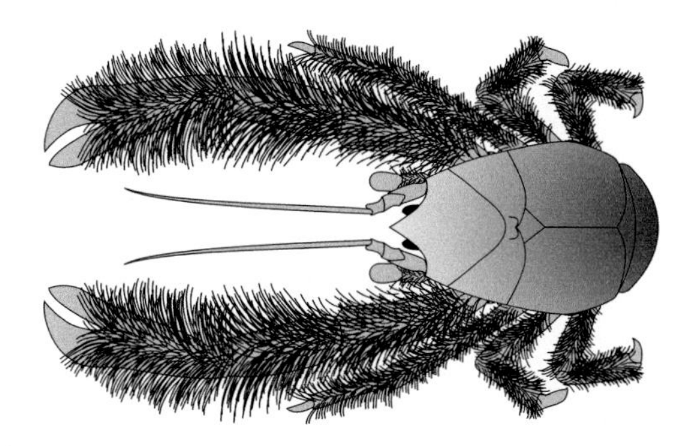

The Yeti is the mythical apeman of the Himalayas, also known as the Abominable Snowman [2332]. The yeti *lobster* is a real animal (*Kiwa hirsuta*) that lives on the floor of the Pacific Ocean [250]. The "fur" of its appendages (whence its name) is made of long, thin bristles. The function of those bristles has baffled researchers ever since the species was described in 2005 [1352]. The riddle was solved in 2011 by studying a related species, *K. puravida*. The bristles, it turns out, provide a huge surface area for the cultivation of bacteria – the lobster's main source of food [2208]. As *K. puravida* waves its arms in the hot water rising from a vent, the bacteria use the upwelling chemicals as fuel for growth. How did the fur arise? Bristle-like setae are common among arthropods, but not at such a high density [1848,2069]. Bees are another exception: their fur keeps their flight muscles warm [1376]. In the fruit fly, furry legs can be induced by disabling the *Notch* pathway [959], so comparable mutations might have created the lobster's mittens. This hypothesis should be testable once the *Kiwa* genome becomes available. Another similarity with flies might lead to further insights. *K. puravida* harvests its bacteria in a bizarre way [2208]. It licks its fur like a cat – albeit with hard mouthparts rather than a tongue – and it uses a comb of stiff bristles on its third maxillipeds (not shown in the sketch here) to scrape the bacteria from the fur. This maxilliped comb bears a striking resemblance to the sex comb of a fruit fly (see Fig. 2.6), so we may be able to use our knowledge of fly genetics to "comb" the *K. puravida* genome for clues to how its combs evolved. The eternally dark, hellishly hot, hydrothermal vents where the yeti monster lurks seem as different from the colorful, solar-powered ecosystems that we inhabit as the alien planets in science-fiction movies [241,2242], but this other-worldly feeling is a delusion of our parochial, post-Cambrian perspective. Those vents were apparently the cradle for earth's first life [1243].

Z

How the zebu got its dewlap

Zebu are Asian cattle that differ from European breeds by the presence of (1) a camel-like hump and (2) a prominent dewlap (a fold of loose skin) hanging from the neck and belly [1350]. In this sketch the dewlap is shaded black. Zebu were domesticated from ancient aurochs – the giant oxen, *Bos primigenius*, depicted in Paleolithic cave paintings [224]. They are a distinct subspecies (*B. p. indicus*) from humpless cattle (*B. p. taurus*), which were tamed separately [376,1320]. The fatty hump may aid survival in an arid climate (see the camel story), but the role – if any – of the dewlap is unclear. How do such areas of loose skin arise genetically? In 2011 the flaccid skin of the Chinese dog breed Shar-Pei was traced to a gene (*HAS2*) that, when mutated, causes excess hyaluronic acid (a component of mucus) to suffuse the skin [1688]. An eerily similar Shar-Pei-like syndrome occurs in humans (don't inspect these photos if you get queasy [1817]!), and people with a less obvious "rubber skin" phenotype were exhibited as freaks in sideshows during the nineteenth century [1214] – for example, James Morris, the "Elastic Skin Man," could stretch his chest skin over his face [567]. The "double chins" that many of us will get as we age are due to similar malfunctions of our connective tissue, so solving this problem might someday spare us the indignity of face-lift surgery.

Epilogue

As we reach the end of our safari, we come to see that this book was really about ourselves all along, albeit as seen through the prism of our colorful cousins across the animal kingdom. By surveying the spectrum of animals living today we have glimpsed what became of those humble little urbilaterians who roamed the Precambrian seas ~600 million years ago. Their external anatomy has diversified enormously, but their genome has not. It persists as the operating system for most of the animal world. We are only beginning to understand the quirky "apps" that make one species different from another.

The winding path from urbilaterians to our own species has been as multifarious as our tour of the animal kingdom. Over the eons, our ancestors were nudged from one ecological niche to another by forces both large (e.g., tectonic plates shifting and meteors hitting the earth) and small (e.g., food availability and sexual selection). These forces reshaped us from fish to amphibian to reptile and eventually to our present form, occupying the place in morphospace that we affectionately call *Homo sapiens*. But all of those forces were beyond our control, and none of them was unique to our lineage. The same can be said, essentially, for every other species on earth. If there is one lesson from this book that stands out above the rest, therefore, it is this: Contingency rules destiny.

> If the motorcade takes a different route through Dallas, if Newton sits under a different tree, or if Ilsa picks another gin joint to walk into, the story is radically altered. [2019]

In this respect Kipling was right. The charming traits that characterize the elephant, the leopard, the camel, and the snake *are* due to their ancestors having been in the right place at the right time to have been shaped in a peculiar way by external factors beyond their control. Darwin's successors in the evo-devo community have merely amended these stories by replacing the fabled factors with the actual ones. The revised stories presented in this book are no less enchanting, and they offer the added virtue of leaving the door open for future revelations. Evo-devo is entering its Golden Age, and its frontier offers endless entertainment. Kipling and Darwin would both be pleased. So would Aristotle, not to mention Alice!

Glossary

Evo-devo terms that are used in this book are defined below, and those that warrant further elaboration are discussed more fully. More terms can be found in *Keywords and Concepts in Evolutionary Developmental Biology* [910], and the major trends of evo-devo are categorized in Wallace Arthur's *Evolution: A Developmental Approach* [75]. Because this glossary is intended primarily as a resource for teachers, references are cited more liberally here than elsewhere in this book.

Adaptive radiation: The diversification of a stem lineage into multiple branches, each of which comes to occupy a particular ecological niche [788,1300,2281]. In some cases, species can carve their own niche, rather than filling an existing one [1515,2359]. Adaptive radiation constitutes an extreme manifestation of **evolvability** that may be aided by genome alterations [314,502]. The cardinal example of adaptive radiation is the Cambrian Explosion of bilaterian phyla about 543 MY ago [632], but that radiation has proven to be slower than once thought [631].

Allometry: A change in body proportions as a function of body size [34,38,2033,2114] due to a fixed ratio in the relative growth rates of body parts [615,825]. One way to think about allometry is to imagine balloons ($\approx$ body parts) being blown up at different rates [1125], with the overall shape of the aggregate ($\approx$ body) changing with the duration of inflation ($\approx$ body size). Reshaping can also be envisioned by D'Arcy Thompson's grids [74,827,2203]. Allometry is a special case of both developmental **constraint** [693] and **morphological integration** [1638]. It is often associated with **heterochrony** [979], but it need not be [1169]. Non-uniform growth rates [613,615,1638] may be caused by genes [2071,2114], nutrients [618,2034], or hormones [616,2071]. Allometry can produce exaggerated traits by either sexual [614,2209] or natural [173] selection. Examples include beetle horns [611,2033], deer antlers [764,765,1312], and the trachea of the trumpeter swan [1961]. Allometric ratios themselves can be altered by selection [617]. The scaling of body size in the absence of allometry poses intriguing problems both anatomically [901] and physiologically [1962].

Atavism: The reappearance of an ancestral trait after many generations of absence [318,904,2122]. Dollo's Law claims that "reverse evolution" [1563,1774,2188] never occurs for complex traits [285,826,1810,2318], but in fact there are exceptions [232,424,1795,2000,2393]. Atavisms can arise suddenly by mutation [742,2360], but only rarely do they spread to fixation in a species [904,2122]. Sporadic throwbacks include hen's teeth [838,935], horse's toes [348,644,838,2103], dew claws (extra digits) in dogs [32,33,2125], premolars in mice [1747], hindlimb flippers in dolphins [2197] and whales [148], and tails or extra nipples in humans [2283]. Possible reversions at the species level include frog teeth [2393], lizard digits [2394], reptile scales (= revived fish scales?) [39], and dorsal fins in ichthyosaurs (= revamped fish fins?) [854]. One confirmed species-level atavism is sex-comb rotation in flies [2000]. One putative species-level atavism [2381] that was disproven involves wings in stick insects [2231].

Character displacement: The divergence of initially similar traits in sympatric subspecies whose fitness is jeopardized by interbreeding [1676,1751]. Because of this hybrid disadvantage, *any* feature that allows one incipient species to tell its members apart from the other's with greater fidelity will be favored [619]. Whichever arbitrary attribute (anatomical or behavioral) happens to be in the right place at the right time could be recruited (**co-opted**) as an arousal cue [51,455,622], a species identifier [1463,1716], and a pre-mating isolating mechanism [617,2033]. Thereafter, it may undergo **runaway selection** to refine its effectiveness [510,585,617,938,2033]. These processes of random trait recruitment and autocatalytic amplification explain why the traits that one gender finds attractive in the other are often ridiculously idiosyncratic [622,986,1133,1244] and frequently maladaptive [612,1195,1428]. Stalk eyes in flies (see Chapter 6) exemplify both trends [2338].

Co-option: The adoption of a new function by an old structure [1513,1535], gene [738,1471,2228], circuit [351,697,1304,2299,2303], or behavior [1440,2469]. Darwin saw it as a ubiquitous trend [489]: "Thus throughout nature almost every part of each living being has probably served, in a slightly modified condition, for diverse purposes, and has acted in the living machinery of many ancient and distinct specific forms." The most famous case is the panda's "thumb" [832], which is a co-opted wrist bone. Other examples include tetrapod toes (= redeployed *Hox* cassette [1526,1650,2233]), octopus arms (= redeployed *Hox* genes [1267]), penguin flippers (= modified wings [676]), and swim bladders in fish (= modified lungs [847]). Possible instances are arthropod legs (= ectopic versions of body axis? [1483,1484,1486,1492]) and butterfly eyespots (= ectopic wing margin? [971]). Part of the problem in assessing co-option is that the bilaterian toolkit contains so few signaling pathways (see Table 1.1) [48,300,396,1103,2458] that they are bound to be reused in many contexts during development [432,1006].

Constraint: A limitation on the kind or amount of hereditary variation [76,912] available to a species for it to respond to selection pressure [830,1414,1995]. Constraints confine evolution to preferred pathways in **morphospace** [1684,2409] – i.e., paths of least resistance [208]. In so doing, they can affect **evolvability** [930,1170,1924,2420]. Constraints can be historical, physical, genetic, or developmental [967,1270,1810,1989,2419]. Developmental constraints include the five digits in tetrapod feet [731], the seven vertebrae in mammal necks [82,273,728], and the phylotypic stage of vertebrate embryos [1118,1755,1793]. Mutations are more likely to yield a selectable phenotype (vs. death) if they occur between (vs. within) **modules** [979,1732,1968,2214]. Constraints can foster **convergence** – for example, two- and three-toed sloths, which evolved separately from different types of digging ancestors [1662].

Convergence: The adoption of similar traits by independent clades as solutions to analogous evolutionary problems [1431,2416]. Textbooks usually depict this trend with placental versus marsupial mammals [12,1123,1410,2097], but other examples are just as colorful, if not more so. For instance, similar ecomorphs also evolved in mammals versus dinosaurs: armadillo ≈ ankylosaur [1014]; bull ≈ *Carnotaurus* [2253]; dolphin ≈ ichthyosaur [849,1400,1548,1549]; and rhinoceros ≈ *Styracosaurus* [1935,2253,2519]. For a greater span of scale (Brobdingnag vs. Lilliput) compare vertebrates versus insects [88,614,880]: giraffe ≈ giraffe weevil, mole ≈ mole cricket, and rhinoceros ≈ rhinoceros beetle, plus the jousting horns of ungulates ≈ those of insects [614,1935] and the recent discovery that the tongues of insectivorous bats ≈ those of bees [932]. If these cases aren't weird enough, then consider (1) the "bear trap" legs of mantids [1518] ≈ those of mantis shrimps [306,1829], (2) the giant eyes of squid ≈ those of ichthyosaurs [1550,1642,1726], and (3) the baleen sieves of whales [359,1939] ≈ those of flamingos [839,1329,1685] as well as those of pterosaurs [382,1246]. The oddest case of all may be octopus arms [1000], which can bend like our arms to handle food [2143,2144] but can also walk like our legs on the ocean floor [1046]. Examples of parallel evolution include the loss of the gas bladder in dozens of fish lineages [1429] and the loss of legs in dozens of lizard clades [2395]. Molecular examples [395,434,561,808] include snake venoms [253], antifreeze proteins [340], and echolocation transducers [1110]. For encyclopedic lists of organs, see McGhee's *Convergent Evolution: Limited Forms Most Beautiful* [1431].

Deep homology: A common ancestry of structures (in distantly related species) that differ overtly but are built by the same conserved genes [341,908]. Examples include the body axes [785,2155], legs [1996,2044,2045], and eyes [5] of divergent animal phyla.

Dissociation: The evolution of overt differences among initially similar structures [158,1599,1812]. Synonymous with **mosaic evolution** among **serially homologous** organs within a **meristic** series.

Emergence: The autonomous elaboration of complexity during development [1317]. Fertilization of the egg launches a chain of events that culminate in the anatomy of the adult [412]. Along the way, cells act like robots [364,2295] – responding to each input with a new output [1501] via hard-wired genetic rules [1230,2476]. Alan Turing showed how noise can create patterns via simple molecular rules (see Fig. 5.4) [2244], and the rule-based nature of emergence is nicely illustrated by John Conway's game of *Life* [747,748].

Evolvability: The potential for adaptive change in a trait when a species is subjected to pressure from either natural [1159,1260,1758,2079] or artificial [228] selection [501,502]. For instance, dogs have been easier to reshape than cats [392,566,844,1969], though cats have been about as evolvable as dogs in coat coloration. The plasticity of pigeons in both shape and color [978] was cited by Darwin as proof of the power of selection [486,2013]. Evolvability can be assessed by the amount of naturally occurring variation [912,1844]. It is typically limited by **constraints** [1170,1924,2420], and it is increased by both **modularity** and redundancy [768,769,1160,1732]. For example, gene duplications allow mutations in one copy (paralog) to explore new phenotypes [428,540,1188] while the other copy continues to serve the old function [1066,1783,1897,2302]. Whole genome duplications thus offer vast reservoirs of potential [314,810,1732,2233,2262]. Other types of genetic reservoirs include (1) dormant ancestral circuits that can be revived by **atavisms** and (2) batteries of genes employed at a juvenile stage, which can be shifted by **heterochrony** to the adult stage or vice versa [44,828]. The evolvability of genomes can also be facilitated by (1) arrays of tandem repeats that grow or shrink [298,674], (2) transposons that catalyze recombination [537], and (3) pleiotropic links [881,1004,2315]. Ecologically, evolvability is aided by phenotypic plasticity [786,1512,1708,2372,2464] and the Baldwin effect [138,472,2354,2373]. We don't yet know why certain clades focus on particular traits [73,75] – for example, gigantism in dinosaurs [626,1025,1938,2010,2448], leg loss in squamates [232,2395], beak shape in birds [1367], pupil shape in felids [1370], and odd tusks in the elephant clan [1790]. Given that mutation rates are uneven within genomes [580,1005], one clever idea is that mutations can be targeted to particular gene loci [319–322]. Indeed, high densities of repetitive elements [537,921] can create recombination hotspots [426,709,982,1686] that could spark **adaptive radiation** [690,1280,1719]. Thus, evolvability itself may be a selectable trait [312,1260,1399,2310,2371].

Heterochrony: A change in the start, end, span, or rate of processes [1443], which alters development and thereby anatomy [34,1091,1409,1841,2501]. Heterochrony has been shown to control individual variation [911], sexual dimorphism [1442], and caste polymorphism [617] within species. Within clades it has been shown to produce digit elongation in dolphins [1854] and digit loss in salamanders [2396], but not jaw reduction in dogs [565]. Its genetic basis has been studied in nematodes [2,2071] and sea urchins [657], but analyzing it is hard because we don't yet understand how embryos measure time [555,1547,1841,2076,2084]. Surprisingly tiny tweaks in timing can cause drastic changes in anatomy [31,34,605,828] that qualify as **saltations** [831,1562]. Pedomorphosis – a type of heterochrony where adults resemble the juveniles of an ancestor – fostered the evolution of birds [190], octopuses [2137], and humans [828,1271,1446,1607,1847].

Heterotopy: The deployment of a gene, circuit, organ, or structure at a new site in the body [1304,1529,2501,2502]. This spatial term and its temporal counterpart (**heterochrony**) were coined by Haeckel [1811]. Examples include the origin of the jaw in vertebrates [427], the relocation of the mouth in nematodes [666], the shift of the anus in irregular sea urchins [268], and possibly the evocation of wing margin to create eyespots in butterflies (see Fig. 3.4). Heterotopy is mediated by genomic rewiring [2156], and it often entails **homeosis** [75,139].

Homeosis: The transformation of one body part to resemble another [1288]. Homeosis can arise from non-genetic (teratogenic) factors [139,2288], but it is typically caused by homeotic mutations [1701]. Most homeotic mutations in animals occur in genes that contain a "homeobox" motif – a 180 base-pair, DNA-binding sequence [353,1432].

Homology: An underlying similarity of organs due to a shared developmental program inherited from a common ancestor [905,2305,2306]. Richard Owen, the famed anatomist who coined this term, used it to denote the same organ under different circumstances [2087], in contrast to analogy, which denoted different organs under similar circumstances [846]. Darwin realized that the epitome of homology is the vertebrate limb [486], whose bone formula (1:2:many:5 from base to tip [2046]) remains constant despite taking forms as diverse as a horse leg, a bat wing, a whale flipper, or a human arm [339,2522]. Homology is easy to define [1973,2309] but can be tricky to apply operationally [215,246,667,2487]. For instance, insect segments and mammal vertebrae look nothing alike but use homologous genes to specify corresponding identities along the head–tail axis [840]. For such cases of conserved circuitry but irreconcilable anatomy the term ***deep*** **homology** is used [2044]. ***Serial*** **homology** is used for homologous organs that develop at iterated sites within a single individual (vs. between species) [1899].

Hopeful monster: An abnormal (typically **homeotic**) mutant which supposedly harbors evolutionary potential [552,2320]. In the late 1800s William Bateson [139] and Francis Galton [734,735] proposed that evolution often relies on discontinuous variation [789], and in the mid twentieth century Richard Goldschmidt promoted this idea [801,803] as an alternative to the Modern Synthesis [547,548,700,829,2112]. Cases do exist where **saltational** events seem to have been instrumental in evolution [389,699,829,1858,2191], but they are the exception rather than the rule [25,277,895,1285,2520].

Hox **Complex:** A conserved array of ~10 homeobox genes that subdivides the head–tail axis of bilaterally symmetric organisms into metameres [457,1045,1738]. The sequence of these genes along the chromosome is typically colinear with the sequence of body zones that they **individuate** [572,1650,1651].

Individuation: The assignment of different identities (area codes) to initially identical members of a **meristic** series during development [2487]. For example, this process governs how our teeth become molars vs. premolars vs. canines vs. incisors [1505,2123]. Related concepts are non-equivalence [439,1290] and Williston's Law [339,828,1485,2418].

Master gene: An executive gene that controls a battery of subordinate genes via the transcription factor that it encodes [762,2384]. *Hox* genes are iconic examples (see Chapter 2) [1650], though they tend to exert control directly as micromanagers rather than through a chain of command [26]. Master genes that specify regional identities are termed **selector genes** [1373,1601,2349].

Merism: The partitioning of an axis into a series of similar parts (**modules**) that share **serial homology** [139]. According to Williston's Law, meristic patterns evolve by reducing the number of parts and increasing the differences (**individuation**) among them [339,828,1485,2418].

Module: A hard-wired genetic circuit (or "app") that governs the development of a particular type of structure [1958,1994,2308] or behavior [1301,2469]. Modules are autonomous subroutines [62,71,578] that can be launched like clicking icons on a computer screen [1529]. Modules that encode anatomy can be accessed from any bodily site (via **heterotopy**) to make stereotyped structures that are inherently

homologous to one another [1160,2314,2417] – for example, mammal teeth [2123], insect sensilla [1230], and fly bristles [965]. Modularity facilitates **evolvability** [769].

Morphogen: A diffusible molecule that controls the spatial organization of cellular differentiation during development [1146,2158] – i.e., a molecule that *gen*erates *morpho*logy. As originally envisioned by Alan Turing [2244], morphogens are chemicals whose distributions become heterogeneous (periodic) as a result of chemical reactions (see Fig. 5.4) [1378]. Lewis Wolpert redefined morphogens as molecules whose concentration diminishes with distance (as a gradient) from the secretion site [2442] so that non-secretory cells can tell how far they are from the source (their "positional information") by measuring the local morphogen level (see Fig. 1.3) [1590,1880,1910].

Morphological integration: The tendency of anatomical features to vary together in the face of evolutionary change [766,1506,2290]. Darwin referred to this yoking of traits as the "correlation of growth" [486,1884]. It is due to the sharing of common genetic circuitry [492,1581,2486], and it is the opposite of **mosaic evolution**, where parts evolve independently [918,2106]. For example, integration ensures a neat fit between the upper and lower teeth in rodents [1257,1995] and in mammals more generally [401,713,1842], but its genetic basis remains a mystery [2078,2123].

Morphospace: The range of possible anatomies that is theoretically available to a particular clade [628,1684]. Some morphospaces are bounded by the algorithms of development [1351,2200] (e.g., snail shells [1430,1824,1825]), while others seem utterly unlimited (e.g., the zany helmets of treehoppers [1474,1514,1796]) – the equivalent of putting putty into the hands of a child [1355,1718,1928] (cf. **evolvability**). Artificial selection allows us to probe the dimensions of a morphospace relative to the subset of morphologies that actually evolved (e.g., butterfly eyespots [152,155,1633]) and thereby to circumscribe regions that were never occupied [830,1351,2417]. Vacant areas could be due to developmental or historical **constraints** [1431,1684,1931] – for example, digit [1013,1849,2124] or tooth [1127,1768] patterns. Where any given species settles in morphospace is mostly a matter of luck [338,1328,1515], because (1) mutations are random [205,1077,1078,1340,2410], (2) sexual preferences are fickle [431,2390], (3) environments are unstable [788,2281], and (4) populations dwindle to bottlenecks [773]. Like it or not, this humbling lesson applies to us too [301,851,1922]. In short, we are a glorious accident, and so is every other species on the planet [858,1922]. Contingency rules destiny [509,627,845].

Mosaic evolution: The independent evolution of different body parts (**modules**) [135,172,918,2106]. Mosaic evolution is the opposite of **morphological integration** [918,2106].

Novelty: A new trait [240,909,1513,2313]. Novelties may arise suddenly (by **saltation**) or gradually [1557]. They typically come from pre-existing structures [1416,1561,1562,1644] by **co-option** [738,1535,2299], in keeping with Darwin's notion of "descent with modification" [486,1514]. Unexplained novelties include the ink sac in octopuses [530], the flapping ears of the Dumbo octopus [425], the lateral fins of the "flying" squid [1202,1665,1703,2279], the tail fin of sea snakes

[1821], and the dorsal or tail fins of ichthyosaurs [849,1548,1549], mosasaurs [1310], and whales [855,1556].

Phenogenetic drift: Evolutionary fluidity in the genetic circuitry that produces a superficially invariant structure [1049,1339,2229,2363,2455]. This phenomenon shows how the connections between "genotype space" and "phenotype space" can often be counterintuitive [365,655,1931,2301].

Preadaptation: A trait that is fortuitously poised to adopt a new function (**co-option**) if and when the opportunity arises [953,1416]. The classic example is feathers [1056], which were first employed for insulation and ornamentation [2521] but later proved useful for flight [404,954,2503]. Gould and Vrba rejected this term [856] and proposed "exaptation" instead [861]. They argued (1) that a new term was needed to denote unselected traits which initially serve no function at all [861], but this need was already satisfied by "**spandrel**" [293,860], and (2) that the prefix "pre-" misleadingly connotes anticipation of future utility, but many authors routinely use "preadaptation" without any teleological overtones [261,279,738]. Other factors militate against adopting "exaptation": (1) the difficulty of ascertaining whether a given trait evolved for its present function [668,2105], (2) the difficulty of determining the ecology of extinct species [610], (3) the need to replace "adaptation" with the neologism "aptation" [1286], and (4) the complication that structures can serve multiple functions at the same time – for example, horns that simultaneously serve as *both* weapons *and* ornaments [614].

Recapitulation: The retracing of ancestral features by an embryo [1419,1937]. This term was immortalized in Haeckel's dictum "Ontogeny recapitulates phylogeny" [859,1023,1853]. Putative examples include gill slits (from fish ancestors) in human embryos [830,1023], internalization of the shell and realignment of body axes (from gastropod ancestors) in octopuses [1202,1267], and venom ducts in centipedes [577].

Runaway selection: A positive feedback loop between two factors that can result in grossly exaggerated traits [510,585,617,938,2033] – effectively an arms race [614,619,671]. Interactions may involve predator–prey contests, male–male rivalry, male–female mate choice [340,500,501,505,2209], or structure–function refinement [1416]. Putative examples include stalked eyes [112,368,2338,2520], bat tongues [1554,1555], beetle horns [610,616], and fly forelegs [2069].

Saltation: A conspicuous departure from the normative anatomy of a species [1285]. Saltations are discontinuous variations [139] that theoretically may have adaptive potential as **hopeful monsters**.

Selector gene: A **master gene** that endows an organ with a unique identity (e.g., eye vs. wing) [1373,1601,2349]. Selector genes encode region-specific transcription factors.

Serial homology: The invisible genetic programming shared by different organs (homologs) in a **meristic** series [1674,1899,2305]. The cardinal example is the arm versus the leg [551,2357]. This common circuitry is attributable to the evolution of these organs by serial repetition of identical elements in an ancestor [1916,2486] and the subsequent **individuation** of those elements as formalized in Williston's Law [339,828,1485,2418].

Spandrel: A feature that arises as an incidental byproduct of some other trait, without being acted upon directly by natural selection [131,668,860,869,1385]. For example, blood is colored red as a side effect of iron oxide having been selected for oxygen transport [926], and bones are colored white as a side effect of calcium phosphate having been selected for tensile strength [293,1128]. Human skull sutures, which allow a baby's head to deform as it squeezes through the mother's birth canal, seem like they *must* have evolved adaptively for this purpose, but, as Darwin stressed, they are actually just lucky spandrels inherited from suture-riddled reptiles that hatched from eggs [486]. Likewise, brain ventricles are spandrels of the hollow neural tube [1333], and brain gyri and sulci are spandrels of the expanding neocortex [2223]. Spandrels can evolve via **allometry** [2033], pleiotropy [1004], or other developmental **constraints** [837]. Although the spandrel term is modern [860], Darwin used the same concept in his *Origin of Species* ([486], p. 196): "Moreover when a modification of structure has primarily arisen from [correlation of growth] or other unknown causes, it may at first have been of no advantage to the species, but may subsequently have been taken advantage of by the descendants of the species under new conditions of life and with newly acquired habits."

Use it or lose it: The tendency of unused structures [894] to deteriorate over time due to the accrual of deleterious mutations [1390,1563]. This decay initially leads to **vestiges** and ultimately to complete loss [1022]. Snake legs and eyes have been discussed in this regard in Chapter 4. Other examples where disuse played a non-trivial role include the eyes of cave animals [219,1020,1199,1778], the rod photoreceptors in geckos [804], the cone photoreceptors in whales [1468,1742], and the lungs of certain frogs [192,1059] and salamanders [1482,2317]. Structures that were lost have occasionally reappeared (see **atavism**). This principle also applies to genes and *cis*-regulatory elements [340,1626,2518].

Vestige: A remnant of a structure that once served a function in a distant ancestor [1723]. Examples include kiwi wings [302,787], horse toes [644,645], manatee fingernails [486], teeth in baleen whale embryos [904], premolars in mice [1165,1747], and the puny arms of *Tyrannosaurus* (http://trextrying.tumblr.com) [1056,1057,2132], *Carnotaurus* [2253], and one-digit dinosaurs [1652,2471]. Vestiges are "ghosts of adaptations past" [668]. They tend to be useless, but some have been **co-opted** – for example, the vestigial hindlimbs of pythons which act as "tickling" spurs during intercourse [2394]. Researchers have tried to restore a few vestiges to their former glory by reawakening their dormant developmental potential [904] – for example, bird teeth [1502,1504] and cavefish eyes [219,1092,1778].

References

1. Abbott, A. and other members of the *C. elegans* Sequencing Consortium (1998). Genome sequence of the nematode *C. elegans*: a platform for investigating biology. *Science* **282**, 2012–2018.
2. Abbott, A.L. (2003). Heterochronic genes. *Curr. Biol.* **13**, R824–R825.
3. Abbott, M.K. and Lengyel, J.A. (1991). Embryonic head involution and rotation of male terminalia require the *Drosophila* locus *head involution defective*. *Genetics* **129**, 783–789.
4. Abmayr, S.M. and Pavlath, G.K. (2012). Myoblast fusion: lessons from flies and mice. *Development* **139**, 641–656.
5. Abouheif, E. (1997). Developmental genetics and homology: a hierarchical approach. *Trends Ecol. Evol.* **12**, 405–408.
6. Abouheif, E. and Wray, G.A. (2002). Evolution of the gene network underlying wing polyphenism in ants. *Science* **297**, 249–252.
7. Abu-Shaar, M. and Mann, R.S. (1998). Generation of multiple antagonistic domains along the proximodistal axis during *Drosophila* leg development. *Development* **125**, 3821–3830.
8. Abzhanov, A. and Kaufman, T.C. (2000). Homologs of *Drosophila* appendage genes in the patterning of arthropod limbs. *Dev. Biol.* **227**, 673–689.
9. Abzhanov, A., Kuo, W.P., Hartmann, C., Grant, B.R., Grant, P.R., and Tabin, C. (2006). The calmodulin pathway and evolution of elongated beak morphology in Darwin's finches. *Nature* **442**, 563–567.
10. Abzhanov, A., Protas, M., Grant, B.R., Grant, P.R., and Tabin, C.J. (2004). Bmp4 and morphological variation of beaks in Darwin's finches. *Science* **305**, 1462–1465.
11. Acampora, D., Annino, A., Tuorto, F., Puelles, E., Lucchesi, W., Papalia, A., and Simeone, A. (2005). Otx genes in the evolution of the vertebrate brain. *Brain Res. Bull.* **66**, 410–420.
12. Achenbach, J. (2010). Lost giants. *Natl. Geogr.* **218** #4, 90–109.
13. Adami, C. (2006). Reducible complexity. *Science* **312**, 61–63.
14. Adams, J.K. (1989). *Pteronotus davyi*. *Mamm. Species* **346**, 1–5.
15. Adams, R.A. and Pedersen, S.C. (1994). Wings on their fingers. *Nat. Hist.* **103** #1, 48–55.
16. Adamska, M., Larroux, C., Adamski, M., Green, K., Lovas, E., Koop, D., Richards, G.S., Zwafink, C., and Degnan, B.M. (2010). Structure and expression of conserved Wnt pathway components in the demosponge *Amphimedon queenslandica*. *Evol. Dev.* **12**, 494–518.
17. Adamska, M., Matus, D.Q., Adamski, M., Green, K., Rokhsar, D.S., Martindale, M.Q., and Degnan, B.M. (2007). The evolutionary origin of hedgehog proteins. *Curr. Biol.* **17**, R836–R837.
18. Adler, P.N. (1992). The genetic control of tissue polarity in *Drosophila*. *BioEssays* **14**, 735–741.

19. Affolter, M., Pyrowolakis, G., Weiss, A., and Basler, K. (2008). Signal-induced repression: the exception or the rule in developmental signaling? *Dev. Cell* **15**, 11–22.

20. Agrawal, S. and Riffell, J.A. (2011). Behavioral neurobiology: The bitter life of male flies. *Curr. Biol.* **21**, R470–R472.

21. Aguinaldo, A.M.A., Turbeville, J.M., Linford, L.S., Rivera, M.C., Garey, J.R., Raff, R.A., and Lake, J.A. (1997). Evidence for a clade of nematodes, arthropods and other moulting animals. *Nature* **387**, 489–493.

22. Ahnesjö, I. and Craig, J.F. (2011). The biology of Syngnathidae: pipefishes, seadragons and seahorses. *J. Fish Biol.* **78**, 1597–1602.

23. Ahuja, A. and Singh, R.S. (2008). Variation and evolution of male sex combs in *Drosophila*: nature of selection response and theories of genetic variation for sexual traits. *Genetics* **179**, 503–509.

24. Aiello, L.C. and Andrews, P. (2006). The australopithecines in review. In *The Human Evolution Source Book* (R.L. Ciochon and J.G. Fleagle, eds.). Advances in Human Evolution series. Pearson Prentice Hall, Upper Saddle River, NJ, pp. 76–89.

25. Akam, M. (1998). *Hox* genes, homeosis and the evolution of segment identity: no need for hopeless monsters. *Int. J. Dev. Biol.* **42**, 445–451.

26. Akam, M. (1998). *Hox* genes: from master genes to micromanagers. *Curr. Biol.* **8**, R676–R678.

27. Akam, M. (1998). The yin and yang of evo/devo. *Cell* **92**, 153–155.

28. Akam, M. (2000). Arthropods: developmental diversity within a (super) phylum. *PNAS* **97** #9, 4438–4441.

29. Akiyama-Oda, Y. and Oda, H. (2006). Axis specification in the spider embryo: *dpp* is required for radial-to-axial symmetry transformation and *sog* for ventral patterning. *Development* **133**, 2347–2357.

30. Albalat, R. (2009). The retinoic acid machinery in invertebrates: ancestral elements and vertebrate innovations. *Mol. Cell. Endocrinol.* **313**, 23–35.

31. Alberch, P. (1980). Ontogenesis and morphological diversification. *Am. Zool.* **20**, 653–667.

32. Alberch, P. (1985). Developmental constraints: why St. Bernards often have an extra digit and poodles never do. *Am. Nat.* **126**, 430–433.

33. Alberch, P. (1986). Possible dogs. *Nat. Hist.* **95** #12, 4–8.

34. Alberch, P., Gould, S.J., Oster, G.F., and Wake, D.B. (1979). Size and shape in ontogeny and phylogeny. *Paleobiology* **5**, 296–317.

35. Alberts, B., Johnson, A., Lewis, J., Raff, M., Roberts, K., and Walter, P. (2002). *Molecular Biology of the Cell*, 4th edn. Garland, New York, NY.

36. Alexander, B., Baggaley, A., Dennis-Bryan, K., McDonald, F., Munsey, E., Preston, P., Tuson, C., Yelland, A., Hamilton, J., Heilman, C., and Perlmutter, J., eds. (2010). *Smithsonian Natural History: The Ultimate Visual Guide to Everything on Earth.* Dorling Kindersley, New York, NY.

37. Alexander, R.M. (1975). *The Chordates*, 2nd edn. Cambridge University Press, New York, NY.

38. Alexander, R.M. (1985). Body support, scaling, and allometry. In *Functional Vertebrate Morphology* (M. Hildebrand, D.M. Bramble, K.F. Liem, and D.B. Wake, eds.). Harvard University Press, Cambridge, MA, pp. 26–37.

39. Alibardi, L. (2012). Perspectives on hair evolution based on some comparative studies on vertebrate cornification. *J. Exp. Zool. (Mol. Dev. Evol.)* **318B**, 325–343.

40. Aliee, M., Röper, J.-C., Landsberg, K.P., Pentzold, C., Widmann, T.J., Jülicher, F., and Dahmann, C. (2012). Physical mechanisms shaping the *Drosophila* dorsovental compartment boundary. *Curr. Biol.* **22**, 967–976.

41. Allen, C.E. (2008). The "eyespot module" and eyespots as modules: development, evolution, and integration of a complex phenotype. *J. Exp. Zool. (Mol. Dev. Evol.)* **310B**, 179–190.

42. Allen, C.E., Beldade, P., Zwaan, B.J., and Brakefield, P.M. (2008). Differences in the selection response of serially repeated color pattern characters: standing variation, development, and evolution. *BMC Evol. Biol.* **8**, Article 94 (13 pp.).

43. Allen, W.L., Cuthill, I.C., Scott-Samuel, N.E., and Baddeley, R. (2011). Why the leopard got its spots: relating pattern development to ecology in felids. *Proc. R. Soc. Lond. B* **278**, 1373–1380.

44. Altig, R. (2006). Tadpoles evolved and frogs are the default. *Herpetologica* **62**, 1–10.

45. Amemiya, C.T., Miyake, T., and Rast, J.P. (2005). Echinoderms. *Curr. Biol.* **15**, R944–R946.

46. Amundson, R. (2005). *The Changing Role of the Embryo in Evolutionary Thought: Roots of Evo-Devo*. Cambridge University Press, New York, NY.

47. Anderson, E., Peluso, S., Lettice, L.A., and Hill, R.E. (2012). Human limb abnormalities caused by disruption of hedgehog signaling. *Trends Genet.* **28**, 364–373.

48. Andersson, E.R., Sandberg, R., and Lendahl, U. (2011). Notch signaling: simplicity in design, versatility in function. *Development* **138**, 3593–3612.

49. Andersson, K., Norman, D., and Werdelin, L. (2011). Sabretoothed carnivores and the killing of large prey. *PLoS ONE* **6** #10, e24971.

50. Andersson, L.S., Larhammar, M., Memic, F., Wootz, H., Schwochow, D., Rubin, C.-J., Patra, K., Arnason, T., Wellbring, L., Hjälm, G., Imsland, F., Petersen, J.L., McCue, M.E., Mickelson, J.R., Cothran, G., Ahituv, N., Roepstorff, L., Mikko, S., Vallstedt, A., Lindgren, G., Andersson, L., and Kullander, K. (2012). Mutations in *DMRT3* affect locomotion in horses and spinal circuit function in mice. *Nature* **488**, 642–646.

51. Andersson, M. and Simmons, L.W. (2006). Sexual selection and mate choice. *Trends Ecol. Evol.* **21**, 296–302.

52. Andersson, O., Reissmann, E., and Ibáñez, C.F. (2006). Growth differentiation factor 11 signals through the transforming growth factor-β receptor ALK5 to regionalize the anterior–posterior axis. *EMBO Reports* **7**, 831–837.

53. Andres, B. (2012). The early evolutionary history and adaptive radiation of the pterosauria. *Acta Geol. Sinica (English Ed.)* **86**, 1356–1365. [*See also* Witton, M.P. (2013). *Pterosaurs: Natural History, Evolution, Anatomy*. Princeton University Press, Princeton, NJ.]

54. Andrew, D.J. and Ewald, A.J. (2010). Morphogenesis of epithelial tubes: insights into tube formation, elongation, and elaboration. *Dev. Biol.* **341**, 34–55.

55. Angelini, D.R. and Kaufman, T.C. (2004). Functional analyses in the hemipteran *Oncopeltus fasciatus* reveal conserved and derived aspects of appendage patterning in insects. *Dev. Biol.* **271**, 306–321.

56. Angelini, D.R. and Kaufman, T.C. (2005). Comparative developmental genetics and the evolution of arthropod body plans. *Annu. Rev. Genet.* **39**, 95–119.

57. Angelini, D.R. and Kaufman, T.C. (2005). Functional analyses in the milkweed bug *Oncopeltus fasciatus* (Hemiptera) support a role for Wnt signaling in body segmentation but not appendage development. *Dev. Biol.* **283**, 409–423.

58. Angelini, D.R. and Kaufman, T.C. (2005). Insect appendages and comparative ontogenetics. *Dev. Biol.* **286**, 57–77.

59. Angelini, D.R., Liu, P.Z., Hughes, C.L., and Kaufman, T.C. (2005). *Hox* gene function and interaction in the milkweed bug *Oncopeltus fasciatus* (Hemiptera). *Dev. Biol.* **287**, 440–455.

60. Aoyama, H. and Asamoto, K. (2000). The developmental fate of the rostral/caudal half of a somite for vertebra and rib formation: experimental confirmation of the resegmentation theory using chick-quail chimeras. *Mechs. Dev.* **99**, 71–82.

61. Apesteguía, S. and Zaher, H. (2006). A Cretaceous terrestrial snake with robust hindlimbs and a sacrum. *Nature* **440**, 1037–1040.

62. Apter, M.J. and Wolpert, L. (1965). Cybernetics and development. I. Information theory. *J. Theor. Biol.* **8**, 244–257.

63. Aragón, J.L., Torres, M., Gil, D., Barrio, R.A., and Maini, P.K. (2002). Turing patterns with pentagonal symmetry. *Phys. Rev. E (Stat. Nonlin. Soft Matter Phys.)* **65**, 051913 (9 pp.).

64. Arakane, Y., Lomakin, J., Gehrke, S.H., Hiromasa, Y., Tomich, J.M., Muthukrishnan, S., Beeman, R.W., Kramer, K.J., and Kanost, M.R. (2012). Formation of rigid, non-flight forewings (elytra) of a beetle requires two major cuticular proteins. *PLoS Genet.* **8** #4, e1002682.

65. Arbib, M.A. (1972). Automata theory in the context of theoretical embryology. In *Foundations of Mathematical Biology*, Vol. 2 (R. Rosen, ed.). Academic Press, New York, NY, pp. 141–215.

66. Arendt, D. and Nübler-Jung, K. (1996). Common ground plans in early brain development in mice and flies. *BioEssays* **18**, 255–259.

67. Arendt, D. and Nübler-Jung, K. (1999). Comparison of early nerve cord development in insects and vertebrates. *Development* **126**, 2309–2325.

68. Arendt, D., Technau, U., and Wittbrodt, J. (2001). Evolution of the bilaterian larval foregut. *Nature* **409**, 81–85.

69. Armstrong, J.R. and Ferguson, M.W.J. (1995). Ontogeny of the skin and the transition from scar-free to scarring phenotype during wound healing in the pouch young of a marsupial, *Monodelphis domestica*. *Dev. Biol.* **169**, 242–260.

70. Arnoult, L., Su, K.F.Y., Manoel, D., Minervino, C., Magriña, J., Gompel, N., and Prud'homme, B. (2013). Emergence and diversification of fly pigmentation through evolution of a gene regulatory module. *Science* **339**, 1423–1426.

71. Arthur, W. (2000). The concept of developmental reprogramming and the quest for an inclusive theory of evolutionary mechanisms. *Evol. Dev.* **2**, 49–57.

72. Arthur, W. (2002). The emerging conceptual framework of evolutionary developmental biology. *Nature* **415**, 757–764.

73. Arthur, W. (2004). *Biased Embryos and Evolution*. Cambridge University Press, New York, NY.

74. Arthur, W. (2006). D'Arcy Thompson and the theory of transformations. *Nature Rev. Genet.* **7**, 401–406.

75. Arthur, W. (2011). *Evolution: A Developmental Approach*. Wiley-Blackwell, Chichester.

76. Arthur, W. (2011). Searching for evo-devo's Holy Grail: the nature of developmental variation. *Evol. Dev.* **13**, 405–407.

77. Arthur, W. and Chipman, A.D. (2005). How does arthropod segment number evolve? Some clues from centipedes. *Evol. Dev.* **7**, 600–607.

78. Arthur, W. and Farrow, M. (1999). The pattern of variation in centipede segment number as an example of developmental constraint in evolution. *J. Theor. Biol.* **200**, 183–191.

79. Asai, R., Taguchi, E., Kume, Y., Saito, M., and Kondo, S. (1999). Zebrafish *Leopard* gene as a component of the putative reaction-diffusion system. *Mechs. Dev.* **89**, 87–92.

80. Ashburner, M. (1989). *Drosophila: A Laboratory Handbook.* CSH Press, Cold Spring Harbor, NY.

81. Ashburner, M. (2006). *Won for All: How the Drosophila Genome Was Sequenced.* Cold Spring Harbor Laboratory Press, Plainview, NY.

82. Asher, R.J., Lin, K.H., Kardjilov, N., and Hautier, L. (2011). Variability and constraint in the mammalian vertebral column. *J. Evol. Biol.* **24**, 1080–1090.

83. Aspiras, A.C., Smith, F.W., and Angelini, D.R. (2011). Sex-specific gene interactions in the patterning of insect genitalia. *Dev. Biol.* **360**, 369–380.

84. Atallah, J., Liu, N.H., Dennis, P., Hon, A., Godt, D., and Larsen, E.W. (2009). Cell dynamics and developmental bias in the ontogeny of a complex sexually dimorphic trait in *Drosophila melanogaster*. *Evol. Dev.* **11**, 191–204.

85. Atallah, J., Liu, N.H., Dennis, P., Hon, A., and Larsen, E.W. (2009). Developmental constraints and convergent evolution in *Drosophila* sex comb formation. *Evol. Dev.* **11**, 205–218.

86. Atallah, J., Watabe, H., and Kopp, A. (2012). Many ways to make a novel structure: a new mode of sex comb development in Drosophilidae. *Evol. Dev.* **14**, 476–483.

87. Attenborough, D. (2002). *The Life of Mammals.* Princeton University Press, Princeton, NJ.

88. Attenborough, D. (2005). *Life in the Undergrowth.* Princeton University Press, Princeton, NJ.

89. Attenborough, D. (2008). *Life in Cold Blood.* Princeton University Press, Princeton, NJ.

90. Attisano, L. and Wrana, J.L. (1998). Mads and Smads in TGFb signalling. *Curr. Opin. Cell Biol.* **10**, 188–194.

91. Aubret, F., Bonnet, X., Pearson, D., and Shine, R. (2005). How can blind tiger snakes (*Notechis scutatus*) forage successfully? *Aust. J. Zool.* **53**, 283–288.

92. Aulehla, A. and Herrmann, B.G. (2004). Segmentation in vertebrates: clock and gradient finally joined. *Genes Dev.* **18**, 2060–2067.

93. Averof, M. (2002). Arthropod *Hox* genes: insights on the evolutionary forces that shape gene functions. *Curr. Opin. Gen. Dev.* **12**, 386–392.

94. Averof, M. and Cohen, S.M. (1997). Evolutionary origin of insect wings from ancestral gills. *Nature* **385**, 627–630.

95. Avise, J.C. (2006). *Evolutionary Pathways in Nature: A Phylogenetic Approach.* Cambridge University Press, New York, NY.

96. Aw, S. and Levin, M. (2008). What's left in asymmetry? *Dev. Dynamics* **237**, 3453–3463.

97. Aw, S. and Levin, M. (2009). Is left-right asymmetry a form of planar cell polarity? *Development* **136**, 355–366.

98. Ayers, K.L., Gallet, A., Staccini-Lavenant, L., and Thérond, P.P. (2010). The long-range activity of Hedgehog is regulated in the apical extracellular space by the glypican Dally and the hydrolase Notum. *Dev. Cell* **18**, 605–620. [*See also* Sanders, T.A., *et al.* (2013). Specialized filopodia direct long-range transport of SHH during vertebrate tissue patterning. *Nature* **497**, 628–632.]

99. Ayers, K.L., Mteirek, R., Cervantes, A., Lavenant-Staccini, L., Thérond, P.P., and Gallet, A. (2012). Dally and Notum regulate the switch between low and high level Hedgehog pathway signalling. *Development* **139**, 3168–3179.

100. Ayres, J.M. (1990). Scarlet faces of the Amazon. *Nat. Hist.* **99** #3, 32–41.

101. Azevedo, A.S., Sousa, S., Jacinto, A., and Saúde, L. (2012). An amputation resets positional information to a proximal identity in the regenerating zebrafish caudal fin. *BMC Dev. Biol.* **12**, Article 24 (10 pp.).

102. Bada, J.L., Mitchell, E., and Kemper, B. (1983). Aspartic acid racemization in narwhal teeth. *Nature* **303**, 418–420.

103. Badyaev, A.V. (2011). How do precise adaptive features arise in development? Examples with evolution of context-specific sex ratios and perfect beaks. *The Auk* **128**, 467–474.

104. Baena-Lopez, L.A., Nojima, H., and Vincent, J.-P. (2012). Integration of morphogen signalling within the growth regulatory network. *Curr. Opin. Cell Biol.* **24**, 166–172.

105. Baer, M.M., Chanut-Delalande, H., and Affolter, M. (2009). Cellular and molecular mechanisms underlying the formation of biological tubes. *Curr. Topics Dev. Biol.* **89**, 137–162.

106. Baguñà, J., Martinez, P., Paps, J., and Riutort, M. (2008). Unravelling body plan and axial evolution in the Bilateria with molecular phylogenetic markers. In *Evolving Pathways: Key Themes in Evolutionary Developmental Biology* (A. Minelli and G. Fusco, eds.). Cambridge University Press, New York, NY, pp. 217–238.

107. Bailly, X., Reichert, H., and Hartenstein, V. (2013). The urbilaterian brain revisited: novel insights into old questions from new flatworm clades. *Dev. Genes Evol.* **223**, 149–157.

108. Bak, P., Chen, K., and Creutz, M. (1989). Self-organized criticality in the "Game of Life". *Nature* **342**, 780–782.

109. Baker, R.E. and Murray, P.J. (2012). Understanding hair follicle cycling: a systems approach. *Curr. Opin. Gen. Dev.* **22**, 607–612.

110. Baker, R.E., Schnell, S., and Maini, P.K. (2006). A clock and wavefront mechanism for somite formation. *Dev. Biol.* **293**, 116–126.

111. Baker, R.E., Schnell, S., and Maini, P.K. (2009). Waves and patterning in developmental biology: vertebrate segmentation and feather bud formation as case studies. *Int. J. Dev. Biol.* **53**, 783–794.

112. Baker, R.H. and Wilkinson, G.S. (2001). Phylogenetic analysis of sexual dimorphism and eye-span allometry in stalk-eyed flies (Diopsidae). *Evolution* **55**, 1373–1385.

113. Ball, E.E., de Jong, D.M., Schierwater, B., Shinzato, C., Hayward, D.C., and Miller, D.J. (2011). Cnidarian gene expression patterns and the origins of bilaterality: are cnidarians reading the same game plan as "higher" animals? In *Key Transitions in Animal Evolution* (R. DeSalle and B. Schierwater, eds.). Science Publishers, Enfield, NH, pp. 197–216.

114. Ball, P. (1999). *The Self-Made Tapestry: Pattern Formation in Nature.* Oxford University Press, New York, NY.

115. Ball, P. (2012). Pattern formation in nature: physical constraints and self-organising characteristics. *Architectural Design* **216** #SI (Special Issue), 22–27.

116. Ball, P. (2013). Celebrate the unknowns. *Nature* **496**, 419–420.

117. Ballard, J.W.O., Olsen, G.J., Faith, D.P., Odgers, W.A., Rowell, D.M., and Atkinson, P.W. (1992). Evidence from 12S ribosomal RNA sequences that onychophorans are modified arthropods. *Science* **258**, 1345–1348.

118. Balter, M. (2013). Dramatic fossils suggest early birds were biplanes. *Science* **339**, 1261.

119. Baratte, S., Andouche, A., and Bonnaud, L. (2007). *Engrailed* in cephalopods: a key gene related to the emergence of morphological novelties. *Dev. Genes Evol.* **217**, 353–362.

120. Bard, J. and Lauder, I. (1974). How well does Turing's theory of morphogenesis work? *J. Theor. Biol.* **45**, 501–531.

121. Bard, J.B.L. (1977). A unity underlying the different zebra striping patterns. *J. Zool. Lond.* **183**, 527–539.

122. Bard, J.B.L. (1981). A model for generating aspects of zebra and other mammalian coat patterns. *J. Theor. Biol.* **93**, 363–385.

123. Bard, J.B.L. (1984). Butterfly wing patterns: how good a determining mechanism is the simple diffusion of a single morphogen? *J. Embryol. Exp. Morph.* **84**, 255–274.

124. Bard, J.B.L. (1990). Traction and the formation of mesenchymal condensations *in vivo*. *BioEssays* **12**, 389–395.

125. Bardach, J.E. (1967). The chemical senses and food intake in the lower vertebrates. In *The Chemical Senses and Nutrition* (M.R. Kare and O. Maller, eds.). Johns Hopkins Press, Baltimore, MD, pp. 19–43.

126. Barmina, O. and Kopp, A. (2007). Sex-specific expression of a HOX gene associated with rapid morphological evolution. *Dev. Biol.* **311**, 277–286.

127. Barnes, J., ed. (1984). *The Complete Works of Aristotle: The Revised Oxford Translation*. Vol. 1. Princeton University Press, Princeton, NJ.

128. Barnett, A.A. and Brandon-Jones, D. (1997). The ecology, biogeography and conservation of the uakaris, *Cacajao* (PItheciinae). *Folia Primatol.* **68**, 223–235.

129. Barnett, R., Barnes, I., Phillips, M.J., Martin, L.D., Harington, C.R., Leonard, J.A., and Cooper, A. (2005). Evolution of the extinct sabretooths and the American cheetah-like cat. *Curr. Biol.* **15**, R589–R590.

130. Barolo, S. and Posakony, J.W. (2002). Three habits of highly effective signaling pathways: principles of transcriptional control by developmental cell signaling. *Genes Dev.* **16**, 1167–1181.

131. Barrett, R.D.H. and Hoekstra, H.E. (2011). Molecular spandrels: tests of adaptation at the genetic level. *Nature Rev. Genet.* **12**, 767–780.

132. Barrio, R.A., Baker, R.E., Vaughan, B., Jr., Tribuzy, K., de Carvalho, M.R., Bassanezi, R., and Maini, P.K. (2009). Modeling the skin pattern of fishes. *Phys. Rev. E* **79**, Article 031908 (11 pp.).

133. Barrio, R.A., Varea, C., Aragón, J.L., and Maini, P.K. (1999). A two-dimensional numerical study of spatial pattern formation in interacting Turing systems. *Bull. Math. Biol.* **61**, 483–505.

134. Barsbold, R., Currie, P.J., Myhrvold, N.P., Osmólska, H., Tsogtbaatar, K., and Watabe, M. (2000). A pygostyle from a non-avian theropod. *Nature* **403**, 155.

135. Barton, R.A. and Harvey, P.H. (2000). Mosaic evolution of brain structure in mammals. *Nature* **405**, 1055–1058.

136. Basler, K. and Struhl, G. (1994). Compartment boundaries and the control of *Drosophila* limb pattern by hedgehog protein. *Nature* **368**, 208–214.

137. Bastida, M.F. and Ros, M.A. (2008). How do we get a perfect complement of digits? *Curr. Opin. Gen. Dev.* **18**, 374–380.

138. Bateson, P. (2004). The active role of behaviour in evolution. *Biol. Philos.* **19**, 283–298.

139. Bateson, W. (1894). *Materials for the Study of Variation Treated with Especial Regard to Discontinuity in the Origin of Species*. Macmillan, London.

140. Baum, B. (2006). Left-right asymmetry: actin-myosin through the looking glass. *Curr. Biol.* **16**, R502–R504.

141. Bautista, A., Mendoza-Degante, M., Coureaud, G., Martínez-Gómez, M., and Hudson, R. (2005). Scramble competition in newborn domestic rabbits for an unusually restricted milk supply. *Anim. Behav.* **70**, 1011–1021.

142. Bavendam, F. (1995). The giant cuttlefish: chameleon of the reef. *Natl. Geogr.* **188** #3, 94–107.

143. Beachy, P.A., Helfand, S.L., and Hogness, D.S. (1985). Segmental distribution of bithorax complex proteins during *Drosophila* development. *Nature* **313**, 545–551.

144. Bechstedt, S. and Howard, J. (2008). Hearing mechanics: a fly in your ear. *Curr. Biol.* **18**, R869–R870.

145. Bechtel, H.B. (1995). *Reptile and Amphibian Variants: Colors, Patterns, and Scales.* Krieger, Malabar, FL.

146. Becker, J.B., Berkley, K.J., Geary, N., Hampson, E., Herman, J.P., and Young, E.A., eds. (2008). *Sex Differences in the Brain: From Genes to Behavior.* Oxford University Press, New York, NY.

147. Beckman, J., Banks, S.C., Sunnucks, P., Lill, A., and Taylor, A.C. (2007). Phylogeography and environmental correlates of a cap on reproduction: teat number in a small marsupial, *Antechinus agilis*. *Mol. Ecol.* **16**, 1069–1083.

148. Bejder, L. and Hall, B.K. (2002). Limbs in whales and limblessness in other vertebrates: mechanisms of evolutionary and developmental transformation and loss. *Evol. Dev.* **4**, 445–458.

149. Belalcazar, A.D., Doyle, K., Hogan, J., Neff, D., and Collier, S. (2013). Insect wing membrane topography is determined by the dorsal wing epithelium. *G3 (Genes, Genomes, Genetics)* **3**, 5–8.

150. Beldade, P. and Brakefield, P.M. (2002). The genetics and evo-devo of butterfly wing patterns. *Nature Rev. Gen.* **3**, 442–452.

151. Beldade, P. and Brakefield, P.M. (2003). Concerted evolution and developmental integration in modular butterfly wing patterns. *Evol. Dev.* **5**, 169–179.

152. Beldade, P. and Brakefield, P.M. (2003). The difficulty of agreeing about constraints. *Evol. Dev.* **5**, 119–120.

153. Beldade, P., Brakefield, P.M., and Long, A.D. (2005). Generating phenotypic variation: prospects from "evo-devo" research on *Bicyclus anynana* wing patterns. *Evol. Dev.* **7**, 101–107.

154. Beldade, P., French, V., and Brakefield, P.M. (2008). Developmental and genetic mechanisms for evolutionary diversification of serial repeats: eyespot size in *Bicyclus anynana* butterflies. *J. Exp. Zool. (Mol. Dev. Evol.)* **310B**, 191–201.

155. Beldade, P., Koops, K., and Brakefield, P.M. (2002). Developmental constraints versus flexibility in morphological evolution. *Nature* **416**, 844–847.

156. Beldade, P., Koops, K., and Brakefield, P.M. (2002). Modularity, individuality, and evo-devo in butterfly wings. *PNAS* **99** #22, 14262–14267.

157. Beldade, P., McMillan, W.O., and Papanicolaou, A. (2008). Butterfly genomics eclosing. *Heredity* **100**, 150–157.

158. Bell, E., Andres, B., and Goswami, A. (2011). Integration and dissociation of limb elements in flying vertebrates: a comparison of pterosaurs, birds and bats. *J. Evol. Biol.* **24**, 2586–2599.

159. Bellairs, A.D. (1948). The eyelids and spectacle in geckos. *Proc. Zool. Soc. Lond.* **118**, 420–425.

160. Bellairs, A.D. and Underwood, G. (1951). The origin of snakes. *Biol. Rev.* **26**, 193–237.

161. Bellone, R.R., Forsyth, G., Leeb, T., Archer, S., Sigurdsson, S., Imsland, F., Mauceli, E., Engensteiner, M., Bailey, E., Sandmeyer, L., Grahn, B., Lindblad-Toh, K., and Wade,

C.M. (2010). Fine-mapping and mutation analysis of *TRPM1*: a candidate gene for leopard complex (*LP*) spotting and congenital stationary night blindness in horses. *Briefings Funct. Genomics* **9**, 193–207.

162. Belote, J.M. and Baker, B.S. (1982). Sex determination in *Drosophila melanogaster*: analysis of transformer-2, a sex-transforming locus. *PNAS* **79** #5, 1568–1572.

163. Beloussov, L.V. (2006). Direct physical formation of anatomical structures by cell traction forces: an interview with Albert Harris. *Int. J. Dev. Biol.* **50**, 93–101.

164. Beloussov, L.V. and Grabovsky, V.I. (2006). Morphomechanics: goals, basic experiments and models. *Int. J. Dev. Biol.* **50**, 81–92.

165. Bely, A.E. (2010). Evolutionary loss of animal regeneration: pattern and process. *Integr. Comp. Biol.* **50**, 515–527.

166. Ben-Jacob, E. and Levine, H. (1998). The artistry of microorganisms. *Sci. Am.* **279** #4, 82–87.

167. Ben-Zvi, D., Shilo, B.-Z., and Barkai, N. (2011). Scaling of morphogen gradients. *Curr. Opin. Gen. Dev.* **21**, 704–710. [*See also* Inomata, H., *et al.* (2013). Scaling of dorsal-ventral patterning by embryo size-dependent degradation of Spemann's organizer signals. *Cell* **153**, 1296–1311.]

168. Bengtson, S., Cunningham, J.A., Yin, C., and Donoghue, P.C.J. (2012). A merciful death for the "earliest bilaterian," *Vernanimalcula*. *Evol. Dev.* **14**, 421–427.

169. Bennett, D.K. (1980). Stripes do not a zebra make, Part 1: A cladistic analysis of *Equus*. *Syst. Zool.* **29**, 272–287.

170. Bensoussan-Trigano, V., Lallemand, Y., Cloment, C.S., and Robert, B. (2012). *Msx1* and *Msx2* in limb mesenchyme modulate digit number and identity. *Dev. Dynamics* **240**, 1190–1202.

171. Benton, M.J. (2010). Studying function and behavior in the fossil record. *PLoS Biol.* **8** #3, e1000321.

172. Berger, L.R. (2013). The mosaic nature of *Australopithecus sediba*. *Science* **340**, 163.

173. Bergmann, P.J. and Berk, C.P. (2012). The evolution of positive allometry of weaponry in horned lizards (*Phrynosoma*). *Evol. Biol.* **39**, 311–323.

174. Bergmann, P.J. and Irschick, D.J. (2010). Alternate pathways of body shape evolution translate into common patterns of locomotor evolution in two clades of lizards. *Evolution* **64**, 1569–1582.

175. Bernard, F., Dutriaux, A., Silber, J., and Lalouette, A. (2006). Notch pathway repression by *vestigial* is required to promote indirect flight muscle differentiation in *Drosophila melanogaster*. *Dev. Biol.* **295**, 164–177.

176. Bernard, F., Lalouette, A., Gullaud, M., Jeantet, A.Y., Cossard, R., Zider, A., Ferveur, J.F., and Silber, J. (2003). Control of *apterous* by *vestigial* drives indirect flight muscle development in *Drosophila*. *Dev. Biol.* **260**, 391–403.

177. Berta, A., Ray, C.E., and Wyss, A.R. (1989). Skeleton of the oldest known pinniped, *Enaliarctos mealsi*. *Science* **244**, 60–62.

178. Berta, A., Sumich, J.L., and Kovacs, K.M. (2006). *Marine Mammals: Evolutionary Biology*, 2nd edn. Elsevier, New York, NY.

179. Bertrand, S., Brunet, F.G., Escriva, H., Parmentier, G., Laudet, V., and Robinson-Rechavi, M. (2004). Evolutionary genomics of nuclear receptors: from 25 ancestral genes to derived endocrine systems. *Mol. Biol. Evol.* **21**, 1923–1937.

180. Bertrand, S. and Escriva, H. (2011). Evolutionary crossroads in developmental biology: amphioxus. *Development* **138**, 4819–4830.

181. Bessho, Y., Hirata, H., Masamizu, Y., and Kageyama, R. (2003). Periodic repression by the bHLH factor Hes7 is an essential mechanism for the somite segmentation clock. *Genes Dev.* **17**, 1451–1456.

182. Bessho, Y., Miyoshi, G., Sakata, R., and Kageyama, R. (2001). *Hes7*: a bHLH-type repressor gene regulated by Notch and expressed in the presomitic mesoderm. *Genes to Cells* **6**, 175–185.

183. Bessodes, N., Haillot, E., Duboc, V., Röttinger, E., Lahaye, F., and Lepage, T. (2012). Reciprocal signaling between the ectoderm and a mesendodermal left-right organizer directs left-right determination in the sea urchin embryo. *PLoS Genet.* **8** #12, e1003121.

184. Best, R.C. (1981). The tusk of the narwhal (*Monodon monoceros* L.): interpretation of its function (Mammalia: Cetacea). *Can. J. Zool.* **59**, 2386–2393.

185. Beverdam, A., Merlo, G.R., Paleari, L., Mantero, S., Genova, F., Barbieri, O., Janvier, P., and Levi, G. (2002). Jaw transformation with gain of symmetry after *Dlx5/Dlx6* inactivation: mirror of the past? *Genesis* **34**, 221–227.

186. Beveridge, W.I.B. (1950). *The Art of Scientific Investigation*. Random House, New York, NY.

187. Bhalla, U.S. and Iyengar, R. (1999). Emergent properties of networks of biological signaling pathways. *Science* **283**, 381–387.

188. Bhattacharyya, R.P., Reményi, A., Yeh, B.J., and Lim, W.A. (2006). Domains, motifs, and scaffolds: the role of modular interactions in the evolution and wiring of cell signaling circuits. *Annu. Rev. Biochem.* **75**, 655–680.

189. Bhullar, B.-A.S. (2012). A phylogenetic approach to ontogeny and heterochrony in the fossil record: cranial evolution and development in anguimorphan lizards (Reptilia: Squamata). *J. Exp. Zool. (Mol. Dev. Evol.)* **318B**, 521–530.

190. Bhullar, B.-A.S., Marugán-Lobón, J., Racimo, F., Bever, G.S., Rowe, T.B., Norell, M.A., and Abzhanov, A. (2012). Birds have paedomorphic dinosaur skulls. *Nature* **487**, 223–226.

191. Bickelmann, C., Mitgutsch, C., Richardson, M.K., Jiménez, R., de Bakker, M.A.G., and Sánchez-Villagra, M.R. (2012). Transcriptional heterochrony in talpid mole autopods. *EvoDevo* **3**, Article 16 (5 pp.).

192. Bickford, D., Iskandar, D., and Barlian, A. (2008). A lungless frog discovered on Borneo. *Curr. Biol.* **18**, R374–R375.

193. Bier, E. (1997). Anti-neural-inhibition: a conserved mechanism for neural induction. *Cell* **89**, 681–684.

194. Bier, E. (2011). Evolution of development: diversified dorsoventral patterning. *Curr. Biol.* **21**, R591–R594.

195. Billet, G., Hautier, L., Asher, R.J., Schwarz, C., Crumpton, N., Martin, T., and Ruf, I. (2012). High morphological variation of vestibular system accompanies slow and infrequent locomotion in three-toed sloths. *Proc. R. Soc. Lond. B* **279**, 3932–3939.

196. Bishop, K.L. (2008). The evolution of flight in bats: narrowing the field of plausible hypotheses. *Q. Rev. Biol.* **83**, 153–169.

197. Bitan, A., Rosenbaum, I., and Abdu, U. (2012). Stable and dynamic microtubules coordinately determine and maintain *Drosophila* bristle shape. *Development* **139**, 1987–1996.

198. Blackstone, N.W. (2006). Charles Manning Child (1869–1954): the past, present, and future of metabolic signaling. *J. Exp. Zool. (Mol. Dev. Evol.)* **306B**, 1–7.

199. Blair, P. (1710). Osteographia elephantina: Or, a full and exact description of all the bones of an elephant, which died near Dundee, April the 27th, 1706. With their several dimensions. Communicated in a letter to Dr. Hans Sloane, R. S. Secr. by Mr Patrick Blair, Surgeon, &c. *Phil. Trans.* **27**, 53–116.

200. Blair, S.S. (2009). Segmentation in animals. *Curr. Biol.* **18**, R991–R995.

201. Blanco, M.J., Misof, B.Y., and Wagner, G.P. (1998). Heterochronic differences of *Hoxa-11* expression in *Xenopus* fore- and hind limb development: Evidence for lower limb identity of the anural ankle bones. *Dev. Genes Evol.* **208**, 175–187.

202. Blaxter, M. and Sunnucks, P. (2011). Velvet worms. *Curr. Biol.* **21**, R238–R240.

203. Boaz, N.T. (1997). *Eco Homo: How the Human Being Emerged from the Cataclysmic History of the Earth.* Basic Books, New York, NY.

204. Bobkova, M.V., Gál, J., Zhukov, V.V., Shepeleva, I.P., and Meyer-Rochow, V.B. (2004). Variations in the retinal designs of pulmonate snails (Mollusca, Gastropoda): squaring phylogenetic background and ecophysiological needs (I). *Invert. Biol.* **123**, 101–115.

205. Bock, G. and Goode, J., eds. (2007). *Tinkering: The Microevolution of Development.* Wiley, Chichester.

206. Bodmann, B.E.J. and Mombach, J.C.M. (2000). On the role of probability amplitudes in cell aggregation: an approach study towards morphogenesis. *Physica A* **278**, 243–259.

207. Bodmer, R. and Venkatesh, T.V. (1998). Heart development in *Drosophila* and vertebrates: conservation of molecular mechanisms. *Dev. Genet.* **22**, 181–186.

208. Boell, L. (2013). Lines of least resistance and genetic architecture of house mouse (*Mus musculus*) mandible shape. *Evol. Dev.* **15**, 197–204.

209. Boero, F. and Piraino, S. (2011). From cnidaria to "higher metazoa" in one step. In *Key Transitions in Animal Evolution* (R. DeSalle and B. Schierwater, eds.). Science Publishers, Enfield, NH, pp. 162–174.

210. Boero, F., Schierwater, B., and Piraino, S. (2007). Cnidarian milestones in metazoan evolution. *Integr. Comp. Biol.* **47**, 693–700.

211. Boisserie, J.-R., Fisher, R.E., Lihoreau, F., and Weston, E.M. (2011). Evolving between land and water: key questions on the emergence and history of the Hippopotamidae (Hippopotamoidea, Cetancodonta, Cetartiodactyla). *Biol. Rev.* **86**, 601–625.

212. Bökel, C. and Brand, M. (2013). Generation and interpretation of FGF morphogen gradients in vertebrates. *Curr. Opin. Gen. Dev.* **23**, 415–422.

213. Boletzky, S.v. (1988). Cephalopod development and evolutionary concepts. In *The Mollusca, Vol. 12: Paleontology and Neontology of Cephalopods* (K.M. Wilbur, M.R. Clarke, and E.R. Trueman, eds.). Academic Press, New York, NY, pp. 185–202.

214. Boletzky, S.v. (1988). Characteristics of cephalopod embryogenesis. In *Cephalopods: Present and Past* (J. Wiedmann and J. Kullmann, eds.). O. H. Schindewolf-Symposium Tübingen 1985 (2nd International Cephalopod Symposium). Schweizerbart, Stuttgart, pp. 167–179.

215. Bolker, J.A. and Raff, R.A. (1996). Developmental genetics and traditional homology. *BioEssays* **18**, 489–494.

216. Bomtorin, A.D., Barchuk, A.R., Moda, L.M., and Simoes, Z.L.P. (2012). Hox gene expression leads to differential hind leg development between honeybee castes. *PLoS ONE* **7** #7, e40111.

217. Bonato, L., Foddai, D., and Minelli, A. (2003). Evolutionary trends and patterns in centipede segment number based on a cladistic analysis of Mecistocephalidae (Chilopoda: Geophilomorpha). *Syst. Entomol.* **28**, 539–579.

218. Bonnet, X., Bradshaw, D., Shine, R., and Pearson, D. (1999). Why do snakes have eyes? The (non-)effect of blindness in island tiger snakes (*Notechis scutatus*). *Behav. Ecol. Sociobiol.* **46**, 267–272.

219. Borowsky, R. (2008). Restoring sight in blind cavefish. *Curr. Biol.* **18**, R23–R24.

220. Bosch, T.C.G. (2003). Ancient signals: peptides and the interpretation of positional information in ancestral metazoans. *Comp. Biochem. Physiol. B* **136**, 185–196.

221. Bourlat, S.J., Juliusdottir, T., Lowe, C.J., Freeman, R., Aronowicz, J., Kirschner, M., Lander, E.S., Thorndyke, M., Nakano, H., Kohn, A.B., Heyland, A., Moroz, L.L., Copley, R.R., and Telford, M.J. (2006). Deuterostome phylogeny reveals monophyletic chordates and the new phylum Xenoturbellida. *Nature* **444**, 85–88.

222. Bouzaffour, M., Dufourcq, P., Lecaudey, V., Haas, P., and Vriz, S. (2009). Fgf and Sdf-1 pathways interact during zebrafish fin regeneration. *PLoS ONE* **4** #6, e5824.

223. Bowsher, J.H. and Nijhout, H.F. (2009). Partial co-option of the appendage patterning pathway in the development of abdominal appendages in the sepsid fly *Themira biloba*. *Dev. Genes Evol.* **219**, 577–587.

224. Bradley, D.G. (2003). Genetic hoofprints. *Nat. Hist.* **112** #1, 36–41.

225. Bragulla, H. and Hirschberg, R.M. (2003). Horse hooves and bird feathers: two model systems for studying the structure and development of highly adapted integumentary accessory organs – the role of the dermo-epidermal interface for the micro-architecture of complex epidermal structures. *J. Exp. Zool. (Mol. Dev. Evol.)* **298B**, 140–151.

226. Bragulla, H.H. and Homberger, D.G. (2009). Structure and functions of keratin proteins in simple, stratified, keratinized and cornified epithelia. *J. Anat.* **214**, 516–559.

227. Braiman, Y., Lindner, J.F., and Ditto, W.L. (1995). Taming spatiotemporal chaos with disorder. *Nature* **378**, 465–467.

228. Brakefield, P.M. (1998). The evolution-development interface and advances with the eyespot patterns of *Bicyclus* butterflies. *Heredity* **80**, 265–272.

229. Brakefield, P.M. (2007). Butterfly eyespot patterns and how evolutionary tinkering yields diversity. In *Tinkering: The Microevolution of Development*. Novartis Foundation Symposium 284. Wiley, Chichester, pp. 90–109.

230. Brakefield, P.M. and French, V. (1999). Butterfly wings: the evolution of development of colour patterns. *BioEssays* **21**, 391–401.

231. Brakefield, P.M., Gates, J., Keys, D., Kesbeke, F., Wijngaarden, P.J., Monteiro, A., French, V., and Carroll, S.B. (1996). Development, plasiticity and evolution of butterfly eyespot patterns. *Nature* **384**, 236–242.

232. Brandley, M.C., Huelsenbeck, J.P., and Wiens, J.J. (2008). Rates and patterns in the evolution of snake-like body form in squamate reptiles: evidence for repeated re-evolution of lost digits and long-term persistence of intermediate body forms. *Evolution* **62**, 2042–2064.

233. Branham, M.A. and Wenzel, J.W. (2003). The origin of photic behavior and the evolution of sexual communication in fireflies (Coleoptera: Lampyridae). *Cladistics* **19**, 1–22.

234. Brear, K., Currey, J.D., Kingsley, M.C.S., and Ramsay, M. (1993). The mechanical design of the tusk of the narwhal (*Monodon monocerous*: Cetacea). *J. Zool. Lond.* **230**, 411–423.

235. Brena, C. and Akam, M. (2012). The embryonic development of the centipede *Strigamia maritima*. *Dev. Biol.* **363**, 290–307.

236. Brena, C., Chipman, A.D., Minelli, A., and Akam, M. (2006). Expression of trunk *Hox* genes in the centipede *Strigamia maritima:* sense and anti-sense transcripts. *Evol. Dev.* **8**, 252–265.

237. Breuker, C.J., Debat, V., and Klingenberg, C.P. (2006). Functional evo-devo. *Trends Ecol. Evol.* **21**, 488–492.

238. Bridges, C.B. and Morgan, T.H. (1923). The third-chromosome group of mutant characters of *Drosophila melanogaster*. *Carnegie Inst. Wash. Publ.* **327**, 1–251.

239. Bridgham, J.T., Carroll, S.B., and Thornton, J.W. (2006). Evolution of hormone-receptor complexity by molecular exploitation. *Science* **312**, 97–101.

240. Brigandt, I. and Love, A.C. (2012). Conceptualizing evolutionary novelty: moving beyond definitional debates. *J. Exp. Zool. (Mol. Dev. Evol.)* **318B**, 417–427.

241. Bright, M., Klose, J., and Nussbaumer, A.D. (2013). Giant tubeworms. *Curr. Biol.* **23**, R224–R225.

242. Brischoux, F., Pizzatto, L., and Shine, R. (2010). Insights into the adaptive significance of vertical pupil shape in snakes. *J. Evol. Biol.* **23**, 1878–1885.

243. Brischoux, F. and Shine, R. (2011). Morphological adaptations to marine life in snakes. *J. Morph.* **272**, 566–572.

244. Briscoe, J., Lawrence, P.A., and Vincent, J.-P., eds. (2010). *Generation and Interpretation of Morphogen Gradients*. Cold Spring Harbor Laboratory Press, Woodbury, NY.

245. Bristow, C.M. (1985). Sugar nannies. *Nat. Hist.* **94** #9, 63–69.

246. Britz, R. and Johnson, G.D. (2011). Comments on the establishment of the one to one relationship between characters as a prerequisite for homology assessment in phylogenetic studies. *Zootaxa* **2946**, 65–72.

247. Britz, R. and Johnson, G.D. (2012). Ontogeny and homology of the skeletal elements that form the sucking disc of remoras (Teleostei, Echeneoidei, Echeneidae). *J. Morph.* **273**, 1353–1366. [*See also* Friedman, M., *et al.* (2013). An early fossil remora (Echeneoideda) reveals the evolutionary assembly of the adhesion disc. *Proc. R. Soc. Lond. B* **280**, doi 20131200.]

248. Brockes, J.P. and Kumar, A. (2005). Appendage regeneration in adult vertebrates and implications for regenerative medicine. *Science* **310**, 1919–1922.

249. Brockes, J.P., Kumar, A., and Velloso, C.P. (2001). Regeneration as an evolutionary variable. *J. Anat.* **199**, 3–11.

250. Brodie, C. (2007). At home in the dark. *Am. Sci.* **95**, 460–461.

251. Brodie, E.D., III (1992). Correlational selection for color pattern and antipredator behavior in the garter snake *Thamnophis ordinoides*. *Evolution* **46**, 1284–1298.

252. Brodie, E.D., III (2009). Toxins and venoms. *Curr. Biol.* **19**, R931–R935.

253. Brodie, E.D., III (2010). Convergent evolution: pick your poison carefully. *Curr. Biol.* **20**, R152–R154.

254. Brodsky, A.K. (1994). *The Evolution of Insect Flight*. Oxford University Press, New York, NY.

255. Brogiolo, W., Stocker, H., Ikeya, T., Rintelen, F., Fernandez, R., and Hafen, E. (2001). An evolutionarily conserved function of the *Drosophila* insulin receptor and insulin-like peptides in growth control. *Curr. Biol.* **11**, 213–221.

256. Bronner, M.E. and LeDouarin, N.M. (2012). Development and evolution of the neural crest: an overview. *Dev. Biol.* **366**, 2–9.

257. Brooke, M.D.L., Hanley, S., and Laughlin, S.B. (1999). The scaling of eye size with body mass in birds. *Proc. R. Soc. Lond. B* **266**, 405–412.

258. Brower, D.L. (1987). *Ultrabithorax* gene expression in *Drosophila* imaginal discs and larval nervous system. *Development* **101**, 83–92.

259. Brown, C., Garwood, M.P., and Williamson, J.E. (2012). It pays to cheat: tactical deception in a cephalopod social signalling system. *Biol. Lett.* **8**, 729–732.

260. Brown, D.M., Brenneman, R.A., Koepfli, K.-P., Pollinger, J.P., Milá, B., Georgiadis, N.J., Louis, E.E., Jr., Grether, G.F., Jacobs, D.K., and Wayne, R.K. (2007). Extensive population genetic structure in the giraffe. *BMC Biol.* **5**, Article 57 (13 pp.).

261. Brown, J.L. (1982). The adaptationist program. *Science* **217**, 884–886.
262. Brown, L. and Rockwood, L.L. (1986). On the dilemma of horns. *Nat. Hist.* **95** #7, 54–61.
263. Brown, N.A. and Wolpert, L. (1990). The development of handedness in left/right asymmetry. *Development* **109**, 1–9.
264. Brown, N.L., Patel, S., Brzezinski, J., and Glaser, T. (2001). *Math5* is required for retinal ganglion cell and optic nerve formation. *Development* **128**, 2497–2508.
265. Bruce, A.E.E. and Shankland, M. (1998). Expression of the head gene *Lox22-Otx* in the leech *Helobdella* and the origin of the bilaterian body plan. *Dev. Biol.* **201**, 101–112.
266. Bruemmer, F. (1993). *The Narwhal: Unicorn of the Sea*. Swan Hill Press, Shrewsbury.
267. Brunetti, C.R., Selegue, J.E., Monteiro, A., French, V., Brakefield, P.M., and Carroll, S.B. (2001). The generation and diversification of butterfly eyespot color patterns. *Curr. Biol.* **11**, 1578–1585.
268. Brusca, R.C. and Brusca, G.J. (1990). *Invertebrates*. Sinauer, Sunderland, MA.
269. Bryant, H.N. and Churcher, C.S. (1987). All sabretoothed carnivores aren't sharks. *Nature* **325**, 488. [*See also* Wroe, S., *et al.* (2013). Comparative biomechanical modeling of metatherian and placental saber-tooths: a different kind of bite for an extreme pouched predator. *PLoS ONE* **8** #6, e66888.]
270. Bryant, P.J. (1978). Pattern formation in imaginal discs. In *The Genetics and Biology of Drosophila*, Vol. 2c (M. Ashburner and T.R.F. Wright, eds.). Academic Press, New York, NY, pp. 229–335.
271. Bryant, P.J. (1993). The Polar Coordinate Model goes molecular. *Science* **259**, 471–472.
272. Bryant, P.J., Bryant, S.V., and French, V. (1977). Biological regeneration and pattern formation. *Sci. Am.* **237** #1, 66–81.
273. Buchholtz, E.A., Bailin, H.G., Laves, S.A., Yang, J.T., Chan, M.-Y., and Drozd, L.E. (2012). Fixed cervical count and the origin of the mammalian diaphram. *Evol. Dev.* **14**, 399–411.
274. Buchholtz, E.A. and Stepien, C.C. (2009). Anatomical transformation in mammals: developmental origin of aberrant cervical anatomy in tree sloths. *Evol. Dev.* **11**, 69–79.
275. Buchtová, M., Handrigan, G.R., Tucker, A.S., Lozanoff, S., Town, L., Fu, K., Diewert, V.M., Wicking, C., and Richman, J.M. (2008). Initiation and patterning of the snake dentition are dependent on Sonic Hedgehog signaling. *Dev. Biol.* **319**, 132–145.
276. Buckley-Beason, V.A., Johnson, W.E., Nash, W.G., Stanyon, R., Menninger, J.C., Driscoll, C.A., Howard, J., Bush, M., Page, J.E., Roelke, M.E., Stone, G., Martelli, P.P., Wen, C., Ling, L., Duraisingam, R.K., Lam, P.V., and O'Brien, S.J. (2006). Molecular evidence for species-level distinctions in clouded leopards. *Curr. Biol.* **16**, 2371–2376.
277. Budd, G.E. (1999). Does evolution in body patterning genes drive morphological change – or vice versa? *BioEssays* **21**, 326–332.
278. Budd, G.E. (2002). A palaeontological solution to the arthropod head problem. *Nature* **417**, 271–275.
279. Budd, G.E. (2006). On the origin and evolution of major morphological characters. *Biol. Rev.* **81**, 609–628.
280. Budd, G.E. (2008). The earliest fossil record of the animals and its significance. *Phil. Trans. R. Soc. Lond. B* **363**, 1425–1434.
281. Budd, G.E. and Telford, M.J. (2009). The origin and evolution of arthropods. *Nature* **457**, 812–817.
282. Budi, E.H., Patterson, L.B., and Parichy, D.M. (2011). Post-embryonic nerve-associated precursors to adult pigment cells: genetic requirements and dynamics of morphogenesis and differentiation. *PLoS Genet.* **7** #5, e1002044.

283. Budrene, E.O. and Berg, H.C. (1991). Complex patterns formed by motile cells of *Escherichia coli*. *Nature* **349**, 630–633.

284. Buecker, C. and Wysocka, J. (2012). Enhancers as information integration hubs in development: lessons from genomics. *Trends Genet.* **28**, 276–284.

285. Bull, J.J. and Charnov, E.L. (1985). On irreversible evolution. *Evolution* **39**, 1149–1155.

286. Burke, A.C., Nelson, C.E., Morgan, B.A., and Tabin, C. (1995). *Hox* genes and the evolution of vertebrate axial morphology. *Development* **121**, 333–346.

287. Burke, R.D. (2011). Deuterostome neuroanatomy and the body plan paradox. *Evol. Dev.* **13**, 110–115.

288. Burnie, D. and Wilson, D.E., eds. (2011). *Animal: The Definitive Visual Guide*. Dorling Kindersley, New York, NY.

289. Burtis, K.C. and Baker, B.S. (1989). *Drosophila doublesex* gene controls somatic sexual differentiation by producing alternatively spliced mRNAs encoding related sex-specific polypeptides. *Cell* **56**, 997–1010.

290. Buschbeck, E.K. and Friedrich, M. (2008). Evolution of insect eyes: tales of ancient heritage, deconstruction, reconstruction, remodeling, and recycling. *Evo. Edu. Outreach* **1**, 448–462.

291. Büschges, A., Scholz, H., and El Manira, A. (2011). New moves in motor control. *Curr. Biol.* **21**, R513–R524.

292. Bush, J.O. and Jiang, R. (2012). Palatogenesis: morphogenetic and molecular mechanisms of secondary palate development. *Development* **139**, 231–243.

293. Buss, D.M., Haselton, M.G., Shackelford, T.K., Bleske, A.L., and Wakefield, J.C. (1998). Adaptations, exaptations, and spandrels. *Am. Psychol.* **53**, 533–548.

294. Butts, T., Holland, P.W.H., and Ferrier, D.E.K. (2008). The Urbilaterian Super-Hox cluster. *Trends Genet.* **24**, 259–262. [*See also* Mallo, M. and Alonso, C.R. (2013). The regulation of *Hox* gene expression during animal development. *Development* **140**, 3951–3963.]

295. Byrnes, G. and Spence, A.J. (2011). Ecological and biomechanical insights into the evolution of gliding in mammals. *Integr. Comp. Biol.* **51**, 991–1001.

296. Caballero, L., Benítez, M., Alvarez-Buylla, E.R., Hernández, S., Arzola, A.V., and Cocho, G. (2012). An epigenetic model for pigment patterning based on mechanical and cellular interactions. *J. Exp. Zool. (Mol. Dev. Evol.)* **318B**, 209–223.

297. Cabrera, C.V., Botas, J., and Garcia-Bellido, A. (1985). Distribution of *Ultrabithorax* proteins in mutants of *Drosophila* bithorax complex and its transregulatory genes. *Nature* **318**, 569–571.

298. Caburet, S., Cocquet, J., Vaiman, D., and Veitia, R.A. (2005). Coding repeats and evolutionary "agility". *BioEssays* **27**, 581–587.

299. Cadigan, K.M. (2012). TCFs and Wnt/â-catenin signaling: more than one way to throw the switch. *Curr. Top. Dev. Biol.* **98**, 1–34.

300. Cadigan, K.M. and Nusse, R. (1997). Wnt signaling: a common theme in animal development. *Genes Dev.* **11**, 3286–3305.

301. Calder, N. (1984). *Timescale: An Atlas of the Fourth Dimension*. Chatto & Windus, London.

302. Calder, W.A., III (1978). The kiwi. *Sci. Am.* **239** #1, 132–142.

303. Calder, W.A., III (1979). The kiwi and egg design: evolution as a package deal. *BioScience* **29** #8, 461–467.

304. Caldwell, M.W. (2003). "Without a leg to stand on": on the evolution and development of axial elongation and limblessness in tetrapods. *Can. J. Earth Sci.* **40**, 573–588.

305. Caldwell, M.W. and Lee, M.S.Y. (1997). A snake with legs from the marine Cretaceous of the Middle East. *Nature* **386**, 705–709.
306. Caldwell, R.L. and Dingle, H. (1976). Stomatopods. *Sci. Am.* **234** #1, 80–89.
307. Camazine, S. (2003). Patterns in nature. *Nat. Hist.* **112** #5, 34–41.
308. Cameron, C.B. (2005). A phylogeny of the hemichordates based on morphological characters. *Can. J. Zool.* **83**, 196–215.
309. Campbell, G. (2002). Distalization of the *Drosophila* leg by graded EGF-receptor activity. *Nature* **418**, 781–785.
310. Campbell, G. and Tomlinson, A. (1995). Initiation of the proximodistal axis in insect legs. *Development* **121**, 619–628.
311. Campbell, G. and Tomlinson, A. (1998). The roles of the homeobox genes *aristaless* and *Distal-less* in patterning the legs and wings of *Drosophila*. *Development* **125**, 4483–4493.
312. Campbell, J.H. (1985). An organizational interpretation of evolution. In *Evolution at a Crossroads: The New Biology and the New Philosophy of Science* (D.J. Depew and B.H. Weber, eds.). MIT Press, Cambridge, MA, pp. 133–167.
313. Campo-Paysaa, F., Marlétaz, F., Laudet, V., and Schubert, M. (2008). Retinoic acid signaling in development: tissue-specific functions and evolutionary origins. *Genesis* **46**, 640–656.
314. Cañestro, C., Albalat, R., Irimia, M., and Garcia-Fernàndez, J. (2013). Impact of gene gains, losses and duplication modes on the origin and diversification of vertebrates. *Semin. Cell Dev. Biol.* **24**, 83–94.
315. Cañestro, C. and Postlethwait, J.H. (2007). Development of a chordate anterior–posterior axis without classical retinoic acid signaling. *Dev. Biol.* **305**, 522–538.
316. Cañestro, C., Postlethwait, J.H., Gonzàlez-Duarte, R., and Albalat, R. (2006). Is retinoic acid genetic machinery a chordate innovation? *Evol. Dev.* **8**, 394–406.
317. Cañestro, C., Yokoi, H., and Postlethwait, J.H. (2007). Evolutionary developmental biology and genomics. *Nature Rev. Genet.* **8**, 932–942.
318. Cantú, J.M. and Ruiz, C. (1985). On atavisms and atavistic genes. *Ann. Génét.* **28** #3, 141–142.
319. Caporale, L. (2005). Darwin in the genome. *BioEssays* **27**, 984.
320. Caporale, L.H. (2003). *Darwin in the Genome: Molecular Strategies in Biological Evolution*. McGraw-Hill, New York, NY.
321. Caporale, L.H. (2003). Foresight in genome evolution. *Am. Sci.* **91**, 234–241.
322. Caporale, L.H. (2008). It's not random anymore. *BioEssays* **30**, 400–402.
323. Cappellari, S.C., Schaefer, H., and Davis, C.C. (2013). Evolution: pollen or pollinators – which came first? *Curr. Biol.* **23**, R316–R318.
324. Caprette, C.L., Lee, M.S.Y., Shine, R., Mokany, A., and Downhower, J.F. (2004). The origin of snakes (Serpentes) as seen through eye anatomy. *Biol. J. Linnean Soc.* **81**, 469–482.
325. Carapuço, M., Nóvoa, A., Bobola, N., and Mallo, M. (2005). *Hox* genes specify vertebral types in the presomitic mesoderm. *Genes Dev.* **19**, 2116–2121.
326. Carlson, B.M. (1994). *Human Embryology and Developmental Biology*. Mosby, St. Louis, MO.
327. Carlson, B.M. (2007). *Principles of Regenerative Biology*. Academic Press, New York, NY.
328. Carmona-Fontaine, C., Matthews, H.K., Kuriyama, S., Moreno, M., Dunn, G.A., Parsons, M., Stern, C.D., and Mayor, R. (2008). Contact inhibition of locomotion *in vivo* controls neural crest directional migration. *Nature* **456**, 957–961.
329. Caro, T. (2009). Contrasting coloration in terrestrial mammals. *Phil. Trans. R. Soc. Lond. B* **364**, 537–548.

330. Caron, J.-B., Conway Morris, S., and Cameron, C.B. (2013). Tubicolous enteropneusts from the Cambrian period. *Nature* **495**, 503–506.

331. Carrier, D.R. (1991). Conflict in the hypaxial musculo-skeletal system: documenting an evolutionary constraint. *Am. Zool.* **31**, 644–654.

332. Carroll, C. (2005). Following the stealth hunter. *Natl. Geogr.* **208** #5, 66–77.

333. Carroll, L. and Gardner, M. (1960). *The Annotated Alice: Alice's Adventures in Wonderland & Through the Looking Glass*. Meridian, New York, NY.

334. Carroll, R.L. and Holmes, R.B. (2007). Evolution of the appendicular skeleton of amphibians. In *Fins into Limbs: Evolution, Development, and Transformation* (B.K. Hall, ed.). University of Chicago Press, Chicago, IL, pp. 185–224.

335. Carroll, S. (1997). Genetics on the wing or how the butterfly got its spots. *Nat. Hist.* **106** #1, 28–32.

336. Carroll, S.B. (1994). Developmental regulatory mechanisms in the evolution of insect diversity. *Development* **1994** Suppl., 217–223.

337. Carroll, S.B. (1995). Homeotic genes and the evolution of arthropods and chordates. *Nature* **376**, 479–485.

338. Carroll, S.B. (2001). Chance and necessity: the evolution of morphological complexity and diversity. *Nature* **409**, 1102–1109.

339. Carroll, S.B. (2005). *Endless Forms Most Beautiful: The New Science of Evo Devo and the Making of the Animal Kingdom*. Norton, New York, NY.

340. Carroll, S.B. (2006). *The Making of the Fittest: DNA and the Ultimate Forensic Record of Evolution*. Norton, New York, NY.

341. Carroll, S.B. (2008). Evo-devo and an expanding evolutionary synthesis: a genetic theory of morphological evolution. *Cell* **134**, 25–36.

342. Carroll, S.B., DiNardo, S., O'Farrell, P.H., White, R.A.H., and Scott, M.P. (1988). Temporal and spatial relationships between segmentation and homeotic gene expression in *Drosophila* embryos: distributions of the fushi tarazu, engrailed, Sex combs reduced, Antennapedia, and Ultrabithorax proteins. *Genes Dev.* **2**, 350–360.

343. Carroll, S.B., Gates, J., Keys, D.N., Paddock, S.W., Panganiban, G.E., Selegue, J.E., and Williams, J.A. (1994). Pattern formation and eyespot determination in butterfly wings. *Science* **265**, 109–114.

344. Carroll, S.B., Grenier, J.K., and Weatherbee, S.D. (2005). *From DNA to Diversity: Molecular Genetics and the Evolution of Animal Design*, 2nd edn. Blackwell, Malden, MA.

345. Carroll, S.B., Weatherbee, S.D., and Langeland, J.A. (1995). Homeotic genes and the regulation and evolution of insect wing number. *Nature* **375**, 58–61.

346. Carroll, S.B. and Whyte, J.S. (1989). The role of the *hairy* gene during *Drosophila* morphogenesis: stripes in imaginal discs. *Genes Dev.* **3**, 905–916.

347. Carson, H.L., Hardy, D.E., Spieth, H.T., and Stone, W.S. (1970). The evolutionary biology of the Hawaiian Drosophilidae. In *Essays in Evolution and Genetics* (M.K. Hecht and W.C. Steere, eds.). Appleton-Century-Crofts, New York, NY, pp. 437–543.

348. Carstanjen, B., Abitbol, M., and Desbois, C. (2007). Bilateral polydactyly in a foal. *J. Vet. Sci.* **8**, 201–203.

349. Casares, F. and Mann, R.S. (2001). The ground state of the ventral appendage in *Drosophila*. *Science* **293**, 1477–1480.

350. Casci, T. (2001). How the butterfly got its spots. *Nature Rev. Gen.* **2**, 911.

351. Cass, A.N., Servetnick, M.D., and McCune, A.R. (2013). Expression of a lung developmental cassette in the adult and developing zebrafish swimbladder. *Evol. Dev.* **15**, 119–132.

352. Castelli-Gair Hombría, J. (2011). Butterfly eyespot serial homology: enter the *Hox* genes. *BMC Biol.* **9**, Article 26 (3 pp.).

353. Castelli-Gair Hombría, J. and Lovegrove, B. (2003). Beyond homeosis: HOX function in morphogenesis and organogenesis. *Differentiation* **71**, 461–476.

354. Castelli-Gair, J. (1998). Implications of the spatial and temporal regulation of *Hox* genes on development and evolution. *Int. J. Dev. Biol.* **42**, 437–444.

355. Castelli-Gair, J. and Akam, M. (1995). How the Hox gene *Ultrabithorax* specifies two different segments: the significance of spatial and temporal regulation within metameres. *Development* **121**, 2973–2982.

356. Castelli-Gair, J., Greig, S., Micklem, G., and Akam, M. (1994). Dissecting the temporal requirements for homeotic gene function. *Development* **120**, 1983–1995.

357. Castro, B., Barolo, S., Bailey, A.M., and Posakony, J.W. (2005). Lateral inhibition in proneural clusters: cis-regulatory logic and default repression by Suppressor of Hairless. *Development* **132**, 3333–3344.

358. Catania, K.C. (2002). Barrels, stripes, and fingerprints in the brain: implications for theories of cortical organization. *J. Neurocytol.* **31**, 347–358.

359. Cavin, L. (2010). On giant filter feeders. *Science* **327**, 968–969.

360. Chadwick, D.H. (2001). Evolution of whales. *Natl. Geogr.* **200** #5, 64–77.

361. Chahda, J.S., Sousa-Neves, R., and Mizutani, C.M. (2013). Variation in the Dorsal gradient distribution is a source for modified scaling of germ layers in *Drosophila*. *Curr. Biol.* **23**, 710–716. [*See also* De Robertis, E.M. and Colozza, G. (2013). Development: scaling to size by protease inhibition. *Curr. Biol.* **23**, R652–R654.]

362. Chan, Y.F., Marks, M.E., Jones, F.C., Villarreal, G., Jr., Shapiro, M.D., Brady, S.D., Southwick, A.M., Absher, D.M., Grimwood, J., Schmutz, J., Myers, R.M., Petrov, D., Jónsson, B., Schluter, D., Bell, M.A., and Kingsley, D.M. (2010). Adaptive evolution of pelvic reduction in sticklebacks by recurrent deletion of a *Pitx1* enhancer. *Science* **327**, 302–305.

363. Chandebois, R. (1980). Cell sociology and the problem of automation in the development of pluricellular animals. *Acta Biotheor.* **29**, 1–35.

364. Chandebois, R. and Faber, J. (1983). Automation in animal development. *Monographs in Developmental Biology*, Vol. 16. Karger, Basel.

365. Chandler, C.H., Chari, S., and Dworkin, I. (2013). Does your gene need a background check? How genetic background impacts the analysis of mutations, genes, and evolution. *Trends Genet.* **29**, 358–366.

366. Chang, C., Wu, P., Baker, R.E., Maini, P.K., Alibardi, L., and Chuong, C.-M. (2009). Reptile scale paradigm: evo-devo, pattern formation and regeneration. *Int. J. Dev. Biol.* **53**, 813–826.

367. Chang, K.C., Wang, C., and Wang, H. (2012). Balancing self-renewal and differentiation by asymmetric division: insights from brain tumor suppressors in *Drosophila* neural stem cells. *BioEssays* **34**, 301–310.

368. Chapman, T., Pomiankowski, A., and Fowler, K. (2005). Stalk-eyed flies. *Curr. Biol.* **15**, R533–R535.

369. Charles, C., Hovorakova, M., Ahn, Y., Lyons, D.B., Marangoni, P., Churava, S., Biehs, B., Jheon, A., Lesot, H., Balooch, G., Krumlauf, R., Viriot, L., Peterkova, R., and Klein, O.D. (2011). Regulation of tooth number by fine-tuning levels of receptor-tyrosine kinase signaling. *Development* **138**, 4063–4073.

370. Chatterjee, S. and Templin, R.J. (2004). Posture, locomotion, and paleoecology of pterosaurs. *Geol. Soc. Am. Spec. Ppr.* **376**, 1–64.

371. Chea, H.K., Wright, C.V., and Swalla, B.J. (2005). Nodal signaling and the evolution of deuterostome gastrulation. *Dev. Dynamics* **234**, 269–278.

372. Chen, H., Xu, Z., Mei, C., Yu, D., and Small, S. (2012). A system of repressor gradients spatially organizes the boundaries of Bicoid-dependent target genes. *Cell* **149**, 618–629.

373. Chen, J. and Chuong, C.-M. (2011). Patterning skin by planar cell polarity: the multi-talented hair designer. *Exp. Dermatol.* **21**, 81–85.

374. Chen, M.-H., Wilson, C.W., and Chuang, P.-T. (2007). Snapshot: Hedgehog signaling pathway. *Cell* **130**, 386.

375. Chen, S., Lee, A.Y., Bowens, N.M., Huber, R., and Kravitz, E.A. (2002). Fighting fruit flies: a model system for the study of aggression. *PNAS* **99** #8, 5664–5668.

376. Chen, S., Lin, B.-Z., Baig, M., Mitra, B., Lopes, R.J., Santos, A.M., Magee, D.A., Azevedo, M., Tarroso, P., Sasazaki, S., Ostrowski, S., Mahgoub, O., Chaudhuri, T.K., Zhang, Y.-P., Costa, V., Royo, L.J., Goyache, F., Luikart, G., Boivin, N., Fuller, D.Q., Mannen, H., Bradley, D.G., and Beja-Pereira, A. (2010). Zebu cattle are an exclusive legacy of the South Asia Neolithic. *Mol. Biol. Evol.* **27** #1, 1–6.

377. Chen, X., McClusky, R., Chen, J., Beaven, S.W., Tontonoz, P., Arnold, A.P., and Reue, K. (2012). The number of X chromosomes causes sex differences in adiposity in mice. *PLoS Genet.* **8** #5, e1002709.

378. Chen, Y., Knezevic, V., Ervin, V., Hutson, R., Ward, Y., and Mackem, S. (2004). Direct interaction with Hoxd proteins reverses Gli3-repressor function to promote digit formation downstream of Shh. *Development* **131**, 2339–2347.

379. Cherry, A.B.C. and Daley, G.Q. (2012). Reprogramming cellular identity for regenerative medicine. *Cell* **148**, 1110–1122.

380. Chesebro, J., Hrycaj, S., Mahfooz, N., and Popadic, A. (2009). Diverging functions of Scr between embryonic and post-embryonic development in a hemimetabolous insect, *Oncopeltus fasciatus*. *Dev. Biol.* **329**, 142–151.

381. Chiappe, L.M., Codorniú, L., Grellet-Tinner, G., and Rivarola, D. (2004). Argentinian unhatched pterosaur fossil. *Nature* **432**, 571.

382. Chiappe, L.M. and Rivarola, D. (1996). Pterodaustro's smile. *Nat. Hist.* **105** #11, 34–35.

383. Child, C.M. (1941). *Patterns and Problems of Development.* University of Chicago Press, Chicago, IL.

384. Chipman, A.D. (2009). On making a snake. *Evol. Dev.* **11**, 3–5.

385. Chipman, A.D. (2010). Parallel evolution of segmentation by co-option of ancestral gene regulatory networks. *BioEssays* **32**, 60–70.

386. Chipman, A.D. and Akam, M. (2008). The segmentation cascade in the centipede *Strigamia maritima*: involvement of the Notch pathway and pair-rule gene homologues. *Dev. Biol.* **319**, 160–169.

387. Chipman, A.D., Arthur, W., and Akam, M. (2004). A double segment periodicity underlies segment generation in centipede development. *Curr. Biol.* **14**, 1250–1255.

388. Cho, E.H. and Nijhout, H.F. (2013). Development of polyploidy of scale-building cells in the wings of *Manduca sexta*. *Arthropod Struct. Dev.* **42**, 37–46.

389. Chouard, T. (2010). Revenge of the hopeful monster. *Nature* **463**, 864–867.

390. Christiaen, L., Jaszczyszyn, Y., Kerfant, M., Kano, S., Thermes, V., and Joly, J.-S. (2007). Evolutionary modification of mouth position in deuterostomes. *Semin. Cell Dev. Biol.* **18**, 502–511.

391. Christiansen, A.E., Keisman, E.L., Ahmad, S.M., and Baker, B.S. (2002). Sex comes in from the cold: the integration of sex and pattern. *Trends Genet.* **18**, 510–516.

392. Christiansen, P. (2008). Evolution of skull and mandible shape in cats (Carnivora: Felidae). *PLoS ONE* **3** #7, e2807.

393. Christiansen, P. (2008). Species distinction and evolutionary differences in the clouded leopard (*Neofelis nebulosa*) and Diard's clouded leopard (*Neofelis diardi*). *J. Mammalogy* **89**, 1435–1446.

394. Christiansen, P. (2012). The making of a monster: postnatal ontogenetic changes in craniomandibular shape in the great sabercat *Smilodon*. *PLoS ONE* **7** #1, e29699.

395. Christin, P.-A., Weinreich, D.M., and Besnard, G. (2010). Causes and evolutionary significance of genetic convergence. *Trends Genet.* **26**, 400–405.

396. Chuong, C.-M., Patel, N., Lin, J., Jung, H.-S., and Widelitz, R.B. (2000). Sonic hedgehog signaling pathway in vertebrate epithelial appendage morphogenesis: perspectives in development and evolution. *Cell. Mol. Life Sci.* **57**, 1672–1681.

397. Chuong, C.-M., Randall, V.A., Widelitz, R.B., Wu, P., and Jiang, T.-X. (2012). Physiological regeneration of skin appendages and implications for regenerative medicine. *Physiology* **27**, 61–72.

398. Chuong, C.M., Hou, L., Chen, P.J., Wu, P., Patel, N., and Chen, Y. (2001). Dinosaur's feather and chicken's tooth? Tissue engineering of the integument. *Eur. J. Dermatol.* **11**, 286–292.

399. Chuong, C.M., Nickoloff, B.J., Elias, P.M., Goldsmith, L.A., Macher, E., Maderson, P.A., Sundberg, J.P., Tagami, H., Plonka, P.M., Thestrup-Pedersen, K., Bernard, B.A., Schröder, J.M., Dotto, P., Chang, C.H., Williams, M.L., Feingold, K.R., King, L.E., Kligman, A.M., Rees, J.L., and Christophers, E. (2002). What is the 'true' function of skin? *Exp. Dermatol.* **11**, 159–187.

400. Cieslak, M., Reissmann, M., Hofreiter, M., and Ludwig, A. (2011). Colours of domestication. *Biol. Rev.* **86**, 885–899.

401. Cisneros, J.C., Abdala, F., Rubidge, B.S., Dentzien-Dias, P.C., and de Oliveira Bueno, A. (2011). Dental occlusion in a 260-million-year-old therapsid with saber canines from the Permian of Brazil. *Science* **331**, 1603–1605.

402. Citerne, H., Jabbour, F., Nadot, S., and Damerval, C. (2010). The evolution of floral symmetry. *Adv. Bot. Res.* **54**, 85–137.

403. Clark, J.W. (1871). On the skeleton of a narwhal (*Monodon monoceros*) with two fully developed tusks. *Proc. Zool. Soc. Lond.* **1871** #4, 42–53.

404. Clarke, J. (2013). Feathers before flight. *Science* **340**, 690–692.

405. Clarke, S.L., VanderMeer, J.E., Wenger, A.M., Schaar, B.T., Ahituv, N., and Bejerano, G. (2012). Human developmental enhancers conserved between deuterostomes and protostomes. *PLoS Genet.* **8** #8, e1002852.

406. Clevers, H. and Nusse, R. (2012). Wnt/β-catenin signaling and disease. *Cell* **149**, 1192–1205.

407. Clyne, J.D. and Miesenböck, G. (2008). Sex-specific control and tuning of the pattern generator for courtship song in *Drosophila*. *Cell* **133**, 354–363.

408. Coates, M. and Ruta, M. (2000). Nice snake, shame about the legs. *Trends Ecol. Evol.* **15**, 503–507.

409. Coates, M.I. and Clack, J.A. (1990). Polydactyly in the earliest known tetrapod limbs. *Nature* **347**, 66–69.

410. Cobb, J. and Duboule, D. (2005). Comparative analysis of genes downstream of the *Hoxd* cluster in developing digits and external genitalia. *Development* **132**, 3055–3067.

411. Cobourne, M.T. and Mitsiadis, T. (2006). Neural crest cells and patterning of the mammalian dentition. *J. Exp. Zool. (Mol. Dev. Evol.)* **306B**, 251–260.

412. Coen, E. (1999). *The Art of Genes: How Organisms Make Themselves*. Oxford University Press, New York, NY.

413. Cohen, M. (2011). The importance of structured noise in the generation of self-organizing tissue patterns through contact-mediated cell-cell signaling. *J. Roy. Soc. Interface* **8**, 787–798.

414. Cohen, M., Georgiou, M., Stevenson, N.L., Miodownik, M., and Baum, B. (2010). Dynamic filopodia transmit intermittent Delta-Notch signaling to drive pattern refinement during lateral inhibition. *Dev. Cell* **19**, 78–89.

415. Cohen, M.M., Jr. (2012). Perspectives on asymmetry: The Erickson lecture. *Am. J. Med. Genet. A* **158A**, 2981–2998.

416. Cohen, S.M. and Jürgens, G. (1989). Proximal-distal pattern formation in *Drosophila*: cell autonomous requirement for *Distal-less* gene activity in limb development. *EMBO J.* **8**, 2045–2055.

417. Cohen, S.M. and Jürgens, G. (1989). Proximal-distal pattern formation in *Drosophila:* graded requirement for *Distal-less* gene activity during limb development. *Roux's Arch. Dev. Biol.* **198**, 157–169.

418. Cohn, M.J. (2011). Development of the external genitalia: conserved and divergent mechanisms of appendage patterning. *Dev. Dynamics* **240**, 1108–1115.

419. Cohn, M.J., Lovejoy, C.O., Wolpert, L., and Coates, M.I. (2002). Branching, segmentation and the metapterygial axis: pattern versus process in the vertebrate limb. *BioEssays* **24**, 460–465.

420. Cohn, M.J., Patel, K., Krumlauf, R., Wilkinson, D.G., Clarke, J.D.W., and Tickle, C. (1997). *Hox9* genes and vertebrate limb specification. *Nature* **387**, 97–101.

421. Cohn, M.J. and Tickle, C. (1999). Developmental basis of limblessness and axial patterning in snakes. *Nature* **399**, 474–479.

422. Colas, J.-F. and Schoenwolf, G.C. (2001). Towards a cellular and molecular understanding of neurulation. *Dev. Dynamics* **221**, 117–145.

423. Cole, E.S. and Palka, J. (1982). The pattern of campaniform sensilla on the wing and haltere of *Drosophila melanogaster* and several of its homeotic mutants. *J. Embryol. Exp. Morph.* **71**, 41–61.

424. Collin, R. and Miglietta, M.P. (2008). Reversing opinions on Dollo's Law. *Trends Ecol. Evol.* **23**, 602–609.

425. Collins, M.A. (2003). The genus *Grimpoteuthis* (Octopoda: Grimpoteuthidae) in the northeast Atlantic, with descriptions of three new species. *Zool. J. Linnean Soc.* **139**, 93–127.

426. Comeron, J.M., Ratnappan, R., and Bailin, S. (2012). The many landscapes of recombination in *Drosophila melanogaster*. *PLoS Genet.* **8** #10, e1002905.

427. Compagnucci, C., Debiais-Thibaud, M., Coolen, M., Fish, J., Griffin, J.N., Bertocchini, F., Minoux, M., Rijli, F.M., Borday-Birraux, V., Casane, D., Mazan, S., and Depew, M.J. (2013). Pattern and polarity in the development and evolution of the gnathostome jaw: both conservation and heterotopy in the branchial arches of the shark, *Scyliorhinus canicula*. *Dev. Biol.* **377**, 428–448.

428. Conant, G.C. and Wolfe, K.H. (2008). Turning a hobby into a job: how duplicated genes find new functions. *Nature Rev. Genet.* **9**, 938–950.

429. Conceição, I.C., Long, A.D., Gruber, J.D., and Beldade, P. (2011). Genomic sequence around butterfly wing development genes: annotation and comparative analysis. *PLoS ONE* **6** #8, e23778.

430. Conniff, R. (1999). Cheetahs: ghosts of the grasslands. *Natl. Geogr.* **196** #6, 2–31.

431. Conway, B.R. and Rehding, A. (2013). Neuroaesthetics and the trouble with beauty. *PLoS Biol.* **11** #3, e1001504.

432. Conway Morris, S. (2000). Evolution: Bringing molecules into the fold. *Cell* **100**, 1–11.

433. Conway-Morris, S. (2003). The Cambrian "explosion" of metazoans and molecular biology: would Darwin be satisfied? *Int. J. Dev. Biol.* **47**, 505–515.

434. Conway-Morris, S. (2006). Evolutionary convergence. *Curr. Biol.* **16**, R826–R827.

435. Cook, O., Biehs, B., and Bier, E. (2004). *brinker* and *optomotor-blind* act coordinately to initiate development of the L5 wing vein primordium in *Drosophila. Development* **131**, 2113–2124.

436. Cook, T. and Desplan, C. (2001). Photoreceptor subtype specification: from flies to humans. *Semin. Cell Dev. Biol.* **12**, 509–518.

437. Cook, T.A. (1914). *The Curves of Life*. Constable, London.

438. Cooke, J. (1975). The emergence and regulation of spatial organization in early animal development. *Annu. Rev. Biophys. Bioeng.* **4**, 185–217.

439. Cooke, J. (1981). The problem of periodic patterns in embryos. *Phil. Trans. Roy. Soc. Lond. B* **295**, 509–524.

440. Cooke, J. (1981). Scale of body pattern adjusts to available cell number in amphibian embryos. *Nature* **290**, 775–778.

441. Cooke, J. (1982). The relation between scale and the completeness of pattern in vertebrate embryogenesis: models and experiments. *Am. Zool.* **22**, 91–104.

442. Cooke, J. (1998). A gene that resuscitates a theory: somitogenesis and a molecular oscillator. *Trends Genet.* **14**, 85–88.

443. Cooke, J. and Zeeman, E.C. (1976). A clock and wavefront model for control of the number of repeated structures during animal morphogenesis. *J. Theor. Biol.* **58**, 455–476.

444. Coombs, W.P., Jr. (1978). Theoretical aspects of cursorial adaptations in dinosaurs. *Q. Rev. Biol.* **53**, 393–418.

445. Cooper, C.D. and Raible, D.W. (2009). Mechanisms for reaching the differentiated state: insights from neural crest-derived melanocytes. *Semin. Cell Dev. Biol.* **20**, 105–110.

446. Cooper, H.M., Herbin, M., and Nevo, E. (1993). Ocular regression conceals adaptive progression of the visual system in a blind subterranean mammal. *Nature* **361**, 156–159.

447. Cooper, K.L. (2011). The lesser Egyptian jerboa, *Jaculus jaculus*: a unique rodent model for evolution and development. *Cold Spr. Harb. Protoc.* **2011** #12, 1451–1456.

448. Cooper, K.L., Oh, S., Sung, Y., Dasari, R.R., Kirschner, M.W., and Tabin, C.J. (2013). Multiple phases of chondrocyte enlargement underlie differences in skeletal proportions. *Nature* **495**, 375–378.

449. Cooper, K.L. and Tabin, C.J. (2008). Understanding of bat wing evolution takes flight. *Genes Dev.* **22**, 121–124.

450. Cooper, L.N., Armfield, B.A., and Thewissen, J.G.M. (2011). Evolution of the apical ectoderm in the developing vertebrate limb. In *Epigenetics: Linking Genotype and Phenotype in Development and Evolution* (B. Hallgrímsson and B.K. Hall, eds.). University of California Press, Berkeley, CA, pp. 238–255.

451. Cooper, L.N., Berta, A., Dawson, S.D., and Reidenberg, J.S. (2007). Evolution of hyperphalangy and digit reduction in the cetacean manus. *Anat. Rec.* **290**, 654–672.

452. Corbet, P.S. (1999). *Dragonflies: Behavior and Ecology of Odonata*. Cornell University Press, Ithaca, NY.

453. Cordes, R., Schuster-Gossler, K., Serth, K., and Gossier, A. (2004). Specification of vertebral identity is coupled to Notch signalling and the segmentation clock. *Development* **131**, 1221–1233.

454. Corfield, J.R., Gsell, A.C., Brunton, D., Heesy, C.P., Hall, M.I., Acosta, M.L., and Iwaniuk, A.N. (2011). Anatomical specializations for nocturnality in a critically endangered parrot, the Kakapo (*Strigops habroptilus*). *PLoS ONE* **6** #8, e22945.

455. Cornwallis, C.K. and Uller, T. (2008). Towards an evolutionary ecology of sexual traits. *Trends Ecol. Evol.* **25**, 145–152.

456. Costanzo, K. and Monteiro, A. (2007). The use of chemical and visual cues in female choice in the butterfly *Bicyclus anynana*. *Proc. Roy. Soc. Lond. B* **274**, 845–851.

457. Couso, J.P. (2009). Segmentation, metamerism and the Cambrian explosion. *Int. J. Dev. Biol.* **53**, 1305–1316.

458. Couso, J.P., Bate, M., and Martínez-Arias, A. (1993). A *wingless*-dependent polar coordinate system in *Drosophila* imaginal discs. *Science* **259**, 484–489.

459. Couso, J.P. and Bishop, S.A. (1998). Proximo-distal development in the legs of *Drosophila*. *Int. J. Dev. Biol.* **42**, 345–352.

460. Coutelis, J.-B., Géminard, C., Spéder, P., Suzanne, M., Petzoldt, A.G., and Noselli, S. (2013). *Drosophila* left/right asymmetry establishment is controlled by the *Hox* gene *Abdominal-B*. *Dev. Cell* **24**, 89–97.

461. Coutelis, J.B., Petzoldt, A.G., Spéder, P., Suzanne, M., and Noselli, S. (2008). Left-right asymmetry in *Drosophila*. *Semin. Cell Dev. Biol.* **19**, 252–262.

462. Couturier, L., Vodovar, N., and Schweisguth, F. (2012). Endocytosis by Numb breaks Notch symmetry at cytokinesis. *Nat. Cell Biol.* **14**, 131–139.

463. Cox, P.G., Rayfield, E.J., Fagan, M.J., Herrel, A., Pataky, T.C., and Jeffery, N. (2012). Functional evolution of the feeding system in rodents. *PLoS ONE* **7** #4, e36299.

464. Crampton, G. (1916). The phylogenetic origin and the nature of the wings of insects according to the paranotal theory. *N. Y. Entomol. Soc.* **24**, 1–39 (plus 2 plates).

465. Cretekos, C.J., Deng, J.-M., Green, E.D., Rasweiler, J.J., IV, and Behringer, R.R. (2007). Isolation, genomic structure and developmental expression of *Fgf8* in the short-tailed fruit bat, *Carollia perspicillata*. *Int. J. Dev. Biol.* **51**, 333–338.

466. Cretekos, C.J., Wang, Y., Green, E.D., Martin, J.F., Rasweiler, J.J., IV, and Behringer, R.R. (2008). Regulatory divergence modifies limb length between mammals. *Genes Dev.* **22**, 141–151.

467. Cretekos, C.J., Weatherbee, S.D., Chen, C.-H., Badwaik, N.K., Niswander, L., Behringer, R.R., and Rasweiler, J.J., IV (2005). Embryonic staging system for the short-tailed fruit bat, *Carollia perspicillata*, a model organism for the mammalian order *Chiroptera*, based upon timed pregnancies in captive-bred animals. *Dev. Dynamics* **233**, 721–738.

468. Crew, B. (2012). *Zombie Birds, Astronaut Fish, and Other Weird Animals*. Adams Media, Avon, MA. [*See also* Henderson, C. (2013). *The Book of Barely Imagined Beings: A 21st Century Bestiary*. University of Chicago Press, Chicago, IL, which contains its own alphabetized listing of exotic animals. N.B.: The author learned of this book after submitting his own manuscript and was shocked to see the same animals for several letters. When he informed Mr. Henderson of this coincidence, the latter kindly dismissed any concern about overlap. [*See also* Piper, R. (2013). *Animal Earth*. Thames & Hudson, New York, NY.]

469. Crickmore, M.A. and Mann, R.S. (2006). *Hox* control of organ size by regulation of morphogen production and mobility. *Science* **313**, 63–68.

470. Crickmore, M.A. and Mann, R.S. (2007). *Hox* control of morphogen mobility and organ development through regulation of glypican expression. *Development* **134**, 327–334.

471. Crickmore, M.A. and Mann, R.S. (2008). The control of size in animals: insights from selector genes. *BioEssays* **30**, 843–853.

472. Crispo, E. (2007). The Baldwin effect and genetic assimilation: revisiting two mechanisms of evolutionary change mediated by phenotypic plasticity. *Evolution* **61**, 2469–2479.

473. Crocker, J., Tamori, Y., and Erives, A. (2008). Evolution acts on enhancer organization to fine-tune gradient threshold readouts. *PLoS Biol.* **6** #11, e263.

474. Cuadrado, M., Martín, J., and López, P. (2001). Camouflage and escape decisions in the common chameleon *Chamaeleo chamaeleon*. *Biol. J. Linnean Soc.* **72**, 547–554.

475. Cullen, K.E. (2011). The neural encoding of self-motion. *Curr. Opin. Neurobiol.* **21**, 587–595.

476. Cully, M. and Downward, J. (2008). Snapshot: Ras signaling. *Cell* **133**, 1292.

477. Cummins, H. and Midlo, C. (1943). *Finger Prints, Palms and Soles. An Introduction to Dermatoglyphics*. Dover, New York, NY.

478. Cundall, D. (2009). Viper fangs: functional limitations of extreme teeth. *Physiol. Biochem. Zool.* **82**, 63–79.

479. Cundall, D. and Deufel, A. (2006). Influence of the venom delivery system on intraoral prey transport in snakes. *Zool. Anz.* **245**, 193–210.

480. Cunningham, S.J., Alley, M.R., and Castro, I. (2011). Facial bristle feather histology and morphology in New Zealand birds: implications for function. *J. Morph.* **272**, 118–128.

481. Cuthill, I.C. (2007). Animal behaviour: strategic signalling by cephalopods. *Curr. Biol.* **17**, R1059–R1060.

482. Dahmann, C., Oates, A.C., and Brand, M. (2011). Boundary formation and maintenance in tissue development. *Nature Rev. Genet.* **12**, 43–55. [*See also* Bard, J. (2013). Driving developmental and evolutionary change: a systems biology view. *Prog. Biophys. Mol. Biol.* **111**, 83–91.]

483. Dalton, S. (1975). *Borne on the Wind: The Extraordinary World of Insects in Flight*. Reader's Digest Press, New York, NY.

484. Damen, W.G.M. (2004). Arthropod segmentation: why centipedes are odd. *Curr. Biol.* **14**, R557–R559.

485. Damen, W.G.M., Saridaki, T., and Averof, M. (2002). Diverse adaptations of an ancestral gill: a common evolutionary origin for wings, breathing organs, and spinnerets. *Curr. Biol.* **12**, 1711–1716.

486. Darwin, C. (1859). *On the Origin of Species by Means of Natural Selection, or the Preservation of Favoured Races in the Struggle for Life*. John Murray, London.

487. Darwin, C. (1868). *The Variation of Animals and Plants Under Domestication*. John Murray, London.

488. Darwin, C. (1871). *The Descent of Man, and Selection in Relation to Sex*. John Murray, London.

489. Darwin, C. (1877). *The Various Contrivances by which Orchids are Fertilised by Insects*, 2nd (revised) edn. John Murray, London.

490. Davidson, E.H. (1990). How embryos work: a comparative view of diverse modes of cell fate specification. *Development* **108**, 365–389.

491. Davidson, E.H., Cameron, R.A., and Ransick, A. (1998). Specification of cell fate in the sea urchin embryo: summary and some proposed mechansisms. *Development* **125**, 3269–3290.

492. Davidson, E.H. and Erwin, D.H. (2006). Gene regulatory networks and the evolution of animal body plans. *Science* **311**, 796–800.

493. Davidson, E.H., Peterson, K.J., and Cameron, R.A. (1995). Origin of bilaterian body plans: evolution of developmental regulatory mechanisms. *Science* **270**, 1319–1325.

494. Davis, B.W., Li, G., and Murphy, W.J. (2010). Supermatrix and species tree methods resolve phylogenetic relationships within the big cats, *Panthera* (Carnivora: Felidae). *Mol. Phylogenet. Evol.* **56**, 64–76.

495. Davis, D.D. (1964). *The Giant Panda: A Morphological Study of Evolutionary Mechanisms.* Fieldiana: Zoology Memoirs, Vol. 3. Chicago Natural History Museum, Chicago, IL.

496. Davis, G.K. and Patel, N.H. (2003). Playing by pair-rules? *BioEssays* **25**, 425–429.

497. Davis, G.K., Srinivasan, D.G., Wittkopp, P.J., and Stern, D.L. (2007). The function and regulation of *Ultrabithorax* in the legs of *Drosophila melanogaster*. *Dev. Biol.* **308**, 621–631.

498. Davis, M.C., Dahn, R.D., and Shubin, N.H. (2007). An autopodial-like pattern of *Hox* expression in the fins of a basal actinopterygian fish. *Nature* **447**, 473–476.

499. Davit-Béal, T., Tucker, A.S., and Sire, J.-Y. (2009). Loss of teeth and enamel in tetrapods: fossil record, genetic data and morphological adaptations. *J. Anat.* **214**, 477–501.

500. Dawkins, R. (1982). *The Extended Phenotype*. Oxford University Press, New York, NY.

501. Dawkins, R. (1986). *The Blind Watchmaker: Why the Evidence of Evolution Reveals a Universe Without Design*. Norton, New York, NY.

502. Dawkins, R. (1989). The evolution of evolvability. In *Artificial Life, Santa Fe Institute Studies in the Sciences of Complexity, Vol. 6* (C.G. Langton, ed.). Addison-Wesley, New York, NY, pp. 201–220.

503. Dawkins, R. (1995). The evolved imagination: animals as models of their world. *Nat. Hist.* **104** #9, 8–11, 22–24.

504. Dawkins, R. (2004). *The Ancestor's Tale: A Pilgrimage to the Dawn of Evolution*. Houghton Mifflin, New York, NY.

505. Dawkins, R. (2009). *The Greatest Show on Earth: The Evidence for Evolution*. Free Press, New York, NY. [N.B.: All of Richard's books are worth reading, especially for evo-devotees, and this book is no exception, but my personal favorite is his *Unweaving the Rainbow: Science, Delusion and the Appetite for Wonder* (1998).]

506. de Bruijn, S., Angenent, G.C., and Kaufmann, K. (2012). Plant "evo-devo" goes genomic: from candidate genes to regulatory networks. *Trends Plant Sci.* **17**, 441–447.

507. de Celis, J.F. (2004). The Notch signaling module. In *Modularity in Development and Evolution* (G. Schlosser and G.P. Wagner, eds.). University of Chicago Press, Chicago, IL, pp. 81–100.

508. de Celis, J.F. and Diaz-Benjumea, J. (2003). Developmental basis for vein pattern variations in insect wings. *Int. J. Dev. Biol.* **47**, 653–663.

509. de Duve, C. (1996). The gospel of contingency. *Nature* **383**, 771–772.

510. De Jong, M.C.M. and Sabelis, M.W. (1991). Limits to runaway sexual selection: the wallflower paradox. *J. Evol. Biol.* **4**, 637–655.

511. de Joussineau, C., Soulé, J., Martin, M., Anguille, C., Montcourrier, P., and Alexandre, D. (2003). Delta-promoted filopodia mediate long-range lateral inhibition in *Drosophila*. *Nature* **426**, 555–559.

512. de Kok-Mercado, F., Habib, M., Phelps, T., Gregg, L., and Gailloud, P. (2013). Adaptations of the owl's cervical and cephalic arteries in relation to extreme neck rotation. *Science* **339**, 514.

513. de Lussanet, M.H.E. (2011). A hexamer origin of the echinoderms' five rays. *Evol. Dev.* **13**, 228–238.
514. de Muizon, C. (1993). Walrus-like feeding adaptation in a new cetacean from the Pliocene of Peru. *Nature* **365**, 745–748.
515. de Navas, L.F., Garaulet, D.L., and Sánchez-Herrero, E. (2006). The *Ultrabithorax* Hox gene of *Drosophila* controls haltere size by regulating the Dpp pathway. *Development* **133**, 4495–4506.
516. De Robertis, E.M. (2008). Evo-devo: variations on ancestral themes. *Cell* **132**, 185–195.
517. De Robertis, E.M. (2008). The molecular ancestry of segmentation mechanisms. *PNAS* **105** #43, 16411–16412.
518. De Robertis, E.M. and Sasai, Y. (1996). A common plan for dorsoventral patterning in Bilateria. *Nature* **380**, 37–40.
519. Decraene, L.P.R. and Smets, E.F. (1994). Merosity in flowers: definition, origin, and taxonomic significance. *Plant Syst. Evol.* **191**, 83–104.
520. Deforest, M.E. and Basrur, P.K. (1979). Malformations and the Manx syndrome in cats. *Can. Vet. J.* **20**, 304–314.
521. del Álamo, D., Terriente, J., and Díaz-Benjumea, F.J. (2002). Spitz/EGFr signalling via the Ras/MAPK pathway mediates the induction of bract cells in *Drosophila* legs. *Development* **129**, 1975–1982.
522. Delanoue, R. and Léopold, P. (2010). Developmental biology: a DOR connecting growth and clocks. *Curr. Biol.* **20**, R884–R886.
523. DeLaurier, A., Schweitzer, R., and Logan, M. (2006). *Pitx1* determines the morphology of muscle, tendon, and bones of the hindlimb. *Dev. Biol.* **299**, 22–34.
524. Delpretti, S., Zakany, J., and Duboule, D. (2012). A function for all posterior *Hoxd* genes during digit development? *Dev. Dynamics* **241**, 792–802.
525. Delsuc, F., Superina, M., Tilak, M.-K., Douzery, E.J.P., and Hassanin, A. (2012). Molecular phylogenetics unveils the ancient evolutionary origins of the enigmatic fairy armadillos. *Mol. Phylogenet. Evol.* **62**, 673–680.
526. Denes, A.S., Jékely, G., Steinmetz, P.R.H., Raible, F., Snyman, H., Prud'homme, B., Ferrier, D.E.K., Balavoine, G., and Arendt, D. (2007). Molecular architecture of annelid nerve cord supports common origin of nervous system centralization in Bilateria. *Cell* **129**, 277–288.
527. Denver, R.J. (2008). Chordate metamorphosis: ancient control by iodothyronines. *Curr. Biol.* **18**, R567–R569.
528. Depew, M.J., Simpson, C.A., Morasso, M., and Rubenstein, J.L.R. (2005). Reassessing the *Dlx* code: the genetic regulation of branchial arch skeletal pattern and development. *J. Anat.* **207**, 501–561.
529. Dequéant, M.-L. and Pourquié, O. (2008). Segmental patterning of the vertebrate embryonic axis. *Nature Rev. Gen.* **9**, 370–382.
530. Derby, C.D. (2007). Escape by inking and secreting: marine molluscs avoid predators through a rich array of chemicals and mechanisms. *Biol. Bull.* **213**, 274–289.
531. Derelle, R., Lopez, P., Le Guyader, H., and Manuel, M. (2007). Homeodomain proteins belong to the ancestral molecular toolkit of eukaryotes. *Evol. Dev.* **9**, 212–219.
532. Derynck, R. and Miyazono, K. (2008). *The TGF-β Family*. Cold Spring Harbor Laboratory Press (Monograph 50), Cold Spring Harbor, NY.
533. DeSalle, R. and Tattersall, I. (2008). *Human Origins: What Bones and Genomes Tell Us About Ourselves*. Texas A&M University Press, College Station, TX.

534. Deschamps, J. (2007). Ancestral and recently recruited global control of the *Hox* genes in development. *Curr. Opin. Gen. Dev.* **17**, 422–427.

535. Deschamps, J. and van Nes, J. (2005). Developmental regulation of the *Hox* genes during axial morphogenesis in the mouse. *Development* **132**, 2931–2942.

536. Deufel, A. and Cundall, D. (2006). Functional plasticity of the venom delivery system in snakes with a focus on the poststrike prey release behavior. *Zool. Anz.* **245**, 249–267.

537. Di-Poï, N., Montoya-Burgos, J.I., Miller, H., Pourquié, O., Milinkovitch, M.C., and Duboule, D. (2010). Changes in *Hox* genes' structure and function during the evolution of the squamate body plan. *Nature* **464**, 99–103.

538. Diamond, J.M. (1996). Competition for brain space. *Nature* **382**, 756–757.

539. Diaz-Benjumea, F.J., Cohen, B., and Cohen, S.M. (1994). Cell interaction between compartments establishes the proximal-distal axis of *Drosophila* legs. *Nature* **372**, 175–179.

540. Dichtel-Danjoy, M.-L. and Félix, M.-A. (2004). Phenotypic neighborhood and micro-evolvability. *Trends Genet.* **20**, 268–276.

541. Dickerson, A.K., Shankles, P.G., Madhavan, N.M., and Hu, D.L. (2012). Mosquitoes survive raindrop collisions by virtue of their low mass. *PNAS* **109** #25, 9822–9827.

542. Dickinson, M.H. (1999). Haltere-mediated equilibrium reflexes of the fruit fly, *Drosophila melanogaster*. *Phil. Trans. R. Soc. Lond. B* **354**, 903–916.

543. Dickinson, M.H. (2005). The initiation and control of rapid flight maneuvers in fruit flies. *Integr. Comp. Biol.* **45**, 274–281.

544. Dickinson, M.H. (2006). Insect flight. *Curr. Biol.* **16**, R309–R314.

545. Dickinson, M.H., Farley, C.T., Full, R.J., Koehl, M.A.R., Kram, R., and Lehman, S. (2000). How animals move: an integrative view. *Science* **288**, 100–106.

546. Dierick, H.A. (2008). Fly fighting: octopamine modulates aggression. *Curr. Biol.* **18**, R161–R163.

547. Dietrich, M.R. (2000). From hopeful monsters to homeotic effects: Richard Goldschmidt's integration of development, evolution, and genetics. *Am. Zool.* **40**, 738–747.

548. Dietrich, M.R. (2003). Richard Goldschmidt: hopeful monsters and other 'heresies'. *Nature Rev. Genet.* **4**, 68–74.

549. Dimitropoulos, A. (1985). First records of Orsini's viper, *Vipera ursinii* (Viperidae), in Greece. *Ann. Mus. Goulandris* **7**, 319–323.

550. DiNardo, S. and O'Farrell, P.H. (1987). Establishment and refinement of segmental pattern in the *Drosophila* embryo: spatial control of *engrailed* expression by pair-rule genes. *Genes Dev.* **1**, 1212–1225.

551. Diogo, R., Linde-Medina, M., Abdala, V., and Ashley-Ross, M.A. (2013). New, puzzling insights from comparative myological studies on the old and unsolved forelimb/hindlimb enigma. *Biol. Rev.* **88**, 196–214.

552. Dittrich-Reed, D.R. and Fitzpatrick, B.M. (2013). Transgressive hybrids as hopeful monsters. *Evol. Biol.* **40**, 310–315.

553. Dixon, D. (1998). *After Man: A Zoology of the Future*. St. Martin's Griffin, New York, NY.

554. Dodson, P. (1991). Life styles of the huge and famous. *Nat. Hist.* **100** #12, 30–35.

555. Doe, C.Q. (2006). Chinmo and neuroblast temporal identity. *Cell* **127**, 254–256.

556. Domning, D.P. (2001). The earliest known fully quadrupedal sirenian. *Nature* **413**, 625–627.

557. Don, E.K., Currie, P.D., and Cole, N.J. (2013). The evolutionary history of the development of the pelvic fin/hindlimb. *J. Anat.* **222**, 114–133.

558. Dong, J., Feldmann, G., Huang, J., Wu, S., Zhang, N., Comerford, S.A., Gayyed, M.F., Anders, R.A., Maitra, A., and Pan, D. (2007). Elucidation of a universal size-control mechanism in *Drosophila* and mammals. *Cell* **130**, 1120–1133.

559. Donlan, C.J. (2007). Restoring America's big, wild animals. *Sci. Am.* **296** #6, 70–77.

560. Donoghue, P.C.J., Graham, A., and Kelsh, R.N. (2008). The origin and evolution of the neural crest. *BioEssays* **30**, 530–541.

561. Doolittle, R.F. (1994). Convergent evolution: the need to be explicit. *TiBS* **19**, 15–18.

562. Dornan, A.J. and Goodwin, S.F. (2008). Fly courtship song: triggering the light fantastic. *Cell* **133**, 210–212.

563. Douglas, R.H., Collett, T.S., and Wagner, H.-J. (1986). Accommodation in anuran Amphibia and its role in depth vision. *J. Comp. Physiol. A* **158**, 133–143.

564. Doupé, D.P. and Jones, P.H. (2013). Cycling progenitors maintain epithelia while diverse cell types contribute to repair. *BioEssays* **35**, 443–451.

565. Drake, A.G. (2011). Dispelling dog dogma: an investigation of heterochrony in dogs using 3D geometric morphometric analysis of skull shape. *Evol.* Dev.**13**, 204–213.

566. Drake, A.G. and Klingenberg, C.P. (2010). Large-scale diversification of skull shape in domestic dogs: disparity and modularity. *Am. Nat.* **175**, 289–301.

567. Drimmer, F. (1973). *Very Special People*. Amjon, New York, NY.

568. Duboc, V. and Lepage, T. (2008). A conserved role for the nodal signaling pathway in the establishment of dorso-ventral and left-right axes in deuterostomes. *J. Exp. Zool. (Mol. Dev. Evol.)* **310B**, 41–53.

569. Duboc, V. and Logan, M.P.O. (2009). Building limb morphology through integration of signaling modules. *Curr. Opin. Gen. Dev.* **19**, 497–503.

570. Duboc, V. and Logan, M.P.O. (2011). Regulation of limb bud initiation and limb-type morphology. *Dev. Dynamics* **240**, 1017–1027.

571. Duboc, V., Röttinger, E., Lapraz, F., Besnardeau, L., and Lepage, T. (2005). Left-right asymmetry in the sea urchin embryo is regulated by Nodal signaling on the right side. *Dev. Cell* **9**, 147–158.

572. Duboule, D. (2007). The rise and fall of *Hox* gene clusters. *Development* **134**, 2549–2560.

573. Dubrulle, J. and Pourquié, O. (2004). Coupling segmentation to axis formation. *Development* **131**, 5783–5793.

574. Dudley, R. (2000). *The Biomechanics of Insect Flight: Form, Function, Evolution*. Princeton University Press, Princeton, NJ.

575. Dudley, R. and Yanoviak, S.P. (2011). Animal aloft: the origins of aerial behavior and flight. *Integr. Comp. Biol.* **51**, 926–936.

576. Dugon, M.M. and Arthur, W. (2012). Comparative studies on the structure and development of the venom-delivery system of centipedes, and a hypothesis on the origin of this evolutionary novelty. *Evol. Dev.* **14**, 128–137.

577. Dugon, M.M., Hayden, L., Black, A., and Arthur, W. (2012). Development of the venom ducts in the centipede *Scolopendra*: an example of recapitulation. *Evol. Dev.* **14**, 515–521.

578. Duncan, E.J., Leask, M.P., and Dearden, P.K. (2013). The pea aphid (*Acyrthosiphon pisum*) genome encodes two divergent early developmental programs. *Dev. Biol.* **377**, 262–274.

579. Dunn, R.R. (2006). Dig it! *Nat. Hist.* **115** #10, 36–41.

580. Duret, L. (2009). Mutation patterns in the human genome: more variable than expected. *PLoS Biol.* **7** #2, e1000028.

581. Dutko, J.A. and Mullins, M.C. (2011). Snapshot: BMP signaling in development. *Cell* **145**, 636.

582. Dworkin, I. (2005). Canalization, cryptic variation, and developmental buffering: a critical examination and analytical perspective. In *Variation: A Central Concept in Biology* (B. Hallgrímsson and B.K. Hall, eds.). Elsevier Academic Press, New York, NY, pp. 131–158.

583. Eames, B.F. (2008). The genesis of cartilage size and shape during development and evolution. *Development* **135**, 3947–3958.

584. Eberhard, W.G. (1980). Horned beetles. *Sci. Am.* **242** #3, 166–181.

585. Eberhard, W.G. (1987). Runaway sexual selection. *Nat. Hist.* **96** #12, 4–8.

586. Eberlein, S. and Russell, M.A. (1983). Effects of deficiencies in the *engrailed* region of *Drosophila melanogaster*. *Dev. Biol.* **100**, 227–237.

587. Eckalbar, W.L., Lasku, E., Infante, C.R., Elsey, R.M., Markov, G.J., Allen, A.N., Corneveaux, J.J., Losos, J.B., DeNardo, D.F., Huentelman, M.J., Wilson-Rawls, J., Rawls, A., and Kusumi, K. (2012). Somitogenesis in the anole lizard and alligator reveals evolutionary convergence and divergence in the amniote segmentation clock. *Dev. Biol.* **363**, 308–319.

588. Economides, K.D., Zeltser, L., and Capecchi, M.R. (2003). *Hoxb13* mutations cause overgrowth of caudal spinal cord and tail vertebrae. *Dev. Biol.* **256**, 317–330.

589. Economou, A.D., Ohazama, A., Porntaveetus, T., Sharpe, P.T., Kondo, S., Basson, M.A., Gritli-Linde, A., Cobourne, M.T., and Green, J.B.A. (2012). Periodic stripe formation by a Turing mechanism operating at growth zones in the mammalian palate. *Nature Genet.* **44**, 348–352.

590. Eddison, M., Le Roux, I., and Lewis, J. (2000). Notch signaling in the development of the inner ear: lessons from *Drosophila*. *PNAS* **97** #22, 11692–11699.

591. Ede, D.A. (1972). Cell behaviour and embryonic development. *Internat. J. Neurosci.* **3**, 165–174.

592. Edelman, G.M. (1987). *Neural Darwinism: The Theory of Neuronal Group Selection*. Basic Books, New York, NY.

593. Edelman, G.M. (1988). *Topobiology: An Introduction to Molecular Embryology*. Basic Books, New York, NY.

594. Edelman, G.M. (1993). Neural Darwinism: selection and reentrant signaling in higher brain function. *Neuron* **10**, 115–125.

595. Edelman, G.M. and Gallin, W.J. (1987). Cell adhesion as a basis of pattern in embryonic development. *Am. Zool.* **27**, 645–656.

596. Edgecombe, G.D. (2009). Palaeontological and molecular evidence linking arthropods, onychophorans, and other ecdysozoa. *Evo. Edu. Outreach* **2**, 178–190.

597. Edwards, J.S. (1997). The evolution of insect flight: implications for the evolution of the nervous system. *Brain Behav. Evol.* **50**, 8–12.

598. Egri, A., Blahó, M., Kriska, G., Farkas, R., Gyurkovszky, M., Åkesson, S., and Horváth, G. (2012). Polarotactic tabanids find striped patterns with brightness and/or polarization modulation least attractive: an advantage of zebra stripes. *J. Exp. Biol.* **215**, 736–745.

599. Eibner, C., Pittlik, S., Meyer, A., and Begemann, G. (2008). An organizer controls the development of the "sword," a sexually selected trait in swordtail fish. *Evol. Dev.* **10**, 403–412.

600. Eilam, D. (1997). Postnatal development of body architecture and gait in several rodent species. *J. Exp. Biol.* **200**, 1339–1350.

601. Eilam, D. and Shefer, G. (1997). The developmental order of bipedal locomotion in the jerboa (*Jaculus orientalis*): pivoting, creeping, quadrupedalism, and bipedalism. *Dev. Psychobiol.* **31**, 137–142.

602. Eisner, T. (2003). *For Love of Insects*. Harvard University Press, Cambridge, MA.

603. Eizirik, E., David, V.A., Buckley-Beason, V., Roelke, M.E., Schäffer, A.A., Hannah, S.S., Narfström, K., O'Brien, S.J., and Menotti-Raymond, M. (2010). Defining and mapping mammalian coat pattern genes: multiple genomic regions implicated in domestic cat stripes and spots. *Genetics* **184**, 267–275.

604. El-Sherif, E., Averof, M., and Brown, S.J. (2012). A segmentation clock operating in blastoderm and germband stages of *Tribolium* development. *Development* **139**, 4341–4346.

605. Eldredge, N. and Eldredge, G. (2012). Editorial. *Evo. Edu. Outreach* **5**, 179–180.

606. Elgin, R.A., Hone, D.W.E., and Frey, E. (2011). The extent of the pterosaur flight membrane. *Acta Palaeontol. Pol.* **56**, 99–111.

607. Elsdale, T. and Wasoff, F. (1976). Fibroblast cultures and dermatoglyphics: the topology of two planar patterns. *Wilhelm Roux's Arch.* **180**, 121–147.

608. Emerald, B.S. and Cohen, S.M. (2001). Limb development: getting down to the ground state. *Curr. Biol.* **11**, R1025–R1027.

609. Emerson, S.B. (1985). Jumping and leaping. In *Functional Vertebrate Morphology* (M. Hildebrand, D.M. Bramble, K.F. Liem, and D.B. Wake, eds.). Harvard University Press, Cambridge, MA, pp. 58–72.

610. Emlen, D. (2011). Diversity in the weapons of sexual selection: horn evolution in dung beetles. In *In the Light of Evolution: Essays from the Laboratory and Field* (J.B. Losos, ed.). Roberts, Greenwood Village, CO, pp. 149–170.

611. Emlen, D.J. (2000). Integrating development with evolution: a case study with beetle horns. *BioScience* **50**, 403–418.

612. Emlen, D.J. (2001). Costs and the diversification of exaggerated animal structures. *Science* **291**, 1534–1536.

613. Emlen, D.J. (2007). On the origin and evolutionary diversification of beetle horns. *PNAS* **104** (Suppl. 1), 8661–8668.

614. Emlen, D.J. (2008). The evolution of animal weapons. *Annu. Rev. Ecol. Evol. Syst.* **39**, 387–413.

615. Emlen, D.J. and Allen, C.E. (2004). Genotype to phenotype: physiological control of trait size and scaling in insects. *Integr. Comp. Biol.* **43**, 617–634.

616. Emlen, D.J., Hunt, J., and Simmons, L.W. (2005). Evolution of sexual dimorphism and male dimorphism in the expression of beetle horns: phylogenetic evidence for modularity, evolutionary lability, and constraint. *Am. Nat.* **166** (Suppl.), S42–S68.

617. Emlen, D.J. and Nijhout, H.F. (2000). The development and evolution of exaggerated morphologies in insects. *Annu. Rev. Entomol.* **45**, 661–708.

618. Emlen, D.J., Szafran, Q., Corley, L.S., and Dworkin, I. (2006). Insulin signaling and limb-patterning: candidate pathways for the origin and evolutionary diversification of beetle 'horns'. *Heredity* **97**, 179–191.

619. Emlen, D.J., Warren, I.A., Johns, A., Dworkin, I., and Lavine, L.C. (2012). A mechanism of extreme growth and reliable signaling in sexually selected ornaments and weapons. *Science* **337**, 860–864. [*See also* Warren, I.A., *et al.* (2013). A general mechanism for conditional expression of exaggerated sexually-selected traits. *BioEssays* **35**, 889–899.]

620. Emsley, M.G. (1975). *Butterfly Magic*. Viking Press, New York, NY.

621. Endo, H., Yamagiwa, D., Hayashi, Y., Koie, H., Yamaya, Y., and Kimura, J. (1999). Role of the giant panda's "pseudo-thumb". *Nature* **397**, 309–310.

622. Enquist, M. and Arak, A. (1993). Selection of exaggerated male traits by female esthetic senses. *Nature* **361**, 446–448.

623. Eom, D.S., Inoue, S., Patterson, L.B., Gordon, T.N., Slingwine, R., Kondo, S., Watanabe, M., and Parichy, D.M. (2012). Melanophore migration and survival during zebrafish adult pigment stripe development require the immunoglobulin superfamily adhesion molecule Igsf11. *PLoS Genet.* **8** #8, e1002899.

624. Erclik, T., Hartenstein, V., Lipshitz, H.D., and McInnes, R.R. (2008). Conserved role of the *Vsx* genes supports a monophyletic origin for bilaterian visual systems. *Curr. Biol.* **18**, 1278–1287.

625. Erickson, G.M., Krick, B.A., Hamilton, M., Bourne, G.R., Norell, M.A., Lilleodden, E., and Sawyer, W.G. (2012). Complex dental structure and wear biomechanics in hadrosaurid dinosaurs. *Science* **338**, 98–101.

626. Erickson, G.M., Makovicky, P.J., Currie, P.J., Norell, M.A., Yerby, S.A., and Brochu, C.A. (2004). Gigantism and comparative life-history parameters of tyrannosaurid dinosaurs. *Nature* **430**, 772–775.

627. Erwin, D.H. (2006). Evolutionary contingency. *Curr. Biol.* **16**, R825–R826.

628. Erwin, D.H. (2007). Disparity: morphological pattern and developmental context. *Palaeontology* **50**, 57–73.

629. Erwin, D.H. (2009). Early origin of the bilaterian developmental toolkit. *Phil. Trans. R. Soc. Lond. B* **364**, 2253–2261. [*See also* Wilkins, A.S. (2014). The genetic tool-kit: the life-history of an important metaphor. In *Advances in Evolutionary Developmental Biology* (J.T. Streelman, ed.). Wiley, New York, NY.]

630. Erwin, D.H. and Davidson, E.H. (2002). The last common bilaterian ancestor. *Development* **129**, 3021–3032.

631. Erwin, D.H., Laflamme, M., Tweedt, S.M., Sperling, E.A., Pisani, D., and Peterson, K.J. (2011). The Cambrian conundrum: early divergence and later ecological success in the early history of animals. *Science* **334**, 1091–1097.

632. Erwin, D.H. and Valentine, J.W. (2013). *The Cambrian Explosion: The Construction of Animal Biodiversity*. Roberts, Greenwood Village, CO.

633. Essalmani, R., Zaid, A., Marcinkiewicz, J., Chamberland, A., Pasquato, A., Seidah, N.G., and Prat, A. (2008). In vivo functions of the proprotein convertase PC5/6 during mouse development: Gdf11 is a likely substrate. *PNAS* **105** #15, 5750–5755.

634. Estella, C. and Mann, R.S. (2008). Logic of Wg and Dpp induction of distal and medial fates in the *Drosophila* leg. *Development* **135**, 627–636.

635. Estella, C., McKay, D.J., and Mann, R.S. (2008). Molecular integration of Wingless, Decapentaplegic, and autoregulatory inputs into *Distalless* during *Drosophila* leg development. *Dev. Cell* **14**, 86–96.

636. Estella, C., Voutev, R., and Mann, R.S. (2012). A dynamic network of morphogens and transcription factors patterns the fly leg. *Curr. Top. Dev. Biol.* **98**, 173–198.

637. Estrada, B., Casares, F., and Sánchez-Herrero, E. (2003). Development of the genitalia in *Drosophila melanogaster*. *Differentiation* **71**, 299–310.

638. Estrada, B. and Sánchez-Herrero, E. (2001). The Hox gene *Abdominal-B* antagonizes appendage development in the genital disc of *Drosophila*. *Development* **128**, 331–339.

639. Evans, C.J., Hartenstein, V., and Banerjee, U. (2003). Thicker than blood: conserved mechanisms in *Drosophila* and vertebrate hematopoiesis. *Dev. Cell* **5**, 673–690.

640. Evans, D.J.R., Valasek, P., Schmidt, C., and Patel, K. (2006). Skeletal muscle translocation in vertebrates. *Anat. Embryol.* **211** (Suppl. 1), S43–S50.

641. Evans, J.W., Borton, A., HIntz, H.F., and Van Vleck, L.D. (1990). *The Horse*, 2nd edn. W. H. Freeman, New York, NY.

642. Evans, S.E. (2003). At the feet of the dinosaurs: the early history and radiation of lizards. *Biol. Rev.* **78**, 513–551.

643. Evans, S.M. (1999). Vertebrate tinman homologues and cardiac differentiation. *Semin. Cell Dev. Biol.* **10**, 73–83.

644. Ewart, J.C. (1894). The development of the skeleton of the limbs of the horse, with observations on polydactyly. *J. Anat. Physiol.* **28**, 236–256 plus 21 figs.

645. Ewart, J.C. (1894). The second and fourth digits in the horse: their development and subsequent degeneration. *Proc. Roy. Soc. Edinburgh* **1894**, 185–191.

646. Fain, G.L., Hardie, R., and Laughlin, S.B. (2010). Phototransduction and the evolution of photoreceptors. *Curr. Biol.* **20**, R114–R124.

647. Fan, J.-Y., Preuss, F., Muskus, M.J., Bjes, E.S., and Price, J.L. (2009). *Drosophila* and vertebrate casein kinase Iδ exhibits evolutionary conservation of circadian function. *Genetics* **181**, 139–152.

648. Farnum, C.E., Tinsley, M., and Hermanson, J.W. (2008). Forelimb versus hindlimb skeletal development in the big brown bat, *Eptesicus fuscus*: functional divergence is reflected in chondrocytic performance in autopodial growth plates. *Cells Tissues Organs* **187**, 35–47.

649. Faucheux, C., Nicholls, B.M., Allen, S., Danks, J.A., Horton, M.A., and Price, J.S. (2004). Recapitulation of the Parathyroid Hormone-related Peptide–Indian Hedgehog pathway in the regenerating deer antler. *Dev. Dynamics* **231**, 88–97.

650. Fausto-Sterling, A. and Smith-Schiess, H. (1982). Interactions between *fused* and *engrailed*, two mutations affecting pattern formation in *Drosophila melanogaster*. *Genetics* **101**, 71–80.

651. Fayyazuddin, A. and Dickinson, M.H. (1996). Haltere afferents provide direct, electrotonic input to a steering motor neuron in the blowfly, *Calliphora*. *J. Neurosci.* **16**, 5225–5232.

652. Fedak, T.J. and Hall, B.K. (2004). Perspectives on hyperphalangy: patterns and processes. *J. Anat.* **204**, 151–163.

653. Fédrigo, O. and Wray, G.A. (2010). Developmental evolution: how beetles evolved their shields. *Curr. Biol.* **20**, R64–R66.

654. Feldhamer, G.A., Drickamer, L.C., Vessey, S.H., Merritt, J.F., and Krajewski, C. (2007). *Mammalogy: Adaptation, Diversity, Ecology*, 3rd edn. Johns Hopkins University Press, Baltimore, MD.

655. Félix, M.-A. (2012). Evolution in developmental phenotype space. *Curr. Opin. Gen. Dev.* **22**, 593–599.

656. Fenton, M.B. and Ratcliffe, J.M. (2010). Bats. *Curr. Biol.* **20**, R1060–R1062.

657. Ferkowicz, M.J. and Raff, R.A. (2001). *Wnt* gene expression in sea urchin development: heterochronies associated with the evolution of developmental mode. *Evol. Dev.* **3**, 24–33.

658. Fernandes, J., Bate, M., and VijayRaghavan, K. (1991). Development of the indirect flight muscles of *Drosophila*. *Development* **113**, 67–77.

659. Feuda, R., Hamilton, S.C., McInerney, J.O., and Pisani, D. (2012). Metazoan opsin evolution reveals a simple route to animal vision. *PNAS* **109** #46, 18868–18872.

660. Fichelson, P. and Gho, M. (2003). The glial cell undergoes apoptosis in the microchaete lineage of *Drosophila*. *Development* **130**, 123–133.

661. Filoramo, N.I. and Schwenk, K. (2009). The mechanism of chemical delivery to the vomeronasal organs in squamate reptiles: a comparative morphological approach. *J. Exp. Zool.* **311A**, 20–34.

662. Fincham, E.F. (1955). The proportion of ciliary muscular force required for accommodation. *J. Physiol.* **128**, 99–112.
663. Findlater, G.S., McDougall, R.D., and Kaufman, M.H. (1993). Eyelid development, fusion and subsequent reopening in the mouse. *J. Anat.* **183**, 121–129.
664. Finnerty, J.R. (2003). The origins of axial patterning in the metazoa: how old is bilateral symmetry? *Int. J. Dev. Biol.* **47**, 523–529.
665. Finnerty, J.R., Pang, K., Burton, P., Paulson, D., and Martindale, M.Q. (2004). Origins of bilateral symmetry: *Hox* and *Dpp* expression in a sea anemone. *Science* **304**, 1335–1337.
666. Fitch, D.H.A. and Sudhaus, W. (2002). One small step for worms, one giant leap for "Bauplan"? *Evol. Dev.* **4**, 243–246.
667. Fitch, W.M. (2000). Homology: a personal view on some of the problems. *Trends Genet.* **16**, 227–231.
668. Fitch, W.T. (2012). Evolutionary developmental biology and human language evolution: constraints on adaptation. *Evol. Biol.* **39**, 613–637.
669. Fjällbrant, T.T., Manger, P.R., and Pettigrew, J.D. (1998). Some related aspects of platypus electroreception: temporal integration behaviour, electroreceptive thresholds and directionality of the bill acting as an antenna. *Phil. Trans. R. Soc. Lond. B* **353**, 1211–1219.
670. Flatt, T. (2005). The evolutionary genetics of canalization. *Q. Rev. Biol.* **80**, 287–316.
671. Flinn, M.V., Geary, D.C., and Ward, C.V. (2005). Ecological dominance, social competition, and coalitionary arms races: why humans evolved extraordinary intelligence. *Evol. Human Behav.* **26**, 10–46.
672. Flynn, J.J. (2009). Splendid isolation. *Nat. Hist.* **118** #5, 26–32.
673. Fomenou, M.D., Scaal, M., Stockdale, F.E., Christ, B., and Huang, R. (2005). Cells of all somitic compartments are determined with respect to segmental identity. *Dev. Dynamics* **233**, 1386–1393.
674. Fondon, J.W., III and Garner, H.R. (2004). Molecular origins of rapid and continuous morphological evolution. *PNAS* **101** #52, 18058–18063.
675. Ford, J. and Ford, D. (1986). Narwhal: unicorn of the Arctic seas. *Natl. Geogr.* **169** #3, 354–363.
676. Fordyce, R.E. and Ksepka, D.T. (2012). The strangest bird. *Sci. Am.* **307** #5, 56–61.
677. Forgacs, G. and Newman, S.A. (2005). *Biological Physics of the Developing Embryo.* Cambridge University Press, New York, NY.
678. Foronda, D., Martin, P., and Sánchez-Herrero, E. (2012). *Drosophila* Hox and sex-determination genes control segment elimination through EGFR and *extramacrochaete* activity. *PLoS Genet.* **8** #8, e1002874.
679. Forsman, A. (1995). Opposing fitness consequences of colour pattern in male and female snakes. *J. Evol. Biol.* **8**, 53–70.
680. Fortey, R. (2012). *Horseshoe Crabs and Velvet Worms: The Story of the Animals and Plants That Time Has Left Behind.* Knopf, New York, NY.
681. Fortey, R. and Thomas, R.H. (1993). The case of the velvet worm. *Nature* **361**, 205–206.
682. Foster, S.J. and Vincent, A.C.J. (2004). Life history and ecology of seahorses: implications for conservation and management. *J. Fish Biol.* **65**, 1–61.
683. Foureaux, G., Egami, M.I., Jared, C., Antoniazzi, M.M., Gutierre, R.C., and Smith, R.L. (2010). Rudimentary eyes of squamate fossorial reptiles (Amphisbaenia and Serpentes). *Anat. Rec. Adv. Integr. Anat. Evol. Biol.* **293**, 351–357.

684. Fox, C.S., Liu, Y., White, C.C., Feitosa, M., Smith, A.V., Heard-Costa, N., Lohman, K., GIANT Consortium, MAGIC Consortium, GLGC Consortium, Johnson, A.D., Foster, M.C., Greenawalt, D.M., Griffin, P., Ding, J., Newman, A.B., Tylavsky, F., Miljkovic, I., Kritchevsky, S.B., Launer, L., Garcia, M., Eiriksdottir, G., Carr, J.J., Gudnason, V., Harris, T.B., Cupples, L.A., and Borecki, I.B. (2012). Genome-wide association for abdominal subcutaneous and visceral adipose reveals a novel locus for visceral fat in women. *PLoS Genet.* **8** #5, e1002695.

685. Fox, C.S., White, C.C., Lohman, K., Heard-Costa, N., Cohen, P., Zhang, Y., Johnson, A.D., Emilsson, V., Liu, C.-T., Chen, Y.-D.I., Taylor, K.D., Allison, M., Budoff, M., CARDIoGRAM Consortium, Rotter, J.I., Carr, J.J., Hoffmann, U., Ding, J., Cupples, L.A., and Liu, Y. (2012). Genome-wide association of pericardial fat identifies a unique locus for ectopic fat. *PLoS Genet.* **8** #5, e1002705.

686. Fox, C.W., Scheibly, K.L., and Reed, D.H. (2008). Experimental evolution of the genetic load and its implications for the genetic basis of inbreeding depression. *Evolution* **62**, 2236–2249.

687. Fox, D.T. and Duronio, R.J. (2012). Endoreplication and polyploidy: insights into development and disease. *Development* **140**, 3–12.

688. Fox, R.C. and Scott, C.S. (2005). First evidence of a venom delivery apparatus in extinct mammals. *Nature* **435**, 1091–1093.

689. Fraenkel, G. and Pringle, J.W.S. (1938). Halteres of flies as gyroscopic organs of equilibrium. *Nature* **141**, 919–920.

690. Francino, M.P. (2005). An adaptive radiation model for the origin of new gene functions. *Nat. Genet.* **37**, 573–577.

691. Francis, V.A., Zorzano, A., and Teleman, A.A. (2010). dDOR is an EcR coactivator that forms a feed-forward loop connecting insulin and ecdysone signaling. *Curr. Biol.* **20**, 1799–1808.

692. Francis-West, P.H., Robertson, K.E., Ede, D.A., C. Rodriguez, C., Izpisúa-Belmonte, J.C., Houston, B., Burt, D.W., Gribbin, C., Brickell, P.M., and Tickle, C. (1995). Expression of genes encoding bone morphogenetic proteins and Sonic hedgehog in talpid (*ta3*) limb buds: their relationships in the signalling cascade involved in limb patterning. *Dev. Dynamics* **203**, 187–197.

693. Frankino, W.A., Zwaan, B.J., Stern, D.L., and Brakefield, P.M. (2005). Natural selection and developmental constraints in the evolution of allometries. *Science* **307**, 718–720.

694. Frantsevich, L. (2012). Indirect closing of elytra by the prothorax in beetles (Coleoptera): general observations and exceptions. *Zoology* **115**, 12–21.

695. Fraser, F.C. (1938). Vestigial teeth in the narwhal. *Proc. Linnean Soc. London* **150**, 155–162.

696. Fraser, G.J., Bloomquist, R.F., and Streelman, J.T. (2013). Common developmental pathways link tooth shape to regeneration. *Dev. Biol.* **377**, 399–414.

697. Fraser, G.J., Hulsey, C.D., Bloomquist, R.F., Uyesugi, K., Manley, N.R., and Streelman, J.T. (2009). An ancient gene network is co-opted for teeth on old and new jaws. *PLoS Biol.* **7** #2, 233–247 (e1000031).

698. Fraser, M.J., Jr. (2012). Insect transgenesis: current applications and future prospects. *Annu. Rev. Entomol.* **57**, 267–289.

699. Frazzetta, T.H. (1970). From hopeful monsters to bolyerine snakes? *Am. Nat.* **104**, 55–72.

700. Frazzetta, T.H. (2012). Flatfishes, turtles, and bolyerine snakes: evolution by small steps or large, or both? *Evol. Biol.* **39**, 30–60.

701. Freitas, R., Gómez-Marín, C., Wilson, J.M., Casares, F., and Gómez-Skarmeta, J.L. (2012). *Hoxd13* contribution to the evolution of vertebrate appendages. *Dev. Cell* **23**, 1219–1229.

702. Freitas, R., Zhang, G., and Cohn, M.J. (2007). Biphasic *Hoxd* gene expression in shark paired fins reveals an ancient origin of the distal limb domain. *PLoS ONE* **2** #8, e754.

703. French, V. (1997). Pattern formation in colour on butterfly wings. *Curr. Opin. Gen. Dev.* **7**, 524–529.

704. French, V. and Brakefield, P.M. (1992). The development of eyespot patterns on butterfly wings: morphogen sources or sinks? *Development* **116**, 103–109.

705. French, V. and Brakefield, P.M. (1995). Eyespot determination on butterfly wings: the focal signal. *Dev. Biol.* **168**, 112–123.

706. French, V. and Brakefield, P.M. (2004). Pattern formation: a focus on Notch in butterfly eyespots. *Curr. Biol.* **14**, R663–R665.

707. French, V., Bryant, P.J., and Bryant, S.V. (1976). Pattern regulation in epimorphic fields. *Science* **193**, 969–981.

708. French, V. and Monteiro, A. (1994). Butterfly wings: colour patterns and now gene expression patterns. *BioEssays* **16**, 789–791.

709. Friberg, U. and Rice, W.R. (2008). Cut thy neighbor: cyclic birth and death of recombination hotspots via genetic conflict. *Genetics* **179**, 2229–2238.

710. Friedman, M. (2008). The evolutionary origin of flatfish asymmetry. *Nature* **454**, 209–212.

711. Fritz, A.E., Ikmi, A., Seidel, C., Paulson, A., and Gibson, M.C. (2013). Mechanisms of tentacle morphogenesis in the sea anemone *Nematostella vectensis*. *Development* **140**, 2212–2223.

712. Fritz, J., Hummel, J., Kienzle, E., Wings, O., Streich, W.J., and Clauss, M. (2011). Gizzard vs. teeth, it's a tie: food-processing efficiency in herbivorous birds and mammals and implications for dinosaur feeding strategies. *Paleobiology* **37**, 577–586.

713. Fröbisch, J. (2011). On dental occlusion and saber teeth. *Science* **331**, 1525–1528.

714. Fry, B.G. (2005). From genome to "venome": Molecular origin and evolution of the snake venom proteome inferred from phylogenetic analysis of toxin sequences and related body proteins. *Genome Res.* **15**, 403–420.

715. Fry, B.G., Vidal, N., Norman, J.A., Vonk, F.J., Scheib, H., Ramjan, S.F.R., Kuruppu, S., Fung, K., Hedges, S.B., Richardson, M.K., Hodgson, W.C., Ignjatovic, V., Summerhayes, R., and Kochva, E. (2006). Early evolution of the venom system in lizards and snakes. *Nature* **439**, 584–588.

716. Fry, C.L. (2006). Juvenile hormone mediates a trade-off between primary and secondary sexual traits in stalk-eyed flies. *Evol. Dev.* **8**, 191–201.

717. Fuchs, E. (2009). The tortoise and the hair: slow-cycling cells in the stem cell race. *Cell* **137**, 811–819.

718. Fuchs, Y. and Steller, H. (2011). Programmed cell death in animal development and disease. *Cell* **147**, 742–758.

719. Fusco, G. (2005). Trunk segment numbers and sequential segmentation in myriapods. *Evol. Dev.* **7**, 608–617.

720. Gad, J.M. and Tam, P.P.L. (1999). Axis development: the mouse becomes a dachshund. *Curr. Biol.* **9**, R783–R786.

721. Galant, R. and Carroll, S.B. (2002). Evolution of a transcriptional repression domain in an insect Hox protein. *Nature* **415**, 910–913.

722. Galant, R., Skeath, J.B., Paddock, S., Lewis, D.L., and Carroll, S.B. (1998). Expression pattern of a butterfly *achaete-scute* homolog reveals the homology of butterfly wing scales and insect sensory bristles. *Curr. Biol.* **8**, 807–813.

723. Galant, R., Walsh, C.M., and Carroll, S.B. (2002). Hox repression of a target gene: extradenticle-independent, additive action through multiple monomer binding sites. *Development* **129**, 3115–3126.

724. Galindo, M.I., Bishop, S.A., and Couso, J.P. (2005). Dynamic EGFR-Ras signalling in *Drosophila* leg development. *Dev. Dynamics* **233**, 1496–1508.

725. Galindo, M.I., Bishop, S.A., Greig, S., and Couso, J.P. (2002). Leg patterning driven by proximal-distal interactions and EGFR signaling. *Science* **297**, 256–259.

726. Galindo, M.I., Fernández-Garza, D., Phillips, R., and Couso, J.P. (2011). Control of *Distal-less* expression in the *Drosophila* appendages by functional 3′ enhancers. *Dev. Biol.* **353**, 396–410.

727. Galindo, M.I., Pueyo, J.I., Fouix, S., Bishop, S.A., and Couso, J.P. (2007). Peptides encoded by short ORFs control development and define a new eukaryotic gene family. *PLoS Biol.* **5** #5, e106. [*See also* Lleras-Forero, L. (2013). Neuropeptides: developmental signals in placode progenitor formation. *Dev. Cell* **26**, 195–203.]

728. Galis, F. (1999). Why do almost all mammals have seven cervical vertebrae? Developmental constraints, *Hox* genes, and cancer. *J. Exp. Zool. (Mol. Dev. Evol.)* **285**, 19–26.

729. Galis, F. and Metz, J.A.J. (2003). Anti-cancer selection as a source of developmental and evolutionary constraints. *BioEssays* **25**, 1035–1039.

730. Galis, F. and Metz, J.A.J. (2007). Evolutionary novelties: the making and breaking of pleiotropic constraints. *Integr. Comp. Biol.* **47**, 409–419.

731. Galis, F., van Alphen, J.J.M., and Metz, J.A.J. (2001). Why five fingers? Evolutionary constraints on digit numbers. *Trends Ecol. Evol.* **16**, 637–646.

732. Galis, F., Van Dooren, T.J.M., Feuth, J.D., Metz, J.A.J., Witkam, A., Ruinard, S., Steigenga, M.J., and Wijnaendts, L.C.D. (2006). Extreme selection in humans against homeotic transformations of cervical vertebrae. *Evolution* **60**, 2643–2654.

733. Galis, F., Wagner, G.P., and Jockusch, E.L. (2003). Why is limb regeneration possible in amphibians but not in reptiles, birds, and mammals? *Evol. Dev.* **5**, 208–220.

734. Galton, F. (1869). *Hereditary Genius: An Inquiry into Its Laws and Consequences*. Macmillan, London.

735. Galton, F. (1889). *Natural Inheritance*. Macmillan, London.

736. Gamble, T. and Zarkower, D. (2012). Sex determination. *Curr. Biol.* **22**, 257–262.

737. Gañan, Y., Macias, D., Basco, R.D., Merino, R., and Hurle, J.M. (1998). Morphological diversity of the avian foot is related with the pattern of *msx* gene expression in the developing autopod. *Dev. Biol.* **196**, 33–41.

738. Ganfornina, M.D. and Sánchez, D. (1999). Generation of evolutionary novelty by functional shift. *BioEssays* **21**, 432–439.

739. Gans, C. (1975). Tetrapod limblessness: Evolution and functional corollaries. *Am. Zool.* **15**, 455–467.

740. Gans, C. (1984). Slide-pushing: A transitional locomotor method of elongate squamates. *Symp. Zool. Soc. Lond.* **52**, 13–26.

741. Garcia-Bellido, A. (1998). The engrailed story. *Genetics* **148**, 539–544.

742. García-Bellido, A. (1977). Homoeotic and atavic mutations in insects. *Am. Zool.* **17**, 613–629.

743. García-Bellido, A. and de Celis, J.F. (2009). The complex tale of the *achaete-scute* complex: a paradigmatic case in the analysis of gene organization and function during development. *Genetics* **182**, 631–639.

744. Garcia-Bellido, A. and Santamaria, P. (1972). Developmental analysis of the wing disc in the mutant *engrailed* of *Drosophila melanogaster*. *Genetics* **72**, 87–104.

745. Garcia-Fernàndez, J. (2005). The genesis and evolution of homeobox gene clusters. *Nature Rev. Genet.* **6**, 881–892.

746. Garcia-Fernàndez, J. and Benito-Gutiérrez, É. (2009). It's a long way from amphioxus: descendants of the earliest chordate. *BioEssays* **31**, 665–675.

747. Gardner, M. (1970). The fantastic combinations of John Conway's new solitare game "life". *Sci. Am.* **223** #4, 120–123.

748. Gardner, M. (1971). On cellular automata, self-reproduction, the Garden of Eden and the game "life". *Sci. Am.* **224** #2, 112–117.

749. Garrouste, R., Clément, G., Nel, P., Engel, M.S., Grandcolas, P., D'Haese, C., Lagebro, L., Denayer, J., Gueriau, P., Lafaite, P., Olive, S., Prestianni, C., and Nel, A. (2012). A complete insect from the Late Devonian period. *Nature* **488**, 82–85.

750. Garvie, C.W. and Wolberger, C. (2001). Recognition of specific DNA sequences. *Mol. Cell* **8**, 937–946.

751. Garza-Garcia, A.A., Driscoll, P.C., and Brockes, J.P. (2010). Evidence for the local evolution of mechanisms underlying limb regeneration in salamanders. *Integr. Comp. Biol.* **50**, 528–535.

752. Gatesy, S.M. and Dial, K.P. (1996). From frond to fan: *Archaeopteryx* and the evolution of short-tailed birds. *Evolution* **50**, 2037–2048.

753. Gaunt, S.J. (1994). Conservation in the Hox code during morphological evolution. *Int. J. Dev. Biol.* **38**, 549–552.

754. Gaviño, M.A. and Reddien, P.W. (2011). A Bmp/Admp regulatory circuit controls maintenance and regeneration of dorsal–ventral polarity in planarians. *Curr. Biol.* **21**, 294–299.

755. Gazave, E., Lapébie, P., Richards, G.S., Brunet, F., Ereskovsky, A.V., Degnan, B.M., Borchiellini, C., Vervoort, M., and Renard, E. (2009). Origin and evolution of the Notch signalling pathway: an overview from eukaryotic genomes. *BMC Evol. Biol.* **9**, Article 249 (27 pp.).

756. Gebelein, B., Culi, J., Ryoo, H.D., Zhang, W., and Mann, R.S. (2002). Specificity of *Distalless* repression and limb primordia development by abdominal Hox proteins. *Dev. Cell* **3**, 487–498.

757. Gebelein, B., McKay, D.J., and Mann, R.S. (2004). Direct integration of *Hox* and segmentation gene inputs during *Drosophila* development. *Nature* **431**, 653–659.

758. Gebo, D.L. (1987). Functional anatomy of the tarsier foot. *Am. J. Phys. Anthrop.* **73**, 9–31.

759. Gee, H. (2008). The amphioxus unleashed. *Nature* **453**, 999–1000.

760. Geertsema, A. (1991). The servals of Gorigor. *Nat. Hist.* **100** #2, 52–61.

761. Gehring, W. (2012). The animal body plan, the prototypic body segment, and eye evolution. *Evol. Dev.* **14**, 34–46.

762. Gehring, W.J. (1998). *Master Control Genes in Development and Evolution: The Homeobox Story*. Yale University Press, New Haven, CT.

763. Gehring, W.J. (2002). The genetic control of eye development and its implications for the evolution of the various eye-types. *Int. J. Dev. Biol.* **46**, 65–73.

764. Geist, V. (1986). The paradox of the great Irish stags. *Nat. Hist.* **95** #3, 54–65.

765. Geist, V. (1994). Why antlers branched out. *Nat. Hist.* **103** #4, 78–83.

766. Gerber, S. (2013). On the relationship between the macroevolutionary trajectories of morphological integration and morphological disparity. *PLoS ONE* **8** #5, e63913.

767. Gerhart, J. (1999). Signaling pathways in development (1998 Warkany lecture). *Teratology* **60**, 226–239.

768. Gerhart, J. and Kirschner, M. (1997). *Cells, Embryos, and Evolution*. Blackwell Science, Malden, MA.

769. Gerhart, J. and Kirschner, M. (2007). The theory of facilitated variation. *PNAS* **104** (Suppl. 1), 8582–8589.

770. Gerhart, J., Lowe, C., and Kirschner, M. (2005). Hemichordates and the origin of chordates. *Curr. Opin. Gen. Dev.* **15**, 461–467.

771. Ghazi, A., Anant, S., and VijayRaghavan, K. (2000). Apterous mediates development of direct flight muscles autonomously and indirect flight muscles through epidermal cues. *Development* **127**, 5309–5318.

772. Ghazi, A., Paul, L., and VijayRaghavan, K. (2003). Prepattern genes and signaling molecules regulate *stripe* expression to specify *Drosophila* flight muscle attachment sites. *Mechs. Dev.* **120**, 519–528.

773. Gherman, A., Chen, P.E., Teslovich, T.M., Stankiewicz, P., Withers, M., Kashuk, C.S., Chakravarti, A., Lupski, J.R., Cutler, D.J., and Katsanis, N. (2007). Population bottlenecks as a potential major shaping force of human genome architecture. *PLoS Genet.* **3** #7, 1223–1231 (e119).

774. Ghiradella, H. (2010). Insect cuticular surface modifications: scales and other structural formations. *Adv. Insect Physiol.* **38**, 135–180.

775. Ghiradella, H. and Schmidt, J.T. (2004). Fireflies at one hundred plus: a new look at flash control. *Integr. Comp. Biol.* **44**, 203–212.

776. Ghiselin, M.T. (1995). A movable feaster. *Nat. Hist.* **94** #9, 54–61.

777. Ghysen, A. (2003). The origin and evolution of the nervous system. *Int. J. Dev. Biol.* **47**, 555–562.

778. Gibson, G. (1999). Developmental evolution: Going beyond the "just so". *Curr. Biol.* **9**, R942–R945.

779. Gibson, G. (2000). Evolution: *Hox* genes and the cellared wine principle. *Curr. Biol.* **10**, R452–R455.

780. Gibson, G. and Honeycutt, E. (2002). The evolution of developmental regulatory pathways. *Curr. Opin. Gen. Dev.* **12**, 695–700.

781. Gierer, A. and Meinhardt, H. (1974). Biological pattern formation involving lateral inhibition. In *Lectures on Mathematics in the Life Sciences*, Vol. 7. American Mathematical Society, Providence, RI, pp. 163–183.

782. Gilbert, A.N. (1986). Mammary number and litter size in Rodentia: the "one-half rule". *PNAS* **83** #13, 4828–4830.

783. Gilbert, S.F. (2003). Opening Darwin's black box: teaching evolution through developmental genetics. *Nature Rev. Genet.* **4**, 735–741.

784. Gilbert, S.F. (2010). *Developmental Biology*, 9th edn. Sinauer, Sunderland, MA.

785. Gilbert, S.F. and Bolker, J.A. (2001). Homologies of process and modular elements of embryonic construction. *J. Exp. Zool. (Mol. Dev. Evol.)* **291**, 1–12.

786. Gilbert, S.F. and Epel, D. (2008). *Ecological Developmental Biology*. Sinauer, Sunderland, MA.

787. Gill, F.B. (2007). *Ornithology*, 3rd edn. W. H. Freeman, New York, NY.

788. Gillespie, R.G. (2013). Adaptive radiation: convergence and non-equilibrium. *Curr. Biol.* **23**, R71–R74. [*See also* Irschick, D.J., *et al.* (2013). Evo-devo beyond morphology: from genes to resource use. *Trends Ecol. Evol.* **28**, 267–273.]

789. Gillham, N.W. (2001). Evolution by jumps: Francis Galton and William Bateson and the mechanism of evolutionary change. *Genetics* **159**, 1383–1392.

790. Gingerich, P.D., ul Haq, M., Zalmout, I.S., Khan, I.H., and Malkani, M.S. (2001). Origin of whales from early artiodactyls: hands and feet of Eocene protocetidae from Pakistan. *Science* **293**, 2239–2242.

791. Giorgianni, M. and Patel, N.H. (2005). Conquering land, air and water: the evolution and development of arthropod appendages. In *Evolving Form and Function: Fossils and Development* (D.E.G. Briggs, ed.). Yale Peabody Museum of Natural History, New Haven, CT, pp. 159–180.

792. Giorgianni, M.W. and Mann, R.S. (2011). Establishment of medial fates along the proximodistal axis of the *Drosophila* leg through direct activation of *dachshund* by Distalless. *Dev. Cell* **20**, 455–468.

793. Gleichauf, R. (1936). Anatomie und Variabilität des Geschlechtsapparates von *Drosophila melanogaster* (Meigen). *Z. wiss. Zool.* **148**, 1–66.

794. Glimm, T., Zhang, J., Shen, Y.-Q., and Newman, S.A. (2012). Reaction-diffusion systems and external morphogen gradients: the two-dimensional case, with an application to skeletal pattern formation. *Bull. Math. Biol.* **74**, 666–687.

795. Gmitro, J.I. and Scriven, L.E. (1969). A physicochemical basis for pattern and rhythm. In *Towards a Theoretical Biology. II. Sketches.* (C.H. Waddington, ed.). Edinburgh University Press, Edinburgh, pp. 184–203.

796. Gnatzy, W., Grünert, U., and Bender, M. (1987). Campaniform sensilla of *Calliphora vicina* (Insecta, Diptera). I. Typography. *Zoomorphology* **106**, 312–319.

797. Goberdhan, D.C.I. and Wilson, C. (2002). Insulin receptor-mediated organ overgrowth in *Drosophila* is not restricted by body size. *Dev. Genes Evol.* **212**, 196–202.

798. Goel, N.S. and Thompson, R.L. (1989). Movable finite automata (MFA): a new tool for computer modeling of living systems. In *Artificial Life* (C.G. Langton, ed.). Addison-Wesley, New York, NY, pp. 317–340.

799. Goldbeter, A., Gonze, D., and Pourquié, O. (2007). Sharp developmental thresholds defined through bistability by antagonistic gradients of retinoic acid and FGF signaling. *Dev. Dynamics* **236**, 1495–1508.

800. Goldschmidt, R. (1938). *Physiological Genetics*. McGraw-Hill, New York, NY.

801. Goldschmidt, R. (1940). *The Material Basis of Evolution*. Yale University Press, New Haven, CT.

802. Goldschmidt, R.B. (1949). Phenocopies. *Sci. Am.* **181** #10, 46–49.

803. Goldschmidt, R.B. (1952). Homoeotic mutants and evolution. *Acta Biotheor.* **10**, 87–104.

804. Goldsmith, T.H. (1990). Optimization, constraint, and history in the evolution of eyes. *Q. Rev. Biol.* **65**, 281–322.

805. Gomez, C., Özbudak, E.M., Wunderlich, J., Baumann, D., Lewis, J., and Pourquié, O. (2008). Control of segment number in vertebrate embryos. *Nature* **454**, 335–339.

806. Gomez, C. and Pourquié, O. (2009). Developmental control of segment numbers in vertebrates. *J. Exp. Zool. (Mol. Dev. Evol.)* **312B**, 533–544.

807. Gómez-Skarmeta, J.L., Rodríguez, I., Martínez, C., Culí, J., Ferrés-Marcó, D., Beamonte, D., and Modolell, J. (1995). *Cis*-regulation of *achaete* and *scute*: shared enhancer-like elements drive their coexpression in proneural clusters of the imaginal discs. *Genes Dev.* **9**, 1869–1882.

808. Gompel, N. and Prud'homme, B. (2009). The causes of repeated genetic evolution. *Dev. Biol.* **332**, 36–47.

809. Goodden, R. (1977). *The Wonderful World of Butterflies and Moths*. Hamlyn, New York, NY.

810. Goode, D.K., Callaway, H.A., Cerda, G.A., Lewis, K.E., and Elgar, G. (2011). Minor change, major difference: divergent functions of highly conserved cis-regulatory elements subsequent to whole genome duplication events. *Development* **138**, 879–884.

811. Goodfield, J. (1974). Changing strategies: a comparison of reductionist attitudes in biological and medical research in the nineteenth and twentieth centuries. In *Studies in the Philosophy of Biology. Reduction and Related Problems* (F.J. Ayala and T. Dobzhansky, eds.). University of California Press, Berkeley, CA, pp. 65–86.

812. Goodrich, L.V. and Strutt, D. (2011). Principles of planar polarity in animal development. *Development* **138**, 1877–1892.

813. Goodwin, B. (1994). *How the Leopard Changed Its Spots*. Charles Scribner's Sons, New York, NY.

814. Goodwin, B.C. (1985). Developing organisms as self-organizing fields. In *Mathematical Essays on Growth and the Emergence of Form* (P.L. Antonelli, ed.). University of Alberta Press, Edmonton, pp. 185–200.

815. Goodwin, B.C. and Cohen, M.H. (1969). A phase-shift model for the spatial and temporal organization of developing systems. *J. Theor. Biol.* **25**, 49–107.

816. Gordo, I. and Campos, P.R.A. (2008). Sex and deleterious mutations. *Genetics* **179**, 621–626.

817. Gordo, I., Navarro, A., and Charlesworth, B. (2002). Muller's ratchet and the pattern of variation at a neutral locus. *Genetics* **161**, 835–848.

818. Gordon, R. (1999). *The Hierarchical Genome and Differentiation Waves: Novel Unification of Development, Genetics and Evolution*. World Scientific, Singapore.

819. Gordon, R. (2001). Making waves: the paradigms of developmental biology and their impact on artificial life and embryonics. *Cybernet. Syst.* **32**, 443–458.

820. Gordon, R. and Beloussov, L. (2006). From observations to paradigms; the importance of theories and models: an interview with Hans Meinhardt. *Int. J. Dev. Biol.* **50**, 103–111.

821. Gorfinkiel, N., Morata, G., and Guerrero, I. (1997). The homeobox gene *Distal-less* induces ventral appendage development in *Drosophila*. *Genes Dev.* **11**, 2259–2271.

822. Gorfinkiel, N., Sánchez, L., and Guerrero, I. (1999). *Drosophila* terminalia as an appendage-like structure. *Mechs. Dev.* **86**, 113–123.

823. Gottlieb, A. (2012). It ain't necessarily so. *The New Yorker* Sept. 17, 2012, 84–89.

824. Gould, G.C. and MacFadden, B.J. (2004). Gigantism, dwarfism, and Cope's Rule: "Nothing in evolution makes sense without a phylogeny". *Bull. Am. Mus. Nat. Hist.* **285**, 219–237.

825. Gould, S.J. (1966). Allometry and size in ontogeny and phylogeny. *Biol. Rev.* **41**, 587–640.

826. Gould, S.J. (1970). Dollo on Dollo's law: irreversibility and the status of evolutionary laws. *J. Hist. Biol.* **3**, 189–212.

827. Gould, S.J. (1971). D'Arcy Thompson and the science of form. *New Lit. Hist.* **2**, 229–258.

828. Gould, S.J. (1977). *Ontogeny and Phylogeny*. Harvard University Press, Cambridge, MA.

829. Gould, S.J. (1977). The return of hopeful monsters. *Nat. Hist.* **86** #6, 22–30.

830. Gould, S.J. (1980). The evolutionary biology of constraint. *Proc. Am. Acad. Arts Sci.* **109** #2, 39–52.

831. Gould, S.J. (1980). Is a new and general theory of evolution emerging? *Paleobiol.* **6**, 119–130.

832. Gould, S.J. (1980). *The Panda's Thumb: More Reflections in Natural History*. Norton, New York, NY.

833. Gould, S.J. (1981). Kingdoms without wheels. *Nat. Hist.* **90** #3, 42–48. [*See also* Burrows, M. and Sutton, G. (2013). Interacting gears synchronize propulsive leg movements in a jumping insect. *Science* **341**, 1254–1256.]

834. Gould, S.J. (1981). Quaggas, coiled oysters, and flimsy facts. *Nat. Hist.* **90** #9, 16–26.

835. Gould, S.J. (1981). What color is a zebra? *Nat. Hist.* **90** #8, 16–22.

836. Gould, S.J. (1981). What, if anything, is a zebra? *Nat. Hist.* **90** #7, 6–12.

837. Gould, S.J. (1982). Darwinism and the expansion of evolutionary theory. *Science* **216**, 380–387.

838. Gould, S.J. (1983). *Hen's Teeth and Horse's Toes*. Norton, New York, NY.

839. Gould, S.J. (1985). The flamingo's smile. *Nat. Hist.* **94** #3, 6–19.

840. Gould, S.J. (1985). Geoffroy and the homeobox. *Nat. Hist.* **94** #11, 12–23.

841. Gould, S.J. (1985). Nasty little facts. *Nat. Hist.* **94** #2, 14–25.

842. Gould, S.J. (1985). To be a platypus. *Nat. Hist.* **94** #8, 10–15.

843. Gould, S.J. (1986). Archetype and adaptation. *Nat. Hist.* **95** #10, 16–27.

844. Gould, S.J. (1986). The egg-a-day barrier. *Nat. Hist.* **95** #7, 16–24.

845. Gould, S.J. (1986). Play it again, life. *Nat. Hist.* **95** #2, 18–26.

846. Gould, S.J. (1988). The heart of terminology. *Nat. Hist.* **97** #2, 24–31.

847. Gould, S.J. (1989). Full of hot air. *Nat. Hist.* **98** #10, 28–38.

848. Gould, S.J. (1989). Through a lens darkly. *Nat. Hist.* **98** #9, 16–24.

849. Gould, S.J. (1990). Bent out of shape. *Nat. Hist.* **99** #5, 12–27.

850. Gould, S.J. (1990). Everlasting legends. *Nat. Hist.* **99** #6, 12–17.

851. Gould, S.J. (1990). *Wonderful Life: The Burgess Shale and the Nature of History*. Norton, New York, NY.

852. Gould, S.J. (1991). Eight (or fewer) little piggies. *Nat. Hist.* **100** #1, 22–29.

853. Gould, S.J. (1991). Exaptation: a crucial tool for an evolutionary psychology. *J. Social Issues* **47** #3, 43–65.

854. Gould, S.J. (1994). Common pathways of illumination. *Nat. Hist.* **103** #12, 10–20.

855. Gould, S.J. (1994). Hooking Leviathan by its past. *Nat. Hist.* **103** #5, 8–15.

856. Gould, S.J. (1995). Not necessarily a wing. *Nat. Hist.* **94** #10, 12–25.

857. Gould, S.J. (1996). The tallest tale: Is the textbook version of giraffe evolution a bit of a stretch? *Nat. Hist.* **105** #5, 18–23, 54–57.

858. Gould, S.J. (1997). Unanswerable questions. In *A Glorious Accident* (W. Kayzer, ed.). W. H. Freeman, New York, NY, pp. 75–104.

859. Gould, S.J. (2000). Abscheulich! (atrocious!): Haeckel's distortions did not help Darwin. *Nat. Hist.* **109** #2, 42–49.

860. Gould, S.J. and Lewontin, R.C. (1979). The spandrels of San Marco and the Panglossian paradigm: a critique of the adaptationist programme. *Proc. Roy. Soc. Lond. B* **205**, 581–598.

861. Gould, S.J. and Vrba, E.S. (1982). Exaptation: a missing term in the science of form. *Paleobiology* **8**, 4–15.

862. Gracheva, E.O., Cordero-Morales, J.F., González-Carcacía, J.A., Ingolia, N.T., Manno, C., Aranguren, C.I., Weissman, J.S., and Julius, D. (2011). Ganglion-specific splicing of TRPV1 underlies infrared sensation in vampire bats. *Nature* **476**, 88–91.

863. Gracheva, E.O., Ingolia, N.T., Kelly, Y.M., Cordero-Morales, J.F., Hollopeter, G., Chesler, A.T., Sánchez, E.E., Perez, J.C., Weissman, J.S., and Julius, D. (2010). Molecular basis of infrared detection by snakes. *Nature* **464**, 1006–1011.

864. Graf, W. and Baker, R. (1983). Adaptive changes of the vestibulo-ocular reflex in flatfish are achieved by reorganization of central nervous pathways. *Science* **221**, 777–779.

865. Graham, A. and McGonnell, I. (1999). Developmental evolution: This side of paradise. *Curr. Biol.* **9**, R630–R632.

866. Graham, A. and McGonnell, I. (1999). Limb development: farewell to arms. *Curr. Biol.* **9**, R368–R370.

867. Grant, P.R., Grant, B.R., and Abzhanov, A. (2006). A developing paradigm for the development of bird beaks. *Biol. J. Linnean Soc.* **88**, 17–22.

868. Grant, R. (2012). How tigers get their stripes. *The Scientist Mag.* **2012** Feb. 22, Article 31728 (1 p.).

869. Grantham, T.A. (2004). Constraints and spandrels in Gould's *Structure of Evolutionary Theory*. *Biol. Philos.* **19**, 29–43.

870. Graves, J.A.M. (2013). How to evolve new vertebrate sex determining genes. *Dev. Dynamics* **242**, 354–359.

871. Gray, R.S., Roszko, I., and Solnica-Krezel, L. (2011). Planar cell polarity: coordinating morphogenetic cell behaviors with embryonic polarity. *Dev. Cell* **21**, 120–133.

872. Greenberg, L. and Hatini, V. (2009). Essential roles for *lines* in mediating leg and antennal proximodistal patterning and generating a stable Notch signaling interface at segment borders. *Dev. Biol.* **330**, 93–104.

873. Greene, H.W. (1997). *Snakes: The Evolution of Mystery in Nature*. University of California Press, Berkeley, CA.

874. Greene, H.W. and Cundall, D. (2000). Limbless tetrapods and snakes with legs. *Science* **287**, 1939–1941.

875. Greer, J.M., Puetz, J., Thomas, K.R., and Capecchi, M.R. (2000). Maintenance of functional equivalence during paralogous *Hox* gene evolution. *Nature* **403**, 661–665.

876. Gregor, T., McGregor, A.P., and Wieschaus, E.F. (2008). Shape and function of the Bicoid morphogen gradient in dipteran species with different sized embryos. *Dev. Biol.* **316**, 350–358.

877. Grenier, J.K. and Carroll, S.B. (2000). Functional evolution of the Ultrabithorax protein. *PNAS* **97** #2, 704–709.

878. Griffin, K.J.P., Stoller, J., Gibson, M., Chen, S., Yelon, D., Stainier, D.Y.R., and Kimelman, D. (2000). A conserved role for *H15*-related T-box transcription factors in zebrafish and *Drosophila* heart formation. *Dev. Biol.* **218**, 235–247.

879. Griffiths, M. (1988). The platypus. *Sci. Am.* **258** #5, 84–91.

880. Grimaldi, D. and Engel, M.S. (2005). *Evolution of the Insects*. Cambridge University Press, New York, NY.

881. Griswold, C.K. (2006). Pleiotropic mutation, modularity and evolvability. *Evol. Dev.* **8**, 81–93.

882. Gross, J.C. and Boutros, M. (2013). Secretion and extracellular space travel of Wnt proteins. *Curr. Opin. Gen. Dev.* **23**, 385–390.

883. Grünbaum, B. and Shephard, G.C. (1987). *Tilings and Patterns*. W.H. Freeman, New York, NY.

884. Grus, W.E. and Zhang, J. (2006). Origin and evolution of the vertebrate vomeronasal system viewed through system-specific genes. *BioEssays* **28**, 709–718.

885. Gubb, D. (1985). Further studies on *engrailed* mutants in *Drosophila melanogaster*. *Wilhelm Roux's Arch.* **194**, 236–246.

886. Guerreiro, I., Nunes, A., Woltering, J.M., Casaca, A., Nóvoa, A., Vinagre, T., Hunter, M.E., Duboule, D., and Mallo, M. (2013). Role of a polymorphism in a Hox/Pax-responsive enhancer in the evolution of the vertebrate spine. *PNAS* **110** #26, 10682–10686.

887. Guichard, C., Harricane, M.-C., Lafitte, J.-J., Godard, P., Zaegel, M., Tack, V., Lalau, G., and Bouvagnet, P. (2001). Axonemal dynein intermediate-chain gene (*DNAI1*) mutations result in situs inversus and primary ciliary diskinesia (Kartagener syndrome). *Am. J. Hum. Genet.* **68**, 1030–1035.

888. Guild, G.M., Connelly, P.S., Ruggiero, L., Vranich, K.A., and Tilney, L.G. (2005). Actin filament bundles in *Drosophila* wing hairs: hairs and bristles use different strategies for assembly. *Mol. Biol. Cell* **16**, 3620–3631.

889. Guo, Z.V. and Mahadevan, L. (2008). Limbless undulatory propulsion on land. *PNAS* **105** #9, 3179–3184.

890. Guruharsha, K.G., Kankel, M.W., and Artavanis-Tsakonas, S. (2012). The Notch signalling system: recent insights into the complexity of a conserved pathway. *Nature Rev. Gen.* **13**, 654–666.

891. Gustavson, E., Goldsborough, A.S., Ali, Z., and Kornberg, T.B. (1996). The *Drosophila engrailed* and *invected* genes: partners in regulation, expression and function. *Genetics* **142**, 893–906.

892. Gutowitz, H., ed. *Cellular Automata: Theory and Experiment.* MIT Press, Cambridge, MA.

893. Gwin, P. (2012). Rhino wars. *Natl. Geogr.* **221** #3, 106–125.

894. Haag, E.S. and Lenski, R.E. (2011). L'enfant terrible at 30: the maturation of evolutionary developmental biology. *Development* **138**, 2633–2637.

895. Haag, E.S. and True, J.R. (2001). From mutants to mechanisms? Assessing the candidate gene paradigm in evolutionary biology. *Evolution* **55**, 1077–1084.

896. Haas, M.S., Brown, S.J., and Beeman, R.W. (2001). Homeotic evidence for the appendicular origin of the labrum in *Tribolium castaneum*. *Dev. Genes Evol.* **211**, 96–102.

897. Haas, M.S., Brown, S.J., and Beeman, R.W. (2001). Pondering the procephalon: the segmental origin of the labrum. *Dev. Genes Evol.* **211**, 89–95.

898. Hack, M.A. and Rubenstein, D.I. (1998). Zebra zones. *Nat. Hist.* **107** #2, 26–33.

899. Hadorn, E. (1961). *Developmental Genetics and Lethal Factors.* Methuen, London (translated from 1955 German original, Thieme Verlag, Stuttgart, by U. Mittwoch).

900. Halanych, K.M. (2004). The new view of animal phylogeny. *Annu. Rev. Ecol. Evol. Syst.* **35**, 229–256.

901. Haldane, J.B.S. (1928). On being the right size. In *Possible Worlds and Other Papers.* Harper, New York, NY, pp. 20–28.

902. Haldane, J.B.S. (1928). *Possible Worlds and Other Papers.* Harper, New York, NY.

903. Halder, G., Callaerts, P., and Gehring, W.J. (1995). Induction of ectopic eyes by targeted expression of the *eyeless* gene in *Drosophila*. *Science* **267**, 1788–1792.

904. Hall, B.K. (1984). Developmental mechanisms underlying the formation of atavisms. *Biol. Rev.* **59**, 89–124.

905. Hall, B.K. (2003). Descent with modification: the unity underlying homology and homoplasy as seen through an analysis of development and evolution. *Biol. Rev.* **78**, 409–433.

906. Hall, B.K., ed. (2007) *Fins into Limbs: Evolution, Development, and Transformation.* University of Chicago Press, Chicago, IL. [*See also* the amazing video of a walking shark by going to the *Los Angeles Times* website (www.latimes.com/science) and typing "walking shark" in the search window.]

907. Hall, B.K. (2009). *The Neural Crest and Neural Crest Cells in Vertebrate Development and Evolution*. Springer, New York, NY.

908. Hall, B.K. (2012). Parallelism, deep homology, and evo-devo. *Evol. Dev.* **14**, 29–33.

909. Hall, B.K. and Kerney, R. (2012). Levels of biological organization and the origin of novelty. *J. Exp. Zool. (Mol. Dev. Evol.)* **318B**, 428–437.

910. Hall, B.K. and Olson, W.M., eds. (2003) *Keywords and Concepts in Evolutionary Developmental Biology*. Harvard University Press, Cambridge, MA.

911. Hallgrímsson, B. and Hall, B.K., eds. (2005). *Variation: A Central Concept in Biology*. Elsevier Academic Press, New York, NY.

912. Hallgrímsson, B., Jamniczky, H.A., Young, N.M., Rolian, C., Schmidt-Ott, U., and Marcucio, R.S. (2012). The generation of variation and the developmental basis for evolutionary novelty. *J. Exp. Zool. (Mol. Dev. Evol.)* **318B**, 501–517.

913. Halloy, J., Bernard, B.A., Loussouarn, G., and Goldbeter, A. (2000). Modeling the dynamics of human hair cycles by a follicular automaton. *PNAS* **97** #15, 8328–8333.

914. Hallsson, J.H., Haflidadóttir, B.S., Stivers, C., Odenwald, W., Arnheiter, H., Pignoni, F., and Steingrímsson, E. (2004). The basic helix-loop-helix leucine zipper transcription factor *Mitf* is conserved in *Drosophila* and functions in eye development. *Genetics* **167**, 233–241.

915. Hamilton, C. (2010). *Requiem for a Species: Why We Resist the Truth about Climate Change*. Earthscan, New York, NY.

916. Hammer, M.F. (2013). Human hybrids. *Sci. Am.* **308** #5, 66–71.

917. Hamrick, M.W. (2001). Development and evolution of the mammalian limb: adaptive diversification of nails, hooves, and claws. *Evol. Dev.* **3**, 355–363.

918. Hamrick, M.W. (2012). The developmental origins of mosaic evolution in the primate limb skeleton. *Evol. Biol.* **39**, 447–455.

919. Han, J., Lee, J.-E., Jin, J., Lim, J.S., Oh, N., Kim, K., Chang, S.-I., Shibuya, M., Kim, H., and Koh, G.Y. (2011). The spatiotemporal development of adipose tissue. *Development* **138**, 5027–5037.

920. Han, K.-A. and Kim, Y.-C. (2010). Courtship behavior: the right touch stimulates the proper song. *Curr. Biol.* **20**, R25–R28.

921. Hancock, J.M. (2005). Gene factories, microfunctionalization and the evolution of gene families. *Trends Genet.* **21**, 591–595.

922. Handrigan, G.R. (2003). *Concordia discors:* duality in the origin of the vertebrate tail. *J. Anat.* **202**, 255–267.

923. Handrigan, G.R. and Richman, J.M. (2011). Unicuspid and bicuspid tooth crown formation in squamates. *J. Exp. Zool. (Mol. Dev. Evol.)* **316B**, 598–608.

924. Handrigan, G.R. and Wassersug, R.J. (2007). The anuran *Bauplan*: a review of the adaptive, developmental, and genetic underpinnings of frog and tadpole morphology. *Biol. Rev.* **82**, 1–25.

925. Hanlon, R. (2007). Cephalopod dynamic camouflage. *Curr. Biol.* **17**, R400–R404.

926. Hansen, T.F. (2011). Epigenetics: adaptation or contingency? In *Epigenetics: Linking Genotype and Phenotype in Development and Evolution* (B. Hallgrímsson and B.K. Hall, eds.). University of California Press, Berkeley, CA, pp. 357–376.

927. Harding, K. and Levine, M. (1988). Gap genes define the limits of Antennapedia and Bithorax gene expression during early development in *Drosophila*. *EMBO J.* **7**, 205–214.

928. Hare, E.E., Peterson, B.K., and Eisen, M.B. (2008). A careful look at binding site reorganization in the *even-skipped* enhancers of *Drosophila* and sepsids. *PLoS Genet.* **4** #11, e1000268.

929. Harima, Y. and Kageyama, R. (2013). Oscillatory links of Fgf signaling and Hes7 in the segmentation clock. *Curr. Opin. Gen. Dev.* **23**, 484–490.

930. Harjunmaa, E., Kallonen, A., Voutilainen, M., Hämäläinen, K., Mikkola, M.L., and Jernvall, J. (2012). On the difficulty of increasing dental complexity. *Nature* **483**, 324–327.

931. Harley, E.H., Knight, M.H., Lardner, C., Wooding, B., and Gregor, M. (2009). The Quagga project: progress over 20 years of selective breeding. *S. Afr. J. Wildlife Res.* **39** #2, 155–163.

932. Harper, C.J., Swartz, S.M., and Brainerd, E.L. (2013). Specialized bat tongue is a hemodynamic nectar mop. *PNAS* **110** #22, 8852–8857.

933. Harris, A.K., Stopak, D., and Warner, P. (1984). Generation of spatially periodic patterns by a mechanical instability: a mechanical alternative to the Turing model. *J. Embryol. Exp. Morph.* **80**, 1–20.

934. Harris, M.L. and Erickson, C.A. (2007). Lineage specification in neural crest cell pathfinding. *Dev. Dynamics* **236**, 1–19.

935. Harris, M.P., Hasso, S.M., Ferguson, M.W.J., and Fallon, J.F. (2006). The development of archosaurian first-generation teeth in a chicken mutant. *Curr. Biol.* **16**, 371–377.

936. Hart, T.B. and Hart, J.A. (1992). Between sun and shadow. *Nat. Hist.* **101** #11, 28–35.

937. Hartenstein, V. (1993). *Atlas of Drosophila Development.* Cold Spring Harbor Laboratory Press, Cold Spring Harbor , NY.

938. Harvey, P.H. and Arnold, S.J. (1982). Female mate choice and runaway sexual selection. *Nature* **297**, 533–534.

939. Hasenfuss, I. (2002). A possible evolutionary pathway to insect flight starting from lepismatid organization. *J. Zool. Syst. Evol. Research* **40**, 65–81.

940. Hashimoto, H., Mizuta, A., Okada, N., Suzuki, T., Tagawa, M., Tabata, K., Yokoyama, Y., Sakaguchi, M., Tanaka, M., and Toyohara, H. (2002). Isolation and characterization of a Japanese flounder clonal line, *reversed*, which exhibits reversal of metamorphic left-right asymmetry. *Mechs. Dev.* **111**, 17–24.

941. Haskel-Ittah, M., Ben-Zvi, D., Branski-Arieli, M., Schejter, E.D., Shilo, B.-Z., and Barkai, N. (2012). Self-organized shuttling: generating sharp dorsoventral polarity in the early *Drosophila* embryo. *Cell* **150**, 1016–1028.

942. Hauswirth, R., Haase, B., Blatter, M., Brooks, S.A., Burger, D., Drögemüller, C., Gerber, V., Henke, D., Janda, J., Jude, R., Magdesian, K.G., Matthews, J.M., Poncet, P.-A., Svansson, V., Tozaki, T., Wilkinson-White, L., Penedo, M.C.T., Rieder, S., and Leeb, T. (2012). Mutations in *MITF* and *PAX3* cause "splashed white" and other white spotting phenotypes in horses. *PLoS Genet.* **8** #4, e1002653.

943. Hautier, L., Weisbecker, V., Sánchez-Villagra, M.R., Goswami, A., and Asher, R.J. (2010). Skeletal development in sloths and the evolution of mammalian vertebral patterning. *PNAS* **107** #44, 18903–18908.

944. Hayashi, T. and Murakami, R. (2001). Left-right asymmetry in *Drosophila melanogaster* gut development. *Develop. Growth Differ.* **43**, 239–246.

945. Hayden, T. (2011). How to hatch a dinosaur. *Wired* **19** #10, 150–157, 186.

946. Hayes, B. (1984). Computer recreations: The cellular automaton offers a model of the world and a world unto itself. *Sci. Am.* **250** #3, 12–21.

947. Haynic, J.L. (1982). Homologies of positional information in thoracic imaginal discs of *Drosophila melanogaster. Wilhelm Roux's Arch.* **191**, 293–300.

948. Hays, R., Buchanan, K.T., Neff, C., and Orenic, T.V. (1999). Patterning of *Drosophila* leg sensory organs through combinatorial signaling by Hedgehog, Decapentaplegic and Wingless. *Development* **126**, 2891–2899.

949. Head, J.J., Bloch, J.I., Hastings, A.K., Bourque, J.R., Cadena, E.A., Herrera, F.A., Polly, P.D., and Jaramillo, C.A. (2009). Giant boid snake from the Palaeocene neotropics reveals hotter past equatorial temperatures. *Nature* **457**, 715–717.

950. Head, J.J. and Polly, P.D. (2007). Dissociation of somatic growth from segmentation drives gigantism in snakes. *Biol. Lett.* **3**, 296–298.

951. Headland, T.N. and Greene, H.W. (2011). Hunter-gatherers and other primates as prey, predators, and competitors of snakes. *PNAS* **108** #52, E1470–E1474.

952. Hedges, S.B. (2012). Amniote phylogeny and the position of turtles. *BMC Biol.* **10**, Article 64 (2 pp.).

953. Heeren, F. (2011). Rise of the titans. *Nature* **475**, 159–161.

954. Heers, A.M. and Dial, K.P. (2012). From extant to extinct: locomotor ontogeny and the evolution of avian flight. *Trends Ecol. Evol.* **27**, 296–305.

955. Heffer, A. and Pick, L. (2013). Conservation and variation in *Hox* genes: how insect models pioneered the evo-devo field. *Annu. Rev. Entomol.* **58**, 161–179.

956. Heine, P., Dohle, E., Bumsted-O'Brien, K., Engelkamp, D., and Schulte, D. (2008). Evidence for an evolutionary conserved role of *homothorax/Meis1/2* during vertebrate retina development. *Development* **135**, 805–811.

957. Hejnol, A. and Martindale, M.Q. (2008). Acoel development indicates the independent evolution of the bilaterian mouth and anus. *Nature* **456**, 382–386.

958. Held, L.I., Jr. (1979). Pattern as a function of cell number and cell size on the second-leg basitarsus of *Drosophila*. *Wilhelm Roux's Arch.* **187**, 105–127.

959. Held, L.I., Jr. (1990). Sensitive periods for abnormal patterning on a leg segment in *Drosophila melanogaster*. *Roux's Arch. Dev. Biol.* **199**, 31–47.

960. Held, L.I., Jr. (1991). Bristle patterning in *Drosophila*. *BioEssays* **13**, 633–640.

961. Held, L.I., Jr. (1992). *Models for Embryonic Periodicity*. Monographs in Developmental Biology, Vol. 24. Karger, Basel.

962. Held, L.I., Jr. (1993). Segment-polarity mutations cause stripes of defects along a leg segment in *Drosophila*. *Dev. Biol.* **157**, 240–250.

963. Held, L.I., Jr. (1995). Axes, boundaries and coordinates: the ABCs of fly leg development. *BioEssays* **17**, 721–732.

964. Held, L.I., Jr. (2002). Bristles induce bracts via the EGFR pathway on *Drosophila* legs. *Mechs. Dev.* **117**, 225–234.

965. Held, L.I., Jr. (2002). *Imaginal Discs: The Genetic and Cellular Logic of Pattern Formation*. Developmental and Cell Biology Series, Vol. 39. Cambridge University Press, New York, NY.

966. Held, L.I., Jr. (2002). Why should transverse rows need the EGFR pathway to align properly on *Drosophila* legs? *Drosophila Info. Serv.* **85**, 17–20.

967. Held, L.I., Jr. (2009). *Quirks of Human Anatomy: An Evo-Devo Look at the Human Body*. Cambridge University Press, New York, NY.

968. Held, L.I., Jr. (2010). The evo-devo puzzle of human hair patterning. *Evol. Biol.* **37**, 113–122.

969. Held, L.I., Jr. (2010). The evolutionary geometry of human anatomy: discovering our inner fly. *Evol. Anthrop.* **19**, 227–235.

970. Held, L.I., Jr. (2010). How does *Scr* cause first legs to deviate from second legs? *Dros. Info. Serv.* **93**, 132–146.

971. Held, L.I., Jr. (2013). Rethinking butterfly eyespots. *Evol. Biol.* **40**, 158–168.

972. Held, L.I., Jr., Duarte, C.M., and Derakhshanian, K. (1986). Extra tarsal joints and abnormal cuticular polarities in various mutants of *Drosophila melanogaster*. *Roux's Arch. Dev. Biol.* **195**, 145–157.

973. Held, L.I., Jr., Grimson, M.J., and Du, Z. (2004). Proving an old prediction: The sex comb rotates at 16 to 24 hours after pupariation. *Drosophila Info. Serv.* **87**, 76–78.

974. Held, L.I., Jr. and Heup, M. (1996). Genetic mosaic analysis of *decapentaplegic* and *wingless* gene function in the *Drosophila* leg. *Dev. Genes Evol.* **206**, 180–194.

975. Held, L.I., Jr., Heup, M.A., Sappington, J.M., and Peters, S.D. (1994). Interactions of *decapentaplegic*, *wingless*, and *Distal-less* in the *Drosophila* leg. *Roux's Arch. Dev. Biol.* **203**, 310–319.

976. Helfman, G.S., Collette, B.B., Facey, D.E., and Bowen, B.W. (2009). *The Diversity of Fishes: Biology, Evolution and Ecology*, 2nd edn. Wiley-Blackwell, Oxford.

977. Heller, K. (2011). How bird necks get naked. *PLoS Biol.* **9** #3, e1001029.

978. Helms, J.A. and Brugmann, S.A. (2007). The origins of species-specific facial morphology: the proof is in the pigeon. *Integr. Comp. Biol.* **47**, 338–342.

979. Hendrikse, J.L., Parsons, T.E., and Hallgrímsson, B. (2007). Evolvability as the proper focus of evolutionary developmental biology. *Evol. Dev.* **9**, 393–401.

980. Hersh, B.M., Nelson, C.E., Stoll, S.J., Norton, J.E., Albert, T.J., and Carroll, S.B. (2007). The UBX-regulated network in the haltere imaginal disc of *D. melanogaster*. *Dev. Biol.* **302**, 717–727.

981. Hershkovitz, P. (1987). Uacaries, New World monkeys of the genus *Cacajao* (Cebidae, Platyrrhini): a preliminary taxonomic review with the description of a new subspecies. *Am. J. Primatol.* **12**, 1–53.

982. Hey, J. (2004). What's so hot about recombination hotspots? *PLoS Biol.* **2** #6, 0730.

983. Hibino, T., Ishii, Y., Levin, M., and Nishino, A. (2006). Ion flow regulates left-right asymmetry in sea urchin development. *Dev. Genes Evol.* **216**, 265–276.

984. Hibino, T., Nishino, A., and Amemiya, S. (2006). Phylogenetic correspondence of the body axes in bliaterians is revealed by the right-sided expression of *Pitx* genes in echinoderm larvae. *Develop. Growth Differ.* **48**, 587–595.

985. Hieronymus, T.L., Witmer, L.M., and Ridgely, R.C. (2006). Structure of white rhinoceros (*Ceratotherium simum*) horn investigated by X-ray computed tomography and histology with implications for growth and external form. *J. Morph.* **267**, 1172–1176.

986. Higgie, M. and Blows, M.W. (2008). The evolution of reproductive character displacement conflicts with how sexual selection operates within a species. *Evolution* **62**, 1192–1203.

987. Higuchi, R., Bowman, B., Freiberger, M., Ryder, O.A., and Wilson, A.C. (1984). DNA sequences from the quagga, an extinct member of the horse family. *Nature* **312**, 282–284.

988. Higuchi, R.G., Wrischnik, L.A., Oakes, E., George, M., Tong, B., and Wilson, A.C. (1987). Mitochondrial DNA of the extinct quagga: relatedness and extent of postmortem change. *J. Mol. Evol.* **25**, 283–287.

989. Hildebrand, J.G. and Shepherd, G.M. (1997). Mechanisms of olfactory discrimination: converging evidence for common principles across phyla. *Annu. Rev. Neurosci.* **20**, 595–631.

990. Hildebrand, M. (1974). *Analysis of Vertebrate Structure*. Wiley, New York, NY.

991. Hildebrand, M. (1985). Walking and running. In *Functional Vertebrate Morphology* (M. Hildebrand, D.M. Bramble, K.F. Liem, and D.B. Wake, eds.). Harvard University Press, Cambridge, MA, pp. 38–57.

992. Hilgetag, C.C. and Barbasthis, H. (2009). Sculpting the brain. *Sci. Am.* **300** #2, 66–71.

993. Hill, R.V. (2006). Comparative anatomy and histology of Xenarthran osteoderms. *J. Morph.* **267**, 1441–1460.

994. Hillis, D.M. (2007). Making evolution relevant and exciting to biology students. *Evolution* **61**, 1261–1264.

995. Hillmer, A.M., Flaquer, A., Hanneken, S., Eigelshoven, S., Kortüm, A.-K., Brockschmidt, F.F., Golla, A., Metzen, C., Thiele, H., Kolberg, S., Reinartz, R., Betz, R.C., Ruzicka, T., Hennies, H.C., Kruse, R., and Nöthen, M.M. (2008). Genome-wide scan and fine-mapping linkage study of androgenetic alopecia reveals a locus on chromosome 3q26. *Am. J. Hum. Genet.* **82**, 737–743.

996. Hirata, M., Nakamura, K.-i., and Kondo, S. (2005). Pigment cell distributions in different tissues of the zebrafish, with special reference to the striped pigment pattern. *Dev. Dynamics* **234**, 293–300.

997. Hironaka, K.-i. and Morishita, Y. (2012). Encoding and decoding of positional information in morphogen-dependent patterning. *Curr. Opin. Gen. Dev.* **22**, 553–561.

998. Hirth, F. (2010). On the origin and evolution of the tripartite brain. *Brain Behav. Evol.* **76**, 3–10.

999. Hirth, F., Kammermeier, L., Frei, E., Walldorf, U., Noll, M., and Reichert, H. (2003). An urbilaterian origin of the tripartite brain: developmental genetic insights from *Drosophila*. *Development* **130**, 2365–2373.

1000. Hochner, B. (2012). An embodied view of octopus neurobiology. *Curr. Biol.* **22**, R887–R892.

1001. Hockman, D., Cretekos, C.J., Mason, M.K., Behringer, R.R., Jacobs, D.S., and Illing, N. (2008). A second wave of *Sonic hedgehog* expression during the development of the bat limb. *PNAS* **105** #44, 16982–16987.

1002. Hockman, D., Mason, M.K., Jacobs, D.K., and Illing, N. (2009). The role of early development in mammalian limb diversification: a descriptive comparison of early limb development between the natal long-fingered bat (*Miniopterus natalensis*) and the mouse (*Mus musculus*). *Dev. Dynamics* **238**, 965–979.

1003. Hodges, A. (1983). *Alan Turing: The Enigma*. Simon & Schuster, New York.

1004. Hodgkin, J. (1998). Seven types of pleiotropy. *Int. J. Dev. Biol.* **42**, 501–505.

1005. Hodgkinson, A. and Eyre-Walker, A. (2011). Variation in the mutation rate across mammalian genomes. *Nature Rev. Genet.* **12**, 756–766.

1006. Hodin, J. (2000). Plasticity and constraints in development and evolution. *J. Exp. Zool. (Mol. Dev. Evol.)* **288**, 1–20.

1007. Hoekstra, H.E. (2006). Genetics, development and evolution of adaptive pigmentation in vertebrates. *Heredity* **97**, 222–234.

1008. Höfer, T., Maini, P.K., Kondo, S., and Asai, R. (1996). Turing patterns in fish skin? *Nature* **380**, 678.

1009. Hoffstetter, R. and Gasc, J.-P. (1969). Vertebrae and ribs of modern reptiles. In *Biology of the Reptilia, Vol. 1: Morphology, Part A* (C. Gans, A.D. Bellairs, and T.S. Parsons, eds.). Academic Press, New York, NY, pp. 201–310.

1010. Hofreiter, M. and Schöneberg, T. (2010). The genetic and evolutionary basis of color variation in vertebrates. *Cell. Mol. Life Sci.* **67**, 2591–2603.

1011. Hofstadter, D. and Sander, E. (2013). *Surfaces and Essences: Analogy as the Fuel and Fire of Thinking*. Basic Books, New York, NY.

1012. Hogan, B.L.M. and Kolodziej, P.A. (2002). Molecular mechanisms of tubulogenesis. *Nature Rev. Genet.* **3**, 513–523.

1013. Holder, N. (1983). Developmental constraints and the evolution of vertebrate digit patterns. *J. Theor. Biol.* **104**, 451–471.

1014. Holland, J.S. (2010). Hard hit. *Natl. Geogr.* **218** #2.

1015. Holland, L.Z., Kene, M., Williams, N.A., and Holland, N.D. (1997). Sequence and embryonic expression of the amphioxus *engrailed* gene (*AmphiEn*): the metameric pattern of transcription resembles that of its segment-polarity homolog in *Drosophila*. *Development* **124**, 1723–1732.

1016. Holland, P.W.H. (1998). Major transitions in animal evolution: a developmental genetic perspective. *Am. Zool.* **38**, 829–842.

1017. Hölldobler, B. and Wilson, E.O. (1990). *The Ants*. Harvard University Press, Cambridge, MA.

1018. Holley, S.A., Jackson, P.D., Sasai, Y., Lu, B., De Robertis, E.M., Hoffmann, F.M., and Ferguson, E.L. (1995). A conserved system for dorsal–ventral patterning in insects and vertebrates involving *sog* and *chordin*. *Nature* **376**, 249–253.

1019. Holloway, M. (2000). Cuttlefish say it with skin. *Nat. Hist.* **109** #3, 70–79.

1020. Holsinger, J.R. (1988). Troglobites: The evolution of cave-dwelling organisms. *Am. Sci.* **76**, 146–153.

1021. Holstein, T.W., Watanabe, H., and Özbek, S. (2011). Signaling pathways and axis formation in the lower metazoa. *Curr. Top. Dev. Biol.* **97**, 137–177.

1022. Holt, R.D. (2000). Use it or lose it. *Nature* **407**, 689–690.

1023. Horder, T.J. (2006). Gavin Rylands de Beer: how embryology foreshadowed the dilemmas of the genome. *Nature Rev. Gen.* **7**, 892–898.

1024. Horne-Badovinac, S. and Munro, E. (2011). Tubular transformations. *Science* **333**, 294–295.

1025. Horner, J.R., Padian, K., and de Ricqlès, A. (2005). How dinosaurs grew so large – and so small. *Sci. Am.* **293** #1, 56–63.

1026. Hosken, D.J. and Stockley, P. (2004). Sexual selection and genital evolution. *Trends Ecol. Evol.* **19**, 87–93.

1027. Hou, L.-h., Zhou, Z., Martin, L.D., and Feduccia, A. (1995). A beaked bird from the Jurassic of China. *Nature* **377**, 616–618.

1028. Houde, P. (1986). Ostrich ancestors found in the Northern Hemisphere suggest new hypothesis of ratite origins. *Nature* **324**, 563–565.

1029. Houssaye, A., Xu, F., Helfen, L., de Buffrénil, V., Baumbach, T., and Tafforeau, P. (2011). Three-dimensional pelvis and limb anatomy of the Cenomanian hind-limbed snake *Eupodophis descouensi* (Squamata, Ophidia) revealed by synchrotron-radiation computed laminography. *J. Vert. Paleo.* **31**, 2–7.

1030. Houston, D.C. (1994). To the vultures belong the spoils. *Nat. Hist.* **103** #9, 34–41.

1031. Howarth, D.G., Martins, T., Chimney, E., and Donoghue, M.J. (2011). Diversification of *CYCLOIDEA* expression in the evolution of bilateral flower symmetry in Caprifoliaceae and *Lonicera* (Dipsacales). *Ann. Bot.* **107**, 1521–1532.

1032. Howland, H.C., Merola, S., and Basarab, J.R. (2004). The allometry and scaling of the size of vertebrate eyes. *Vision Res.* **44**, 2043–2065.

1033. Hoyer, S.C., Eckart, A., Herrel, A., Zars, T., Fischer, S.A., Hardie, S.L., and Heisenberg, M. (2008). Octopamine in male aggression of *Drosophila*. *Curr. Biol.* **18**, 159–167.

1034. Hozumi, S., Maeda, R., Taniguchi, K., Kanai, M., Shirakabe, S., Sasamura, T., Spéder, P., Noselli, S., Aigaki, T., Murakami, R., and Matsuno, K. (2006). An unconventional myosin in *Drosophila* reverses the default handedness in visceral organs. *Nature* **440**, 798–802.

1035. Hrycaj, S., Chesebro, J., and Popadic, A. (2010). Functional analysis of *Scr* during embryonic and post-embryonic development in the cockroach, *Periplaneta americana*. *Dev. Biol.* **341**, 324–334.

1036. Hrycaj, S., Mihajlovic, M., Mahfooz, N., Couso, J.P., and Popadic, A. (2008). RNAi analysis of *nubbin* embryonic functions in a hemimetabolous insect, *Oncopeltus fasciatus*. *Evol. Dev.* **10**, 705–716.

1037. Hsia, C.C. and McGinnis, W. (2003). Evolution of transcription factor function. *Curr. Opin. Gen. Dev.* **13**, 199–206.

1038. Hsia, C.C., Paré, A.C., Hannon, M., Ronshaugen, M., and McGinnis, W. (2010). Silencing of an abdominal *Hox* gene during early development is correlated with limb development in a crustacean trunk. *Evol. Dev.* **12**, 131–143. [*See also* Mallo, M. and Alonso, C.R. (2013). The regulation of *Hox* gene expression during animal development. *Development* **140**, 3951–3963.]

1039. Hu, D. and Marcucio, R.S. (2009). Unique organization of the frontonasal ectodermal zone in birds and mammals. *Dev. Biol.* **325**, 200–210.

1040. Hu, D.L., Nirody, J., Scott, T., and Shelley, M.J. (2009). The mechanics of slithering locomotion. *PNAS* **106** #25, 10081–10085.

1041. Hu, J. and He, L. (2008). Patterning mechanisms controlling digit development. *J. Genet. Genomics* **35**, 517–524.

1042. Huang, R., Zhi, Q., Schmidt, C., Wilting, J., Brand-Saberi, B., and Christ, B. (2000). Sclerotomal origin of the ribs. *Development* **127**, 527–532.

1043. Hubbard, J.K., Uy, J.A.C., Hauber, M.E., Hoekstra, H.E., and Safran, R.J. (2010). Vertebrate pigmentation: from underlying genes to adaptive function. *Trends Genet.* **26**, 231–239.

1044. Huber, B.A., Sinclair, B.J., and Schmitt, M. (2007). The evolution of asymmetric genitalia in spiders and insects. *Biol. Rev.* **82**, 647–698.

1045. Hueber, S.D., Weiller, G.F., Djordjevic, M.A., and Frickey, T. (2010). Improving Hox protein classification across the major model organisms. *PLoS ONE* **5** #5, e10820.

1046. Huffard, C.L., Boneka, F., and Full, R.J. (2005). Underwater bipedal locomotion by octopuses in disguise. *Science* **307**, 1927.

1047. Hughes, A.L. and Friedman, R. (2005). Loss of ancestral genes in the genomic evolution of *Ciona intestinalis*. *Evol. Dev.* **7**, 196–200.

1048. Hughes, C.L. and Kaufman, T.C. (2002). Exploring the myriapod body plan: expression patterns of the ten *Hox* genes in a centipede. *Development* **129**, 1225–1238.

1049. Hughes, C.L. and Kaufman, T.C. (2002). *Hox* genes and the evolution of the arthropod body plan. *Evol. Dev.* **4**, 459–499.

1050. Hughes, M.W., Wu, P., Jiang, T.-X., Lin, S.-J., Dong, C.-Y., Li, A., Hsieh, F.-J., Widelitz, R.B., and Chuong, C.M. (2011). In search of the Golden Fleece: unraveling principles of morphogenesis by studying the integrative biology of skin appendages. *Integr. Biol.* **3**, 388–407.

1051. Hunt, P. and Krumlauf, R. (1992). Hox codes and positional specification in vertebrate embryonic axes. *Annu. Rev. Cell Biol.* **8**, 227–256.

1052. Hunter, C.M., Caswell, H., Runge, M.C., Regehr, E.V., Amstrup, S.C., and Stirling, I. (2010). Climate change threatens polar bear populations: a stochastic demographic analysis. *Ecology* **91**, 2883–2897.

1053. Hunter, L. (2011). *Carnivores of the World*. Princeton University Press, Princeton, NJ.

1054. Hurle, J.M. and Fernandezteran, M.A. (1984). Fine-structure of the interdigital membranes during the morphogenesis of the digits of the webbed foot of the duck embryo. *J. Morph.* **79**, 201–210.

1055. Huszar, D., Sharpe, A., Hashmi, S., Bouchard, B., Houghton, A., and Jaenisch, R. (1991). Generation of pigmented stripes in albino mice by retroviral marking of neural crest melanoblasts. *Development* **113**, 653–660.

1056. Hutchinson, J.R. and Allen, V. (2009). The evolutionary continuum of limb function from early theropods to birds. *Naturwissenschaften* **96**, 423–448.

1057. Hutchinson, J.R., Bates, K.T., Molnar, J., Allen, V., and Makovicky, P.J. (2011). A computational analysis of limb and body dimensions in *Tyrannosaurus rex* with implications for locomotion, ontogeny, and growth. *PLoS ONE* **6** #10, e26037.

1058. Hutchinson, J.R., Delmer, C., Miller, C.E., Hildebrandt, T., Pitsillides, A.A., and Boyde, A. (2011). From flat foot to fat foot: structure, ontogeny, function, and evolution of elephant "sixth toes". *Science* **334**, 1699–1703.

1059. Hutchison, V.H. (2008). Amphibians: lungs' lift lost. *Curr. Biol.* **18**, R392–R393.

1060. Ide, H. (2012). Bone pattern formation in mouse limbs after amputation at the forearm level. *Dev. Dynamics* **241**, 435–441.

1061. Inaba, M., Yamanaka, H., and Kondo, S. (2012). Pigment pattern formation by contact-dependent depolarization. *Science* **335**, 677.

1062. Indrebo, A., Langeland, M., Juul, H.M., Skogmo, H.K., Rengmark, A.H., and Lingaas, F. (2008). A study of inherited short tail and taillessness in Pembroke Welsh corgi. *J. Small Anim. Pract.* **49**, 220–224.

1063. Infante, C.R., Park, S., Mihala, A.G., Kingsley, D.M., and Menke, D.B. (2013). Pitx broadly associates with limb enhancers and is enriched on hindlimb *cis*-regulatory elements. *Dev. Biol.* **374**, 234–244.

1064. Ingber, D.E. and Levin, M. (2007). What lies at the interface of regenerative medicine and developmental biology? *Development* **134**, 2541–2547.

1065. Ingham, P.W., Nakano, Y., and Seger, C. (2011). Mechanisms and functions of Hedgehog signalling across the metazoa. *Nature Rev. Gen.* **12**, 393–406.

1066. Innan, H. and Kondrashov, F. (2010). The evolution of gene duplications: classifying and distinguishing between models. *Nature Rev. Genet.* **11**, 97–108.

1067. Irish, F.J. (1989). The role of heterochrony in the origin of a novel bauplan: evolution of the ophidian skull. *Geobios* **22** (Suppl. 2, Ontogenèse et Evolution), 227–233.

1068. Irish, V.F., Martinez-Arias, A., and Akam, M. (1989). Spatial regulation of the *Antennapedia* and *Ultrabithorax* homeotic genes during *Drosophila* early development. *EMBO J.* **8**, 1527–1537.

1069. Iulianella, A., Melton, K.R., and Trainor, P.A. (2003). Somitogenesis: breaking new boundaries. *Neuron* **40**, 11–14.

1070. Iwasato, T., Katoh, H., Nishimaru, H., Ishikawa, Y., Inoue, H., Saito, Y.M., Ando, R., Iwama, M., Takahashi, R., Negishi, M., and Itohara, S. (2007). Rac-GAP a-chimerin regulates motor-circuit formation as a key mediator of ephrinB3/EphA4 forward signaling.

Cell **130**, 742–753. [*See also* Talpalar, A.E., *et al.* (2013). Dual-mode operation of neuronal networks involved in left-right alternation. *Nature* **500**, 85–88.]

1071. Iwashita, M., Watanabe, M., Ishii, M., Chen, T., Johnson, S.L., Kurachi, Y., Okada, N., and Kondo, S. (2006). Pigment pattern in *jaguar/obelix* zebrafish is caused by a Kir7.1 mutation: implications for the regulation of melanosome movement. *PLoS Genet.* **2** #11, e197.

1072. Jablonski, N.G. (2006). *Skin: A Natural History*. University of California Press, Berkeley, CA.

1073. Jackman, W.R., Davies, S.H., Lyons, D.B., Stauder, C.K., Denton-Schneider, B.R., Jowdry, A., Aigler, S.R., Vogel, S.A., and Stock, D.W. (2013). Manipulation of Fgf and Bmp signaling in teleost fishes suggests potential pathways for the evolutionary origin of multicuspid teeth. *Evol. Dev.* **15**, 107–118.

1074. Jackson, D.J., Meyer, N.P., Seaver, E., Pang, K., McDougall, C., Moy, V.N., Gordon, K., Degnan, B.M., Martindale, M.Q., Burke, R.D., and Peterson, K.J. (2010). Developmental expression of *COE* across the Metazoa supports a conserved role in neuronal cell-type specification and mesodermal development. *Dev. Genes Evol.* **220**, 221–234.

1075. Jackson, K. (2003). The evolution of venom-delivery systems in snakes. *Zool. J. Linnean Soc.* **137**, 337–354.

1076. Jackson, K. (2007). The evolution of venom-conducting fangs: insights from developmental biology. *Toxicon* **49**, 975–981.

1077. Jacob, F. (1977). Evolution and tinkering. *Science* **196**, 1161–1166.

1078. Jacob, F. (1982). *The Possible and the Actual*. University of Washington Press, Seattle, WA.

1079. Jacobs, D.K., Hughes, N.C., Fitz-Gibbon, S.T., and Winchell, C.J. (2005). Terminal addition, the Cambrian radiation and the Phanerozoic evolution of bilaterian form. *Evol. Dev.* **7**, 498–514.

1080. Jaeger, J., Manu, and Reinitz, J. (2012). *Drosophila* blastoderm patterning. *Curr. Opin. Gen. Dev.* **22**, 533–541.

1081. Jain, A.K., Prabhakar, S., and Pankanti, S. (2002). On the similarity of identical twin fingerprints. *Patt. Recog.* **35**, 2653–2663.

1082. Jameson, N.M., Xu, K., Yi, S.V., and Wildman, D.E. (2012). Development and annotation of shotgun sequence libraries from New World monkeys. *Mol. Ecol. Resources* **12**, 950–955.

1083. Janda, C.Y., Waghray, D., Levin, A.M., Thomas, C., and García, K.C. (2012). Structural basis of Wnt recognition by Frizzled. *Science* **337**, 59–64.

1084. Jane, S.M., Ting, S.B., and Cunningham, J.M. (2005). Epidermal impermeable barriers in mouse and fly. *Curr. Opin. Gen. Dev.* **15**, 447–453.

1085. Janecka, J.E., Helgen, K.M., Lim, N.T.-L., Baba, M., Izawa, M., Boeadi, N.i., and Murphy, W.J. (2009). Evidence for multiple species of Sunda colugo. *Curr. Biol.* **18**, R1001–R1002.

1086. Janis, C. (1994). The sabertooth's repeat performances. *Nat. Hist.* **103** #4, 78–83.

1087. Janssen, J.M., Monteiro, A., and Brakefield, P.M. (2001). Correlations between scale structure and pigmentation in butterfly wings. *Evol. Dev.* **3**, 415–423.

1088. Janssen, R., Eriksson, B.J., Budd, G.E., Akam, M., and Prpic, N.-M. (2010). Gene expression patterns in an onychophoran reveal that regionalization predates limb segmentation in pan-arthropods. *Evol. Dev.* **12**, 363–372.

1089. Janvier, P. (2008). Squint of the fossil flatfish. *Nature* **454**, 169–170.

1090. Jarman, A.P. (2000). Developmental genetics: vertebrates and insects see eye to eye. *Curr. Biol.* **10**, R857–R859.

1091. Jeffery, J.E., Bininda-Emonds, O.R.P., Coates, M.I., and Richardson, M.K. (2002). Analyzing evolutionary patterns in amniote embryonic development. *Evol. Dev.* **4**, 292–302.

1092. Jeffery, W.R. (2009). Regressive evolution in *Astyanax* cavefish. *Annu. Rev. Genet.* **43**, 25–47.

1093. Jegalian, B.G. and De Robertis, E.M. (1992). Homeotic transformation in the mouse induced by overexpression of a human *Hox3.3* transgene. *Cell* **71**, 901–910.

1094. Jegalian, B.G., Miller, R.W., Wright, C.V.E., Blum, M., and De Robertis, E.M. (1992). A *Hox 3.3*-lacZ transgene expressed in developing limbs. *Mechs. Dev.* **39**, 171–180.

1095. Jenkins, F.A., Jr. and Walsh, D.M. (1993). An early Jurassic caecilian with limbs. *Nature* **365**, 246–250.

1096. Jenner, R.A. (2000). Evolution of animal body plans: the role of metazoan phylogeny at the interface between pattern and process. *Evol. Dev.* **2**, 208–221.

1097. Jenner, R.A. (2006). Unburdening evo-devo: ancestral attractions, model organisms, and basal baloney. *Dev. Genes Evol.* **216**, 385–394.

1098. Jernvall, J. (2000). Linking development with generation of novelty in mammalian teeth. *PNAS* **97** #6, 2641–2645.

1099. Jernvall, J. and Thesleff, I. (2012). Tooth shape formation and tooth renewal: evolving with the same signals. *Development* **139**, 3487–3497.

1100. Jheon, A.H. and Schneider, R.A. (2009). The cells that fill the bill: neural crest and the evolution of craniofacial development. *J. Dent. Res.* **88**, 12–21.

1101. Ji, C., Wu, L., Zhao, W., Wang, S., and Lv, J. (2012). Echinoderms have bilateral tendencies. *PLoS ONE* **7** #1, e28978.

1102. Ji, Q., Ji, S.-A., Cheng, Y.-N., You, H.-L., Lü, J.-C., Liu, Y.-Q., and Yuan, C.-X. (2004). Pterosaur egg with a leathery shell. *Nature* **432**, 572.

1103. Jiang, J. and Chi-chung, H. (2008). Hedgehog signaling in development and cancer. *Dev. Cell* **15**, 801–812.

1104. Jiménez-Guri, E., Philippe, H., Okamura, B., and Holland, P.W.H. (2007). *Buddenbrockia* is a cnidarian worm. *Science* **317**, 116–118.

1105. Jockusch, E.L., Nulsen, C., Newfeld, S.J., and Nagy, L.M. (2000). Leg development in flies versus grasshoppers: differences in *dpp* expression do not lead to differences in the expression of downstream components of the leg patterning pathway. *Development* **127**, 1617–1626.

1106. Jockusch, E.L. and Ober, K.A. (2004). Hypothesis testing in evolutionary developmental biology: a case study from insect wings. *J. Hered.* **95**, 382–396.

1107. Jockusch, E.L., Williams, T.A., and Nagy, L. (2004). The evolution of patterning of serially homologous appendages in insects. *Dev. Genes Evol.* **214**, 324–338.

1108. Johanson, Z., Joss, J., Boisvert, C.A., Ericsson, R., Sutija, M., and Ahlberg, P.E. (2007). Fish fingers: Digit homologues in Sarcopterygian fish fins. *J. Exp. Zool. (Mol. Dev. Evol.)* **308B**, 757–768.

1109. Jones, F.C., Grabherr, M.G., Chan, Y.F., Russell, P., Mauceli, E., Johnson, J., Swofford, R., Pirun, M., Zody, M.C., White, S., Birney, E., Searle, S., Schmutz, J., Grimwood, J., Dickson, M.C., Myers, R.M., Miller, C.T., Summers, B.R., Knecht, A.K., Brady, S.D., Zhang, H., Pollen, A.A., Howes, T., Amemiya, C., Broad Institutre Genome Sequencing Platform & Whole Genome Assembly Team, Lander, E.S., Di Palma, F., Lindblad-Toh, K., and Kingsley, D.M. (2012). The genomic basis of adaptive evolution in threespine sticklebacks. *Nature* **484**, 55–61.

1110. Jones, G. (2010). Molecular evolution: gene convergence in echolocating mammals. *Curr. Biol.* **20**, R62–R64. [*See also* Parker, J., *et al.* (2013). Genome-wide signatures of convergent evolution in echolocating mammals. *Nature* **502**, 228–231.]

1111. Jordan, B., Vercammen, P., and Cooper, K.L. (2011). Husbandry and breeding of the lesser Egyptian jerboa, *Jaculus jaculus*. *Cold Spr. Harb. Protoc.* **2011** #12, 1457–1461.

1112. Jordan, S.A. and Jackson, I.J. (2000). MGF (KIT ligand) is a chemokinetic factor for melanoblast migration into hair follicles. *Dev. Biol.* **225**, 424–436.

1113. Josef, N., Amodio, P., Fiorito, G., and Shashar, N. (2012). Camouflaging in a complex environment – octopuses use specific features of their surroundings for background matching. *PLoS ONE* **7** #5, e37579. [*See also* Courage, K.H. (2013). *Octopus*! Penguin, New York, NY.]

1114. Joshi, M., Buchanan, K.T., Shroff, S., and Orenic, T.V. (2006). Delta and Hairy establish a periodic prepattern that positions sensory bristles in *Drosophila* legs. *Dev. Biol.* **293**, 64–76.

1115. Kaelin, C.B., Xu, X., Hong, L.Z., David, V.A., McGowan, K.A., Schmitdt-Küntzel, A., Roelke, M.E., Pino, J., Pontius, J., Cooper, G.M., Manuel, H., Swanson, W.F., Marker, L., Harper, C.K., van Dyk, A., Yue, B., Mullikin, J.C., Warren, W.C., Eizirik, E., Kos, L., O'Brien, S.J., Barsh, G.S., and Menotti-Raymond, M. (2012). Specifiying and sustaining pigmentation patterns in domestic and wild cats. *Science* **337**, 1536–1541.

1116. Kajiura, S.M. and Holland, K.N. (2002). Electroreception in juvenile scalloped hammerhead and sandbar sharks. *J. Exp. Biol.* **205**, 3609–3621.

1117. Kalay, G. and Wittkopp, P.J. (2010). Nomadic enhancers: Tissue-specific *cis*-reguatory elements of *yellow* have divergent genomic positions among *Drosophila* species. *PLoS Genet.* **6** #11, e1001222.

1118. Kalinka, A.T., Varga, K.M., Gerrard, D.T., Preibisch, S., Corcoran, D.L., Jarrells, J., Ohler, U., Bergman, C.M., and Tomancak, P. (2010). Gene expression divergence recapitulates the developmental hourglass model. *Nature* **468**, 811–818.

1119. Kallman, K.D. and Kazianis, S. (2006). The genus *Xiphophorus* in Mexico and Central America. *Zebrafish* **3**, 271–285.

1120. Kang, K., Pulver, S.R., Panzano, V.C., Chang, E.C., Griffith, L.C., Theobald, D.L., and Garrity, P.A. (2010). Analysis of *Drosophila* TRPA1 reveals an ancient origin for human chemical nociception. *Nature* **464**, 597–600.

1121. Kankel, D.R., Ferrús, A., Garen, S.H., Harte, P.J., and Lewis, P.E. (1980). The structure and development of the nervous system. In *The Genetics and Biology of Drosophila*, Vol. 2d (M. Ashburner and T.R.F. Wright, eds.). Academic Press, New York, NY, pp. 295–368.

1122. Kardong, K.V. (1979). "Protovipers" and the evolution of snake fangs. *Evolution* **33**, 433–443.

1123. Kardong, K.V. (2005). *An Introduction to Biological Evolution*. McGraw-Hill, New York, NY.

1124. Karleskint, G., Jr., Turner, R., and Small, J.W., Jr. (1999). *Introduction to Marine Biology*, 3rd edn. Brooks/Cole, Belmont, CA. [*See also* Loxton, D. (2013). Mermaids. *Skeptic* **18** #3, 65–71.]

1125. Kauffman, S.A. (1983). Developmental constraints: internal factors in evolution. In *Development and Evolution, Symp. Brit. Soc. Dev. Biol., Vol. 6* (B.C. Goodwin, N. Holder, and C.C. Wylie, eds.). Cambridge University Press, Cambridge, pp. 195–225.

1126. Kauffman, S.A. (1993). *The Origins of Order: Self-Organization and Selection in Evolution*. Oxford University Press, Oxford.

1127. Kavanagh, K.D., Evans, A.R., and Jernvall, J. (2007). Predicting evolutionary patterns of mammalian teeth from development. *Nature* **449**, 427–432.

1128. Kawasaki, K. and Weiss, K.M. (2006). Evolutionary genetics of vertebrate tissue mineralization: the origin and evolution of the secretory calcium-binding phosphoprotein family. *J. Exp. Zool. (Mol. Dev. Evol.)* **306B**, 295–316.

1129. Kay, E.H. and Hoekstra, H.E. (2008). Rodents. *Curr. Biol.* **18**, R406–R410.

1130. Kearney, M. and Stuart, B.L. (2004). Repeated evolution of limblessness and digging heads in worm lizards revealed by DNA from old bones. *Proc. R. Soc. Lond. B* **271**, 1677–1683.

1131. Kearny, M. and Stuart, B.L. (2004). Repeated evolution of limblessness and digging heads in worm lizards revealed by DNA from old bones. *Proc. R. Soc. Lond. B* **271**, 1677–1683.

1132. Keays, D.A., Tian, G., Poirier, K., Huang, G.-J., Siebold, C., Cleak, J., Oliver, P.L., Fray, M., Harvey, R.J., Molnár, Z., Piñon, M.C., Dear, N., Valdar, W., Brown, S.D.M., Davies, K.E., Rawlins, J.N.P., Cowan, N.J., Nolan, P., Chelly, J., and Flint, J. (2007). Mutations in α-tubulin cause abnormal neuronal migration in mice and lissencephaly in humans. *Cell* **128**, 45–57.

1133. Keddy-Hector, A.C. (1992). Mate choice in non-human primates. *Am. Zool.* **32**, 62–70.

1134. Keil, T.A. and Steinbrecht, R.A. (1984). Mechanosensitive and olfactory sensilla of insects. In *Insect Ultrastructure*, Vol. 2 (R.C. King and H. Akai, eds.). Plenum, New York, NY, pp. 477–516.

1135. Kelley, J.L., Fitzpatrick, J.L., and Merilaita, S. (2013). Spots and stripes: ecology and colour pattern evolution in butterflyfishes. *Proc. R. Soc. Lond. B* **280**, in press. doi 20122730.

1136. Kellogg, V.L. and Bell, R.G. (1904). Studies of variation in insects. *Proc. Wash. Acad. Sci.* **6**, 203–332.

1137. Kelsh, R.N. and Barsh, G.S. (2011). A nervous origin for fish stripes. *PLoS Genet.* **7** #5, e1002081.

1138. Kelsh, R.N., Harris, M.L., Colanesi, S., and Erickson, C.A. (2009). Stripes and belly-spots: a review of pigment cell morphogenesis in vertebrates. *Semin. Cell Dev. Biol.* **20**, 90–104.

1139. Kenward, B., Wachtmeister, C.-A., Ghirlanda, S., and Enquist, M. (2004). Spots and stripes: the evolution of repetition in visual signal form. *J. Theor. Biol.* **230**, 407–419.

1140. Keogh, J.S., Scott, I.A.W., and Hayes, C. (2005). Rapid and repeated origin of insular gigantism and dwarfism in Australian tiger snakes. *Evolution* **59**, 226–233.

1141. Kerschensteiner, D. (2011). Circuit assembly: the repulsive side of lamination. *Curr. Biol.* **21**, R163–R166.

1142. Keynes, R.J. and Stern, C.D. (1988). Mechanisms of vertebrate segmentation. *Development* **103**, 413–429.

1143. Keys, D.N., Lewis, D.L., Selegue, J.E., Pearson, B.J., Goodrich, L.V., Johnson, R.L., Gates, J., Scott, M.P., and Carroll, S.B. (1999). Recruitment of a *hedgehog* regulatory circuit in butterfly eyespot evolution. *Science* **283**, 532–534.

1144. Khadjeh, S., Turetzek, N., Pechmann, M., Schwager, E.E., Wimmer, E.A., Damen, W.G.M., and Prpic, N.-M. (2012). Divergent role of the Hox gene *Antennapedia* in spiders is responsible for the convergent evolution of abdominal limb repression. *PNAS* **109** #13, 4921–4926.

1145. Khila, A., Abouheif, E., and Rowe, L. (2012). Function, developmental genetics, and fitness consequences of a sexually antagonistic trait. *Science* **336**, 585–589.

1146. Kicheva, A., Bollenbach, T., Wartlick, O., Jülicher, F., and Gonzalez-Gaitan, M. (2012). Investigating the principles of morphogen gradient formation: from tissues to cells. *Curr. Opin. Gen. Dev.* **22**, 527–532.

1147. Kiefer, J.C. (2006). Emerging developmental model systems. *Dev. Dynamics* **235**, 2895–2899.

1148. Kijimoto, T., Andrews, J., and Moczek, A.P. (2010). Programmed cell death shapes the expression of horns within and between species of horned beetles. *Evol. Dev.*12, 449–458.

1149. Kim, J., Sebring, A., Esch, J.J., Kraus, M.E., Vorwerk, K., Magee, J., and Carroll, S.B. (1996). Integration of positional signals and regulation of wing formation and identity by *Drosophila vestigial* gene. *Nature* **382**, 133–138.

1150. Kim, J.S., Jin, D.I., Lee, J.H., Son, D.S., Lee, S.H., Yi, Y.J., and Park, C.S. (2005). Effects of teat number on litter size in gilts. *Anim. Reprod. Sci.* **90**, 111–116.

1151. Kim, S.Y., Paylor, S.W., Magnuson, T., and Schumacher, A. (2007). Juxtaposed Polycomb complexes co-regulate vertebral identity. *Development* **133**, 4957–4968.

1152. Kimm, M.A. and Prpic, N.-M. (2006). Formation of the arthropod labrum by fusion of paired and rotated limb-bud-like primordia. *Zoomorphology* **125**, 147–155.

1153. King, N. and Carroll, S.B. (2001). A receptor tyrosine kinase from choanoflagellates: molecular insights into early animal evolution. *PNAS* **98** #26, 15032–15037.

1154. King, N., Hittinger, C.T., and Carroll, S.B. (2003). Evolution of key cell signaling and adhesion protein families predates animal origins. *Science* **301**, 361–363.

1155. Kingdon, J. (1977). *East African Mammals: An Atlas of Evolution in Africa: Carnivores*, Vol. IIIA. University of Chicago Press, Chicago, IL.

1156. Kingsley, M.C.S. and Ramsay, M.A. (1988). The spiral in the tusk of the narwhal. *Artic* **41**, 236–238.

1157. Kingsolver, J.G. and Koehl, M.A.R. (1985). Aerodynamics, thermoregulation, and the evolution of insect wings: differential scaling and evolutionary change. *Evolution* **39**, 488–504.

1158. Kingsolver, J.G. and Koehl, M.A.R. (1994). Selective factors in the evolution of insect wings. *Annu. Rev. Entomol.* **39**, 425–451.

1159. Kirschner, M. and Gerhart, J. (1998). Evolvability. *PNAS* **95** #15, 8420–8427.

1160. Kirschner, M.W. and Gerhart, J.C. (2005). *The Plausibility of Life: Resolving Darwin's Dilemma*. Yale University Press, New Haven, CT.

1161. Kitchener, A.C., Beaumont, M.A., and Richardson, D. (2006). Geographical variation in the clouded leopard, *Neofelis nebulosa*, reveals two species. *Curr. Biol.* **16**, 2377–2383.

1162. Klauber, L.M. (1997). *Rattlesnakes: Their Habits, Life Histories, and Infuence on Mankind*. 2nd edn., Vol. 2. University of California Press, Berkeley, CA.

1163. Klauer, G., Burda, H., and Nevo, E. (1997). Adaptive differentiations of the skin of the head in a subterranean rodent, *Spalax ehrenbergi*. *J. Morph.* **233**, 53–66.

1164. Klein, O.D., Lyons, D.B., Balooch, G., Marshall, G.W., Basson, M.A., Peterka, M., Boran, T., Peterkova, R., and Martin, G.R. (2008). An FGF signaling loop sustains the generation of differentiated progeny from stem cells in mouse incisors. *Development* **135**, 377–385.

1165. Klein, O.D., Minowada, G., Peterkova, R., Kangas, A., Yu, B.D., Lesot, H., Peterka, M., Jernvall, J., and Martin, G. (2006). Sprouty genes control diastema tooth development via bidirectional antagonism of epithelial-mesenchymal FGF signaling. *Dev. Cell* **11**, 181–190.

1166. Klein, R. (2012). Eph/ephrin signalling during development. *Development* **139**, 4105–4109.

1167. Kley, N.J. and Brainerd, E.L. (1999). Feeding by mandibular raking in a snake. *Nature* **402**, 369–370.

1168. Kley, N.J. and Kearney, M. (2007). Adaptations for digging and burrowing. *In Fins into Limbs: Evolution, Development, and Transformation* (B.K. Hall, ed.). University of Chicago Press, Chicago, IL, pp. 284–309.

1169. Klingenberg, C.P. (1998). Heterochrony and allometry: the analysis of evolutionary change in ontogeny. *Biol. Rev.* **73**, 79–123.

1170. Klingenberg, C.P. (2005). Developmental constraints, modules, and evolvability. In *Variation: A Central Concept in Biology* (B. Hallgrímsson and B.K. Hall, eds.). Elsevier Academic Press, New York, NY, pp. 219–247.

1171. Kodandaramaiah, U. (2009). Eyespot evolution: phylogenetic insights from *Junonia* and related butterfly genera (Nymphalidae: Junoniini). *Evol. Dev.* **11**, 489–497.

1172. Kodandaramaiah, U. (2009). Fixed eyespot display in a butterfly thwarts attacking birds. *Anim. Behav.* **77**, 1415–1419.

1173. Kodandaramaiah, U. (2011). The evolutionary significance of butterfly eyespots. *Behav. Ecol.* **22**, 1264–1271.

1174. Koestler, A. (1969). Beyond atomism and holism: the concept of the holon. In *Beyond Reductionism. New Perspectives in the Life Sciences* (A. Koestler and J.R. Smythies, eds.). Macmillan, New York, NY, pp. 192–227.

1175. Koganezawa, M., Haba, D., Matsuo, T., and Yamamoto, D. (2010). The shaping of male courtship posture by lateralized gustatory inputs to male-specific interneurons. *Curr. Biol.* **20**, 1–8.

1176. Koh, T.-W. and Carlson, J.R. (2011). Chemoreception: Identifying friends and foes. *Curr. Biol.* **21**, R998–R999.

1177. Kohlsdorf, T., Cummings, M.P., Lynch, V.J., Stopper, G.F., Takahashi, K., and Wagner, G.P. (2008). A molecular footprint of limb loss: sequence variation of the autopodial identity gene *Hoxa-13*. *J. Mol. Evol.* **67**, 581–593.

1178. Kojima, T. (2004). The mechanism of *Drosophila* leg development along the proximodistal axis. *Develop. Growth Differ.* **46**, 115–129.

1179. Kojima, T., Sato, M., and Saigo, K. (2000). Formation and specification of distal leg segments in *Drosophila* by dual *Bar* homeobox genes, *BarH1*and *BarH2*. *Development* **127**, 769–778.

1180. Kolbe, J.J., Leal, M., Schoener, T.W., Spiller, D.A., and Losos, J.B. (2012). Founder effects persist despite adaptive differentiation: a field experiment with lizards. *Science* **335**, 1086–1089.

1181. Kollar, E.J. and Fisher, C. (1980). Tooth induction in chick epithelium: expression of quiescent genes for enamel synthesis. *Science* **207**, 993–995.

1182. Kondo, S. (2002). The reaction-diffusion system: a mechanism for autonomous pattern formation in the animal skin. *Genes to Cells* **7**, 535–541.

1183. Kondo, S. (2005). Cell-cell interaction network that generates the skin pattern of animal. *Genome Informatics* **16**, 287–291.

1184. Kondo, S. and Asai, R. (1995). A reaction-diffusion wave on the skin of the marine angelfish *Pomacanthus*. *Nature* **376**, 765–768.

1185. Kondo, S., Iwashita, M., and Yamaguchi, M. (2009). How animals get their skin patterns: fish pigment pattern as a live Turing wave. *Int. J. Dev. Biol.* **53**, 851–856.

1186. Kondo, S. and Miura, T. (2010). Reaction-diffusion model as a framework for understanding biological pattern formation. *Science* **329**, 1616–1620.

1187. Kondo, S. and Shirota, H. (2009). Theoretical analysis of mechanisms that generate the pigmentation pattern of animals. *Semin. Cell Dev. Biol.* **20**, 82–89.

1188. Koonin, E.V. (2005). Orthologs, paralogs, and evolutionary genomics. *Annu. Rev. Genet.* **39**, 309–338.

1189. Koop, D., Holland, N.D., Sémon, M., Alvarez, S., Rodriguez de Lera, A., Laudet, V., Holland, L.Z., and Schubert, M. (2010). Retinoic acid signaling targets *Hox* genes during the amphioxus gastrula stage: Insights into early anterior–posterior patterning of the chordate body plan. *Dev. Biol.* **338**, 98–106.

1190. Kopp, A. (2009). Metamodels and phylogenetic replication: a systematic approach to the evolution of developmental pathways. *Evolution* **63**, 2771–2789.

1191. Kopp, A. (2011). *Drosophila* sex combs as a model of evolutionary innovations. *Evol. Dev.* **13**, 504–522.

1192. Kopp, A. (2012). *Dmrt* genes in the development and evolution of sexual dimorphism. *Trends Genet.* **28**, 175–184.

1193. Kopp, A. and True, J.R. (2002). Evolution of male sexual characters in the Oriental *Drosophila melanogaster* species group. *Evol. Dev.* **4**, 278–291.

1194. Kornberg, T.B. and Guha, A. (2007). Understanding morphogen gradients: a problem of dispersion and containment. *Curr. Opin. Gen. Dev.* **17**, 264–271.

1195. Kotiaho, J.S. (2001). Costs of sexual traits: a mismatch between theoretical and empirical evidence. *Biol. Rev.* **76**, 365–376.

1196. Koyanagi, M., Kubokawa, K., Tsukamoto, H., Shichida, Y., and Terakita, A. (2005). Cephalochordate melanopsin: evolutionary linkage between invertebrate visual cells and vertebrate photosensitive retinal ganglion cells. *Curr. Biol.* **15**, 1065–1069.

1197. Kozmikova, I., Smolikova, J., Vlcek, C., and Kozmik, Z. (2011). Conservation and diversification of an ancestral chordate gene regulatory network for dorsoventral patterning. *PLoS One* **6** #2, e14650.

1198. Kozopas, K.M. and Nusse, R. (2002). Direct flight muscles in *Drosophila* develop from cells with characteristics of founders and depend on *DWnt-2*for their correct patterning. *Dev. Biol.* **243**, 312–325.

1199. Krajick, K. (2007). Discoveries in the dark. *Natl. Geogr.* **212** #3, 134–147.

1200. Krauss, G. (2008). *Biochemistry of Signal Transduction and Regulation*, 4th edn. Wiley-VCH, Weinheim.

1201. Krochmal, A.R., Bakken, G.S., and LaDuc, T.J. (2004). Heat in evolution's kitchen: evolutionary perspectives on the functions and origin of the facial pit of pitvipers (Viperidae: Crotalinae). *J. Exp. Biol.* **207**, 4231–4238.

1202. Kröger, B., Vinther, J., and Fuchs, D. (2011). Cephalopod origin and evolution: a congruent picture emerging from fossils, development and molecules. *BioEssays* **33**, 602–613.

1203. Krol, A.J., Roellig, D., Dequéant, M.-L., Tassy, O., Glynn, E., Hattem, G., Mushegian, A., Oates, A.C., and Pourquié, O. (2011). Evolutionary plasticity of segmentation clock networks. *Development* **138**, 2783–2792.

1204. Kronforst, M.R., Barsh, G.S., Kopp, A., Mallet, J., Monteiro, A., Mullen, S.P., Protas, M., Rosenblum, E.B., Schneider, C.J., and Hoekstra, H.E. (2012). Unraveling the thread of nature's tapestry: the genetics of diversity and convergence in animal pigmentation. *Pigment Cell Melanoma Res.* **25**, 411–433.

1205. Krumlauf, R. (1994). *Hox* genes in vertebrate development. *Cell* **78**, 191–201.

1206. Kuch, U., Müller, J., Mödden, C., and Mebs, D. (2006). Snake fangs from the Lower Miocene of Germany: evolutionary stability of perfect weapons. *Naturwissenschaften* **93**, 84–87.

1207. Kühn, A. (1971). *Lectures on Developmental Physiology*, 2nd edn. Springer-Verlag, Berlin.

1208. Kukalová-Peck, J. (1978). Origin and evolution of insect wings and their relation to metamorphosis, as documented by the fossil record. *J. Morph.* **156**, 53–125.

1209. Kukalová-Peck, J. (1985). Ephemeroid wing venation based upon new gigantic Carboniferous mayflies and basic morphology, phylogeny, and metamorphosis of pterygote insects (Insecta, Ephemerida). *Can. J. Zool.* **63**, 933–955.

1210. Kukalová-Peck, J. (1987). New Carboniferous Diplura, Monura, and Thysanura, the hexapod ground plan, and the role of thoracic side lobes in the origin of wings (Insecta). *Can. J. Zool.* **65**, 2327–2345.

1211. Kukalová-Peck, J. (2008). Phylogeny of higher taxa in insecta: finding synapomorphies in the extant fauna and separating them from homoplasies. *Evol. Biol.* **35**, 4–51.

1212. Kulesa, P.M. and Fraser, S.E. (2002). Cell dynamics during somite boundary formation revealed by time-lapse analysis. *Science* **298**, 991–995.

1213. Kundrát, M. (2009). Heterochronic shift between early organogenesis and migration of cephalic neural crest cells in two divergent evolutionary phenotypes of archosaurs: crocodile and ostrich. *Evol. Dev.* **11**, 535–546.

1214. Kunhardt, P.B., Jr., Kunhardt, P.B., III, and Kunhardt, P.W. (1995). *P. T. Barnum: America's Greatest Showman*. Knopf, New York, NY.

1215. Kuo, D.-H. and Weisblat, D.A. (2011). A new molecular logic for BMP-mediated dorsoventral patterning in the leech *Helobdella*. *Curr. Biol.* **21**, 1282–1288.

1216. Kuranaga, E., Matsunuma, T., Kanuka, H., Takemoto, K., Koto, A., Kimura, K.-i., and Miura, M. (2011). Apoptosis controls the speed of looping morphogenesis in *Drosophila* male terminalia. *Development* **138**, 1493–1499.

1217. Kuratani, S. (2012). Evolution of the vertebrate jaw from developmental perspectives. *Evol. Dev.* **14**, 76–92.

1218. Kuratani, S., Kuraku, S., and Nagashima, H. (2011). Evolutionary developmental perspective for the origin of turtles: the folding theory for the shell based on the developmental nature of the carapacial ridge. *Evol. Dev.* **13**, 1–14.

1219. Kurshan, P.T. and Shen, K. (2012). Dendritic patterning: three-dimensional position determines dendritic avoidance capability. *Curr. Biol.* **22**, R192–R194.

1220. Kusumi, K., May, C.M., and Eckalbar, W.L. (2013). A large-scale view of the evolution of amniote development: insights from somitogenesis in reptiles. *Curr. Opin. Gen. Dev.* **23**, 491–497.

1221. Kuzniar, A., van Ham, R.C.H.J., Pongor, S., and Leunissen, J.A.M. (2008). The quest for orthologs: finding the corresponding gene across genomes. *Trends Genet.* **24**, 539–551.

1222. Kwon, C., Hays, R., Fetting, J., and Orenic, T.V. (2004). Opposing inputs by Hedgehog and Brinker define a stripe of *hairy* expression in the *Drosophila* leg imaginal disc. *Development* **131**, 2681–2692.

1223. Lacalli, T. (1996). Dorsoventral axis inversion: a phylogenetic perspective. *BioEssays* **18**, 251–254.

1224. Lacalli, T. (2008). Head organization and the head/trunk relationship in protochordates: problems and prospects. *Integr. Comp. Biol.* **48**, 620–629.

1225. Lacalli, T. (2010). The emergence of the chordate body plan: some puzzles and problems. *Acta Zool. (Stockholm)* **91**, 4–10.

1226. Lacalli, T. (2012). The middle Cambrian fossil *Pikaia* and the evolution of chordate swimming. *EvoDevo* **3**, Article 12 (6 pp.).

1227. Lacalli, T.C. (2005). Protochordate body plan and the evolutionary role of larvae: old controversies resolved? *Can. J. Zool.* **83**, 216–224.

1228. Lacalli, T.C. (2008). Mucus secretion and transport in amphioxus larvae: organization and ultrastructure of the food trapping system, and implications for head evolution. *Acta Zool. (Stockholm)* **89**, 219–230.

1229. Lai, E.C. (2004). Notch signaling: control of cell communication and cell fate. *Development* **131**, 965–973.

1230. Lai, E.C. and Orgogozo, V. (2004). A hidden program in *Drosophila* peripheral neurogenesis revealed: fundamental principles underlying sensory organ diversity. *Dev. Biol.* **269**, 1–17.

1231. Lakoff, G. and Johnson, M. (2003). *Metaphors We Live By*, 2nd edn. University of Chicago Press, Chicago, IL.

1232. Laman, T. (2000). Wild gliders: the creatures of Borneo's rain forest go airborne. *Natl. Geogr.* **198** #4, 68–85.

1233. Lamar, W.W., Carmichael, P., and Shumway, G. (2002). *The World's Most Spectacular Reptiles & Amphibians*. World Publications, Tampa, FL.

1234. Lamb, T.D. (2011). Evolution of the eye. *Sci. Am.* **305** #1, 64–69.

1235. Lanctôt, C., Moreau, A., Chamberland, M., Tremblay, M.L., and Drouin, J. (1999). Hindlimb patterning and mandible development require the *Ptx1* gene. *Development* **126**, 1805–1810.

1236. Land, M.F. (2006). Visual optics: the shapes of pupils. *Curr. Biol.* **16**, R167–R168.

1237. Land, M.F. and Nilsson, D.-E. (2002). *Animal Eyes*. Oxford University Press, New York, NY.

1238. Landberg, T., Mailhot, J.D., and Brainerd, E.L. (2003). Lung ventilation during treadmill locomotion in a terrestrial turtle, *Terrapene carolina*. *J. Exp. Biol.* **206**, 3391–3404.

1239. Lande, R. (1978). Evolutionary mechanisms of limb loss in tetrapods. *Evolution* **32**, 73–92.

1240. Lander, A.D. (2007). Morpheus unbound: Reimagining the morphogen gradient. *Cell* **128**, 245–256.

1241. Lander, A.D. (2011). Pattern, growth, and control. *Cell* **144**, 955–969.

1242. Lander, A.D. (2013). How cells know where they are. *Science* **339**, 923–927.

1243. Lane, N. and Martin, W.F. (2012). The origin of membrane bioenergetics. *Cell* **151**, 1406–1416.

1244. Langdon, J.H. (2005). *The Human Strategy: An Evolutionary Perspective on Human Anatomy*. Oxford University Press, New York, NY.

1245. Langridge, K.V., Broom, M., and Osorio, D. (2007). Selective signalling by cuttlefish to predators. *Curr. Biol.* **17**, R1044–R1045.

1246. Langston, W., Jr. (1981). Pterosaurs. *Sci. Am.* **244** #2, 122–136.

1247. Langton, C.G., ed. *Artificial Life*. Addison-Wesley, New York, NY.

1248. LaPolla, J.S., Dlussky, G.M., and Perrichot, V. (2013). Ants and the fossil record. *Annu. Rev. Entomol.* **58**, 609–630. [*See also* Johnson, B.R., *et al.* (2013). Phylogenomics resolves evolutionary relationships among ants, bees, and wasps. *Curr. Biol.* **23**, 2058–2062.]

1249. Lapraz, F., Besnardeau, L., and Lepage, T. (2009). Patterning of the dorsal–ventral axis in echinoderms: insights into the evolution of the BMP-Chordin signaling network. *PLoS Biol.* **7** #11, e1000248.

1250. Larsen, C., Bardet, P.-L., Vincent, J.-P., and Alexandre, C. (2008). Specification and positioning of parasegment grooves in *Drosophila*. *Dev. Biol.* **321**, 310–318.

1251. Larsen, E. and McLaughlin, H.M.G. (1987). The morphogenetic alphabet: lessons for simple-minded genes. *BioEssays* **7**, 130–132.

1252. Larson, G. (1991). *Unnatural Selections: A Far Side Collection*. Andrews & McMeel (Universal Press Syndicate), Kansas City, MO.

1253. Lau, F.H., Xia, F., Kaplan, A., Cerrato, F., Greene, A.K., Taghinia, A., Cowan, C.A., and Labow, B.I. (2012). Expression analysis of macrodactyly identifies pleiotrophin upregulation. *PLoS ONE* **7** #7, e40423.

1254. Lawrence, P.A. (1966). Development and determination of hairs and bristles in the milkweed bug *Oncopeltus fasciatus* (Lygaeidae, Hemiptera). *J. Cell Sci.* **1**, 475–498.

1255. Lawrence, P.A. (1984). Homoeotic selector genes: a working definition. *BioEssays* **1**, 227–229.

1256. Lawrence, P.A. and Struhl, G. (1982). Further studies of the *engrailed* phenotype in *Drosophila*. *EMBO J.* **1**, 827–833.

1257. Lazzari, V., Charles, C., Tafforeau, P., Vianey-Liaud, M., Aguilar, J.-P., Jaeger, J.-J., Michaux, J., and Viriot, L. (2008). Mosaic convergence of rodent dentitions. *PLoS ONE* **3** #10, e3607.

1258. Le Douarin, N.M., Creuzet, S., Couly, G., and Dupin, E. (2004). Neural crest plasticity and its limits. *Development* **131**, 4637–4650.

1259. Le Gros Clark, W.E. (1945). Deformation patterns in the cerebral cortex. In *Essays on Growth and Form* (W.E. Le Gros Clark and P.B. Medawar, eds.). Clarendon Press, Oxford, pp. 1–22.

1260. Le Rouzic, A., Álvarez-Castro, J.M., and Hansen, T.F. (2013). The evolution of canalization and evolvability in stable and fluctuating environments. *Evol. Biol.* **40**, 317–340.

1261. Lecuit, T. and Cohen, S.M. (1997). Proximal-distal axis formation in the *Drosophila* leg. *Nature* **388**, 139–145.

1262. Lee, H.-G., Kim, Y.-C., Dunning, J.S., and Han, K.-A. (2008). Recurring ethanol exposure induces disinhibited courtship in *Drosophila*. *PLoS ONE* #1, e1391.

1263. Lee, J. and Tumbar, T. (2012). Hairy tale of signaling in hair follicle development and cycling. *Semin. Cell Dev. Biol.* **23**, 906–916.

1264. Lee, M.S.Y. (2009). Hidden support from unpromising data sets strongly unites snakes with anguimorph 'lizards'. *J. Exp. Biol.* **22**, 1308–1316.

1265. Lee, M.S.Y., Bell, G.L., Jr., and Caldwell, M.W. (1999). The origin of snake feeding. *Nature* **400**, 656–659.

1266. Lee, M.S.Y., Jago, J.B., García-Bellido, D.C., Edgecombe, G.D., Gehling, J.G., and Paterson, J.R. (2011). Modern optics in exceptionally preserved eyes of early Cambrian arthropods from Australia. *Nature* **474**, 631–634.

1267. Lee, P.N., Callaerts, P., de Couet, H.G., and Martindale, M.Q. (2003). Cephalopod *Hox* genes and the origin of morphological novelties. *Nature* **424**, 1061–1065. [*See also* Hoving, H.J.T., *et al.* (2013). First *in situ* observations of the deep-sea squid *Grimalditeuthis bonplandi* reveal unique use of tentacles. *Proc.R.Soc.Lond.B* **280**, doi 20131463.]

1268. Lee, R.T.H., Thiery, J.P., and Carney, T.J. (2013). Dermal fin rays and scales derive from mesoderm, not neural crest. *Curr. Biol.* **23**, R336–R337.

1269. Legué, E. and Nicolas, J.-F. (2005). Hair follicle renewal: organization of stem cells in the matrix and the role of stereotyped lineages and behaviors. *Development* **132**, 4143–4154.

1270. Leichty, A.R., Pfennig, D.W., Jones, C.D., and Pfennig, K.S. (2012). Relaxed genetic constraint is ancestral to the evolution of phenotypic plasiticity. *Integr. Comp. Biol.* **52**, 16–30.

1271. Leigh, S.R. (2012). Brain size growth and life history in human evolution. *Evol. Biol.* **39**, 587–599.

1272. Lelli, K.M., Slattery, M., and Mann, R.S. (2012). Disentangling the many layers of eukaryotic transcriptional regulation. *Annu. Rev. Genet.* **46**, 43–68.

1273. Lemaire, P. (2011). Evolutionary crossroads in developmental biology: the tunicates. *Development* **138**, 2143–2152.

1274. Lemaire, P., Smith, W.C., and Nishida, H. (2008). Ascidians and the plasticity of the chordate developmental program. *Curr. Biol.* **18**, R620–R631.

1275. Lemmon, M.A. and Schlessinger, J. (2010). Cell signaling by receptor tyrosine kinases. *Cell* **141**, 1117–1134.

1276. Lemons, D. and McGinnis, W. (2006). Genomic evolution of *Hox* gene clusters. *Science* **313**, 1918–1922.

1277. Lemus, D. (1995). Contributions of heterospecific tissue recombinations to odontogenesis. *Int. J. Dev. Biol.* **39**, 291–297.

1278. Lennox, J.G. (1991). Darwinian thought experiments: a function for just-so stories. In *Thought Experiments in Science and Philosophy* (T. Horowitz and G.J. Massey, eds.). Rowman & Littlefield, Savage, MD, pp. 223–245.

1279. Leonard, J.A., Rohland, N., Glaberman, S., Fleischer, R.C., Caccone, A., and Hofreiter, M. (2005). A rapid loss of stripes: the evolutionary history of the extint quagga. *Biol. Lett.* **1**, 291–295.

1280. Lerner, H.R.L., Meyer, M., James, H.F., Hofreiter, M., and Fleischer, R.C. (2011). Multilocus resolution of phylogeny and timescale in the extant adaptive radiation of Hawaiian honeycreepers. *Curr. Biol.* **21**, 1838–1844.

1281. Lettice, L.A., Hill, A.E., Devenney, P.S., and Hill, R.E. (2008). Point mutations in a distant sonic hedgehog *cis*-regulator generate a variable regulatory output responsible for preaxial polydactyly. *Hum. Mol. Genet.* **17**, 978–985.

1282. Levin, M. (2005). Left-right asymmetry in embryonic development: a comprehensive review. *Mechs. Dev.* **122**, 3–25.

1283. Levine, M. (2002). How insects lose their limbs. *Nature* **415**, 848–849.

1284. Levine, M. (2010). Transcriptional enhancers in animal development and evolution. *Curr. Biol.* **20**, R754–R763.

1285. Levinton, J.S. (1986). Developmental constraints and evolutionary saltations: a discussion and critique. In *Genetics, Development, and Evolution* (J.P. Gustafson, G.L. Stebbins, and F.J. Ayala, eds.). 17th Stadler Genetics Symposium. Plenum, New York, NY, pp. 253–288.

1286. Lewin, R. (1982). Adaptation can be a problem for evolutionists. *Science* **216**, 1212–1213.

1287. Lewis, D.L., DeCamillis, M., and Bennett, R.L. (2000). Distinct roles of the homeotic genes *Ubx* and *abd-A* in beetle embryonic abdominal appendage development. *PNAS* **97** #9, 4504–4509.

1288. Lewis, E.B. (1994). Homeosis: the first 100 years. *Trends Genet.* **10**, 341–343.

1289. Lewis, J. (2003). Autoinhibition with transcriptional delay: a simple mechanism for the zebrafish somitogenesis oscillator. *Curr. Biol.* **13**, 1398–1408.

1290. Lewis, J.H. and Wolpert, L. (1976). The principle of non-equivalence in development. *J. Theor. Biol.* **62**, 479–490.

1291. Lewis, S.M. and Cratsley, C.K. (2008). Flash signal evolution, mate choice, and predation in fireflies. *Annu. Rev. Entomol.* **53**, 293–321.

1292. Leysen, H., Roos, G., and Adriaens, D. (2011). Morphological variation in head shape of pipefishes and seahorses in relation to snout length and developmental growth. *J. Morph.* **272**, 1259–1270.

1293. Li, B.-W., Zhao, H.-P., and Feng, X.-Q. (2011). Static and dynamic mechanical properties of cattle horns. *Materials Sci. Eng. C* **31**, 179–183.

1294. Li, C., Wu, X.-C., Rieppel, O., Wang, L.-T., and Zhao, L.-J. (2008). An ancestral turtle from the Late Triassic of southwestern China. *Nature* **456**, 497–501.

1295. Li, C., Yang, F., Haines, S., Zhao, H., Wang, W., Xing, X., Sun, H., Chu, W., Lu, X., Liu, L., and McMahon, C. (2012). Stem cells responsible for deer antler regeneration are unable to recapitulate the process of first antler development – revealed through intradermal and subcutaneous tissue transplantation. *J. Exp. Zool. (Mol. Dev. Evol.)* **314B**, 552–570.

1296. Li, C.-Y., Cha, W., Luder, H.-U., Charles, R.-P., McMahon, M., Mitsiadis, T.A., and Klein, O.D. (2012). E-cadherin regulates the behavior and fate of epithelial stem cells and their progeny in the mouse incisor. *Dev. Biol.* **366**, 357–366.

1297. Li, H. and Popadic, A. (2004). Analysis of *nubbin* expression patterns in insects. *Evol. Dev.* **6**, 310–324.

1298. Liang, H.-L., Xu, M., Chuang, Y.-C., and Rushlow, C. (2012). Response to the BMP gradient requires highly combinatorial inputs from multiple patterning systems in the *Drosophila* embryo. *Development* **139**, 1956–1964.

1299. Lichtneckert, R. and Reichert, H. (2005). Insights into the urbilaterian brain: conserved genetic patterning mechanisms in insect and vertebrate brain development. *Heredity* **94**, 465–477.

1300. Lieberman, B.S. (2012). Adaptive radiations in the context of macroevolutionary theory: a paleontological perspective. *Evol. Biol.* **39**, 181–191.

1301. Lim, M.M., Wang, Z., Olazábal, D.E., Ren, X., Terwilliger, E.F., and Young, L.J. (2004). Enhanced partner preference in a promiscuous species by manipulating the expression of a single gene. *Nature* **429**, 754–757.

1302. Lim, W.A. and Pawson, T. (2010). Phosphotyrosine signaling: evolving a new cellular communication system. *Cell* **142**, 661–667.

1303. Lin, B., Wang, S.W., and Masland, R.H. (2004). Retinal ganglion cell type, size, and spacing can be specified independent of homotypic dendritic contacts. *Neuron* **43**, 475–485.

1304. Lin, C., Yin, Y., Bell, S.M., Veith, G.M., Chen, H., Huh, S.-H., Ornitz, D.M., and Ma, L. (2013). Delineating a conserved genetic cassette promoting outgrowth of body appendages. *PLoS Genet.* **9** #1, e1003231.

1305. Lin, C.-M., Jiang, T.X., Baker, R.E., Maini, P.K., Widelitz, R.B., and Chuong, C.-M. (2009). Spots and stripes: pleomorphic patterning of stem cells via p-ERK-dependent cell chemotaxis shown by feather morphogenesis and mathematical simulation. *Dev. Biol.* **334**, 369–382.

1306. Lin, C.H., Liu, J.H., Osterburg, J.W., and Nicol, J.D. (1982). Fingerprint comparison. I. Similarity of fingerprints. *J. Forensic Sci.* **27**, 290–304.

1307. Lin, C.H. and Rankin, C.H. (2012). Alcohol addiction: chronic ethanol leads to cognitive dependence in *Drosophila*. *Curr. Biol.* **22**, R1043–R1044.

1308. Lin, G., Chen, Y., and Slack, J.M.W. (2013). Imparting regenerative capacity to limbs by progenitor cell transplantation. *Dev. Cell* **24**, 41–51.

1309. Lin, J.Y. and Fisher, D.E. (2007). Melanocyte biology and skin pigmentation. *Nature* **445**, 843–850.

1310. Lindgren, J., Caldwell, M.W., Konishi, T., and Chiappe, L. (2010). Convergent evolution in aquatic tetrapods: insights from an exceptional fossil mosasaur. *PLoS ONE* **5** #8, e11998.

1311. Lindsley, D.L. and Zimm, G.G. (1992). *The Genome of* Drosophila melanogaster. Academic Press, New York, NY.

1312. Lister, A.M., Edwards, C.J., Nock, D.A.W., Bunce, M., van Pijlen, I.A., Bradley, D.G., Thomas, M.G., and Barnes, I. (2005). The phylogenetic position of the "giant deer" *Megalocerous giganteus*. *Nature* **438**, 850–853.

1313. Litingtung, Y., Dahn, R.D., Li, Y., Fallon, J.F., and Chiang, C. (2002). *Shh* and *Gli3* are dispensable for limb skeleton formation but regulate digit number and identity. *Nature* **418**, 979–983.

1314. Little, J.W., Byrd, C.A., and Brower, D.L. (1990). Effect of *abx*, *bx* and *pbx* mutations on expression of homeotic genes in *Drosophila* larvae. *Genetics* **124**, 899–908.

1315. Liu, B., Rooker, S.M., and Helms, J.A. (2010). Molecular control of facial morphology. *Semin. Cell Dev. Biol.* **21**, 309–313.

1316. Liu, C., Fu, X., Liu, L., Ren, X., Chau, C.K.L., Li, S., Xiang, L., Zeng, H., Chen, G., Tang, L.-H., Lenz, P., Cui, X., Huang, W., Hwa, T., and Huang, J.-D. (2011). Sequential establishment of stripe patterns in an expanding cell population. *Science* **334**, 238–241.

1317. Lloyd, S. (1996). Complexity simplified. *Sci. Am.* **274** #5, 104–108.

1318. Locke, M. (2008). Structure of ivory. *J. Morph.* **269**, 423–450.

1319. Loehlin, D.W. and Werren, J.H. (2012). Evolution of shape by multiple regulatory changes to a growth gene. *Science* **335**, 943–947.

1320. Loftus, R.T., MacHugh, D.E., Bradley, D.G., Sharp, P.M., and Cunningham, P. (1994). Evidence for two independent domestications of cattle. *PNAS* **91** #7, 2757–2761.

1321. Logan, M. (2003). Finger or toe: the molecular basis of limb identity. *Development* **130**, 6401–6410.

1322. Logan, M. and Tabin, C.J. (1999). Role of Pitx1 upstream of Tbx4 in specification of hindlimb identity. *Science* **283**, 1736–1739.

1323. Logan, M.A. and Vetter, M.L. (2004). Do-it-yourself tiling: dendritic growth in the absence of homotypic contacts. *Neuron* **43**, 439–440.

1324. Lohmann, I., McGinnis, N., Bodmer, M., and McGinnis, W. (2002). The *Drosophila Hox* gene *Deformed* sculpts head morphology via direct regulation of the apoptosis activator *reaper*. *Cell* **110**, 457–466.

1325. Lohmann, I. and McGinnis, W. (2002). *Hox* genes: it's all a matter of context. *Curr. Biol.* **12**, R514–R516.

1326. Longrich, N.R., Bhullar, B.-A.S., and Gauthier, J.A. (2012). A transitional snake from the Late Cretaceous period of North America. *Nature* **488**, 205–208.

1327. Looso, M., Preussner, J., Sousounis, K., Bruckskotten, M., Michel, C.S., Lignelli, E., Reinhardt, R., Hoeffner, S., Krueger, M., Tsonis, P.A., Borchardt, T., and Braun, T. (2013). A de novo assembly of the newt transcriptome combined with proteomic validation identifies new protein families expressed during tissue regeneration. *Genome Biol.* **14**, R16.

1328. Losos, J.B. and Ricklefs, R.E. (2009). Adaptation and diversification on islands. *Nature* **457**, 830–836.

1329. Louchart, A. and Viriot, L. (2011). From snout to beak: the loss of teeth in birds. *Trends Ecol. Evol.* **26**, 663–673.

1330. Love, A.C. (2010). Darwin's "imaginary illustrations": Creatively teaching evolutionary concepts & the nature of science. *Am. Biol. Teacher* **72**, 82–89.

1331. Lowe, C.J., Terasaki, M., Wu, M., Freeman, R.M., Jr., Runft, L., Kwan, K., Haigo, S., Aronowicz, J., Lander, E., Gruber, C., Smith, M., Kirschner, M., and Gerhart, J. (2006). Dorsoventral patterning in hemichordates: insights into early chordate evolution. *PLoS Biol.* **4** #9, 1603–1619 (e291).

1332. Lowenstein, J.M. and Ryder, O.A. (1985). Immunological systematics of the extinct quagga (Equidae). *Experientia* **41**, 1192–1193.

1333. Lowery, L.A. and Sive, H. (2009). Totally tubular: the mystery behind function and origin of the brain ventricular system. *BioEssays* **31**, 446–458.

1334. Lu, B., LaMora, A., Sun, Y., Welsh, M.J., and Ben-Shahar, Y. (2012). *ppk23*-dependent chemosensory functions contribute to courtship behavior in *Drosophila melanogaster*. *PLoS Genet.* **8** #3, e1002587.

1335. Lu, C.-H., Rincón-Limas, D.E., and Botas, J. (2000). Conserved overlapping and reciprocal expression of *msh/Msx1* and *apterous/Lhx2* in *Drosophila* and mice. *Mechs. Dev.* **99**, 177–181.

1336. Lubarsky, B. and Krasnow, M.A. (2003). Tube morphogenesis: making and shaping biological tubes. *Cell* **112**, 19–28.

1337. Luo, Y.-J. and Su, Y.-H. (2012). Opposing Nodal and BMP signals regulate left-right asymmetry in the sea urchin larva. *PLoS Biol.* **10** #10, e1001402.

1338. Lynch, L.J., Robinson, V., and Anderson, C.A. (1973). A scanning electron microscope study of the morphology of rhinoceros horn. *Aust. J. Biol. Sci.* **26**, 395–399.

1339. Lynch, M. (2007). The evolution of genetic networks by non-adaptive processes. *Nature Rev. Genet.* **8**, 803–813.

1340. Lynch, M. (2007). The frailty of adaptive hypotheses for the origins of organismal complexity. *PNAS* **104** (Suppl. 1), 8597–8604.

1341. Lyons, M.J. and Harrison, L.G. (1992). Stripe selection: an intrinsic property of some pattern-formation models with nonlinear dynamics. *Dev. Dynamics* **195**, 201–215.

1342. Lyson, T.R., Bever, G.S., Scheyer, T.M., Hsiang, A.Y., and Gauthier, J.A. (2013). Evolutionary origin of the turtle shell. *Curr. Biol.* **23**, 1113–1119.

1343. Lyson, T.R. and Joyce, W.G. (2012). Evolution of the turtle bauplan: the topological relationship of the scapula relative to the ribcage. *Biol. Lett.* **8**, 1028–1031.

1344. Ma, X., Hou, X., Edgecombe, G.D., and Strausfeld, N.J. (2012). Complex brain and optic lobes in an early Cambrian arthropod. *Nature* **490**, 258–261.

1345. Ma, Y., Li, A., Faller, W.J., Libertini, S., Fiorito, F., Gillespie, D.A., Sansom, O.J., Yamashiro, S., and Machesky, L.M. (2013). Fascin 1 is transiently expressed in mouse melanoblasts during development and promotes migration and proliferation. *Development* **140**, 2203–2211.

1346. Maas, A.-H. (1948). Über die Auslösbarkeit von Temperatur-Modifikationen während der Embryonal-Entwicklung von *Drosophila melanogaster* Meigen. *W. Roux's Arch. Entw.-Mech. Org.* **143**, 515–572.

1347. MacArthur, J.W. and Ford, N. (1937). *A Biological Study of the Dionne Quintuplets: An Identical Set.* University of Toronto Press, Toronto.

1348. MacDonald, B.T., Semenov, M.V., and He, X. (2007). Snapshot: Wnt/β-catenin signaling. *Cell* **131**, 1204.

1349. Macdonald, W.P., Martin, A., and Reed, R.D. (2010). Butterfly wings shaped by a molecular cookie cutter: evolutionary radiation of lepidopteran wing shapes associated with a derived Cut/*wingless* wing margin boundary system. *Evol. Dev.* **12**, 296–304.

1350. MacHugh, D.E., Shriver, M.D., Loftus, R.T., Cunningham, P., and Bradley, D.G. (1997). Microsatellite DNA variation and the evolution, domestication and phylogeography of taurine and zebu cattle (*Bos taurus* and *Bos indicus*). *Genetics* **146**, 1071–1086.

1351. MacLaurin, J. (2003). The good, the bad and the impossible. *Biol. Philos.* **18**, 463–476.

1352. Macpherson, E., Jones, W., and Segonzac, M. (2005). A new squat lobster family of Galatheoidea (Crustacea, Decapoda, Anomura) from the hydrothermal vents of the Pacific-Antarctic Ridge. *Zoosystema* **27**, 709–723.

1353. Maderspacher, F. (2012). Colour patterns: channelling Turing. *Curr. Biol.* **22**, R266–R268.

1354. Maderspacher, F. and Nüsslein-Volhard, C. (2003). Formation of the adult pigment pattern in zebrafish requires *leopard* and *obelix* dependent cell interactions. *Development* **130**, 3447–3457.

1355. Maderspacher, F. and Stensmyr, M. (2011). Myrmecomorphomania. *Curr. Biol.* **21**, R291–R293.

1356. Madore, B.F. and Freedman, W.L. (1983). Computer simulations of the Belousov-Zhabotinsky reaction. *Science* **222**, 437–438.

1357. Madore, B.F. and Freedman, W.L. (1987). Self-organizing structures. *Am. Sci.* **75**, 252–259.

1358. Mahfooz, N., Turchyn, N., Mihajlovic, M., Hrycaj, S., and Popadic, A. (2007). *Ubx* regulates differential enlargement and diversification of insect hind legs. *PLoS ONE* **2** #9, e866.

1359. Mahler, D.L. and Kearney, M. (2006). The palatal dentition in squamate reptiles: morphology, development, attachment, and replacement. *Fieldiana, Zoology* New Series, No. **108**, 1–61.

1360. Mainguy, G., In der Rieden, P.M.J., Berezikov, E., Woltering, J.M., Plasterk, R.H.A., and Durston, A.J. (2003). A position-dependent organisation of retinoid response elements is conserved in the vertebrate *Hox* clusters. *Trends Genet.* **19**, 476–479.

1361. Maini, P.K. (2003). How the mouse got its stripes. *PNAS* **100** #17, 9656–9657.

1362. Maini, P.K., Baker, R.E., and Chuong, C.-M. (2006). The Turing Model comes of molecular age. *Science* **314**, 1397–1398.

1363. Maître, J.-L., Berthoumieux, H., Krens, S.F.G., Salbreux, G., Jülicher, F., Paluch, E., and Heisenberg, C.-P. (2012). Adhesion functions in cell sorting by mechanically coupling the cortices of adhering cells. *Science* **338**, 253–256.

1364. Makanya, A.N. and Mortola, J.P. (2007). The structural design of the bat wing web and its possible role in gas exchange. *J. Anat.* **211**, 687–697.

1365. Makhijani, K., Kalyani, C., Srividya, T., and Shashidhara, L.S. (2007). Modulation of Decapentaplegic gradient during haltere specification in *Drosophila*. *Dev. Biol.* **302**, 243–255.

1366. Maladen, R.D., Ding, Y., Li, C., and Goldman, D.I. (2009). Undulatory swimming in sand: subsurface locomotion of the sandfish lizard. *Science* **325** #314–318.

1367. Mallarino, R. and Abzhanov, A. (2012). Paths less traveled: Evo-devo approaches to investigating animal morphological evolution. *Annu. Rev. Cell Dev. Biol.* **28**, 743–763.

1368. Mallo, M., Vinagre, T., and Carapuço, M. (2009). The road to the vertebral formula. *Int. J. Dev. Biol.* **53**, 1469–1481.

1369. Mallo, M., Wellik, D.M., and Deschamps, J. (2010). *Hox* genes and regional patterning of the vertebrate body plan. *Dev. Biol.* **344**, 7–15. [*See also* Mallo, M. and Alonso, C.R. (2013). The regulation of *Hox* gene expression during animal development. *Development* **140**, 3951–3963.]

1370. Malmström, T. and Kröger, R.H.H. (2006). Pupil shapes and lens optics in the eyes of terrestrial vertebrates. *J. Exp. Biol.* **209**, 18–25.

1371. Mandal, L., Banerjee, U., and Hartenstein, V. (2004). Evidence for a fruit fly hemangioblast and similarities between lymph-gland hematopoiesis in fruit fly and mammal aorta-gonadal-mesonephros mesoderm. *Nat. Genet.* **36**, 1019–1023.

1372. Manger, P.R., Hall, L.S., and Pettigrew, J.D. (1998). The development of the external features of the platypus (*Ornithorhynchus anatinus*). *Phil. Trans. R. Soc. Lond. B* **353**, 1115–1125.

1373. Mann, R.S. and Carroll, S.B. (2002). Molecular mechanisms of selector gene function and evolution. *Curr. Opin. Gen. Dev.* **12**, 592–600.

1374. Manry, D.E. (1985). Birds of fire. *Nat. Hist.* **94** #1, 38–44.

1375. Manuel, M. and Forêt, S. (2012). Searching for Eve: basal metazoans and the evolution of multicellular complexity. *BioEssays* **34**, 247–251.

1376. Mapalad, K.S., Leu, D., and Nieh, J.C. (2008). Bumble bees heat up for high quality pollen. *J. Exp. Biol.* **211**, 2239–2242.

1377. Marcil, A., Dumontier, É., Chamberland, M., Camper, S.A., and Drouin, J. (2003). *Pitx1* and *Pitx2* are required for development of hindlimb buds. *Development* **130**, 45–55.

1378. Marcon, L. and Sharpe, J. (2012). Turing patterns in development: what about the horse part? *Curr. Opin. Gen. Dev.* **22**, 578–584.

1379. Marcus, J.M. (2001). The development and evolution of crossveins in insect wings. *J. Anat.* **199**, 211–216.

1380. Marcus, J.M., Ramos, D.M., and Monteiro, A. (2004). Germline transformation of the butterfly *Bicyclus anynana*. *Proc. Roy. Soc. Lond. B (Suppl.)* **271**, S263–S265.

1381. Marden, J.H. (1995). Flying lessons from a flightless insect. *Nat. Hist.* **104** #2, 4–8.

1382. Marden, J.H. and Kramer, M.G. (1995). Locomotor performance of insects with rudimentary wings. *Nature* **377**, 332–334.

1383. Marek, P.E. and Bond, J.E. (2006). Rediscovery of the world's leggiest animal. *Nature* **441**, 707.

1384. Maricich, S.M. and Zoghbi, H.Y. (2006). Getting back to basics. *Cell* **126**, 11–15.

1385. Mark, R. (1996). Architecture and evolution. *Am. Sci.* **84**, 383–389.

1386. Markow, T.A. and O'Grady, P.M. (2005). Evolutionary genetics of reproductive behavior in *Drosophila*: Connecting the dots. *Annu. Rev. Genet.* **39**, 263–291.

1387. Marlétaz, F., Holland, L.Z., Laudet, V., and Schubert, M. (2006). Retinoic acid signaling and the evolution of chordates. *Int. J. Biol. Sci.* **2**, 38–47.

1388. Marquardt, T., Ashery-Padan, R., Andrejewski, N., Scardigli, R., Guillemot, F., and Gruss, P. (2001). Pax6 is required for the multipotent state of retinal progenitor cells. *Cell* **105**, 43–55.

1389. Marsh, O.C. (1892). Recent polydactyle horses. *Am J. Sci.* **43** (3rd Series) or **143** (Combined Series), 339–355.

1390. Marshall, C.R., Raff, E.C., and Raff, R.A. (1994). Dollo's law and the death and resurrection of genes. *PNAS* **91** #25, 12283–12287.

1391. Marshall, C.R. and Valentine, J.W. (2010). The importance of preadapted genomes in the origin of the animal bodyplans and the Cambrian Explosion. *Evolution* **64**, 1189–1201.

1392. Marshall, L.G. (1994). The terror birds of South America. *Sci. Am.* **270** #2, 90–95.

1393. Martin, A. and Reed, R.D. (2010). *wingless* and *aristaless2* define a developmental ground plan for moth and butterfly wing pattern evolution. *Mol. Biol. Evol.* **27**, 2864–2878.

1394. Martin, G.R., Wilson, K.-J., Wild, J.M., Parsons, S., Kubke, M.F., and Corfield, J. (2007). Kiwi forego vision in the guidance of their nocturnal activities. *PLoS ONE* **2** #2, e198.

1395. Martindale, M.Q. (2005). The evolution of metazoan axial properties. *Nature Rev. Genet.* **6**, 917–927.

1396. Martindale, M.Q., Finnerty, J.R., and Henry, J.Q. (2002). The Radiata and the evolutionary origins of the bilaterian body plan. *Mol. Phylogenet. Evol.* **24**, 358–365.

1397. Martindale, M.Q. and Hejnol, A. (2009). A developmental perspective: changes in the position of the blastopore during bilaterian evolution. *Dev. Cell* **17**, 162–174.

1398. Martindale, M.Q. and Henry, J.Q. (1998). The development of radial and biradial symmetry: the evolution of bilaterality. *Am. Zool.* **38**, 672–684.

1399. Masel, J. and Trotter, M.V. (2010). Robustness and evolvability. *Trends Genet.* **26**, 406–414.

1400. Massare, J.A. (1992). Ancient mariners. *Nat. Hist.* **101** #9, 48–53.

1401. Masumoto, M., Yaginuma, T., and Niimi, T. (2009). Functional analysis of *Ultrabithorax* in the silkworm, *Bombyx mori*, using RNAi. *Dev. Genes Evol.* **219**, 437–444.

1402. Mateus, O., Maidment, S.C.R., and Christiansen, N.A. (2009). A new long-necked 'sauropod-mimic' stegosaur and the evolution of the plated dinosaurs. *Proc. Roy. Soc. Lond. B* **276**, 1815–1821.

1403. Mäthger, L.M., Bell, G.R.R., Kuzirian, A.M., Allen, J.J., and Hanlon, R.T. (2012). How does the blue-ringed octopus (*Hapalochlaena lunulata*) flash its blue rings? *J. Exp. Biol.* **215**, 3752–3757.

1404. Matson, C.K. and Zarkower, D. (2012). Sex and the singular DM domain: insights into sexual regulation, evolution and plasiticity. *Nature Rev. Gen.* **13**, 163–174.

1405. Matsuda, S. and Shimmi, O. (2012). Directional transport and active retention of Dpp/BMP create wing vein patterns in *Drosophila*. *Dev. Biol.* **366**, 153–162.

1406. Matsuoka, T., Ahlberg, P.E., Kessaris, N., Iannarelli, P., Dennehy, U., Richardson, W.D., McMahon, A.P., and Koentges, G. (2005). Neural crest origins of the neck and shoulder. *Nature* **436**, 347–355.

1407. Mattison, C. (2007). *The New Encyclopedia of Snakes*. Princeton University Press, Princeton, NJ.

1408. Maxmen, A. (2011). A can of worms. *Nature* **470**, 161–162.

1409. Maxwell, E.E. and Harrison, L.B. (2009). Methods for the analysis of developmental sequence data. *Evol. Dev.* **11**, 109–119.

1410. May, R.M. (1978). The evolution of ecological systems. *Sci. Am.* **239** #3, 160–175.

1411. Mayer, J.A., Foley, J., De la Cruz, D., Chuong, C.-M., and Widelitz, R. (2008). Conversion of the nipple to hair-bearing epithelia by lowering bone morphogenetic protein pathway activity at the dermal-epidermal interface. *Am. J. Pathol.* **173**, 1339–1348.

1412. Maynard Smith, J. (1960). Continuous, quantized and modal variation. *Proc. Roy. Soc. Lond. B* **152**, 397–409.

1413. Maynard Smith, J. (1968). The counting problem. In *Towards a Theoretical Biology. I. Prolegomena* (C.H. Waddington, ed.). Aldine, Chicago, IL, pp. 120–124.

1414. Maynard Smith, J., Burian, R., Kauffman, S., Alberch, P., Campbell, J., Goodwin, B., Lande, R., Raup, D., and Wolpert, L. (1985). Developmental constraints and evolution. *Q. Rev. Biol.* **60**, 265–287.

1415. Mayor, R. and Theveneau, E. (2013). The neural crest. *Development* **140**, 2247–2251.

1416. Mayr, E. (1976). *Evolution and the Diversity of Life*. Harvard University Press, Cambridge, MA.

1417. Mayr, E. (1985). How biology differs from the physical sciences. In *Evolution at a Crossroads: The New Biology and the New Philosophy of Science* (D.J. Depew and B.H. Weber, eds.). MIT Press, Cambridge, MA, pp. 43–63.

1418. Mayr, E. (1991). *One Long Argument: Charles Darwin and the Genesis of Modern Evolutionary Thought*. Harvard University Press, Cambridge, MA.

1419. Mayr, E. (1994). Recapitulation reinterpreted: the somatic program. *Q. Rev. Biol.* **69**, 223–232.

1420. Mayr, E. and Provine, W. (1980). *The Evolutionary Synthesis*. Harvard University Press, Cambridge, MA.

1421. Mazák, J.H., Christiansen, P., and Kitchener, A.C. (2011). Oldest known pantherine skull and evolution of the tiger. *PLoS ONE* **6** #10, e25483. [*Note added in proof*: The tiger genome has just been sequenced! *See* Cho, Y.S., *et al.* (2013). The tiger genome and comparative analysis with lion and snow leopard genomes. *Nature Comm.* doi 10.1038/ncomms3433.]

1422. McAlpine, J.F. (1981). Morphology and terminology: adults. In *Manual of Nearctic Diptera*, Vol. 1 (J.F. McAlpine, B.V. Peterson, G.E. Shewell, H.J. Teskey, J.R. Vockeroth, and D.M. Wood, eds.). Canadian Government Publishing Centre, Hull, Quebec, pp. 9–63.

1423. McClure, M. and McCune, A.R. (2003). Evidence for developmental linkage of pigment patterns with body size and shape in Danios (Teleostei: Cyprinidae). *Evolution* **57**, 1863–1875.

1424. McCollum, M. and Sharpe, P.T. (2001). Evolution and development of teeth. *J. Anat.* **199**, 153–159.

1425. McConnell, S.K. and Kaznowski, C.E. (1991). Cell cycle dependence of laminar determination in developing neocortex. *Science* **254**, 282–285.

1426. McCue, M.D. (2007). Prey envenomation does not improve digestive performance in western diamondback rattlesnakes (*Crotalus atrox*). *J. Exp. Zool.* **307A**, 568–577.

1427. McCue, M.E., Bannasch, D.L., Petersen, J.L., Gurr, J., Bailey, E., Binns, M.M., Distl, O., Guérin, G., Hasegawa, T., Hill, E.W., Leeb, T., Lindgren, G., Penedo, M.C.T., Røed, K.H., Ryder, O.A., Swinburne, J.E., Tozaki, T., Valberg, S.J., Vaudin, M., Lindblad-Toh, K., Wade, C.M., and Mickelson, J.R. (2012). A high density SNP array for the domestic horse and extant perissodactyla: utility for association mapping, genetic diversity, and phylogeny studies. *PLoS Genet.* **8** #1, e1002451.

1428. McCullough, E.L., Weingarden, P.R., and Emlen, D.J. (2012). Costs of elaborate weapons in a rhinocerous beetle: how difficult is it to fly with a big horn? *Behav. Ecol.* **23**, 1042–1048.

1429. McCune, A.R. and Carlson, R.L. (2004). Twenty ways to lose your bladder: common natural mutants in zebrafish and widespread convergence of swim bladder loss among teleost fishes. *Evol. Dev.* **6**, 246–259.

1430. McGhee, G.R., Jr. (1999). *Theoretical Morphology: The Concept and Its Applications.* Columbia University Press, New York, NY.

1431. McGhee, G.R., Jr. (2011). Convergent Evolution: Limited Forms Most Beautiful. *Vienna Series in Theoretical Biology*. MIT Press, Cambridge, MA.

1432. McGinnis, W. (1994). A century of homeosis, a decade of homeoboxes. *Genetics* **137**, 607–611.

1433. McGregor, A.P. (2005). How to get ahead: the origin, evolution and function of *bicoid. BioEssays* **27**, 904–913.

1434. McHenry, M.J. (2005). The morphology, behavior, and biomechanics of swimming in ascidian larvae. *Can. J. Zool.* **83**, 62–74.

1435. McIntyre, D.C., Rakshit, S., Yallowitz, A.R., Loken, L., Jeannotte, L., Capecchi, M.R., and Wellik, D.M. (2007). Hox patterning of the vertebrate rib cage. *Development* **134**, 2981–2989.

1436. McKay, D.J., Estella, C., and Mann, R.S. (2009). The origins of the *Drosophila* leg revealed by the cis-regulatory architecture of the *Distalless* gene. *Development* **136**, 61–71.

1437. McLean, W.H.I. (2008). Combing the genome for the root cause of baldness. *Nature Genet.* **11**, 1270–1271.

1438. McLeish, T. (2013). *Narwhals: Arctic Whales in a Melting World.* University of Washington Press, Seattle, WA.

1439. McLellan, J.S., Zheng, X., Hauk, G., Ghirlando, R., Beachy, P.A., and Leahy, D.J. (2008). The mode of Hedgehog binding to Ihog homologues is not conserved across different phyla. *Nature* **455**, 979–983.

1440. McLennan, D.A. (2008). The concept of co-option: why evolution often looks miraculous. *Evo. Edu. Outreach* **1**, 247–258.

1441. McMillan, W.O., Monteiro, A., and Kapan, D.D. (2002). Development and evolution on the wing. *Trends Ecol. Evol.* **17**, 125–133.

1442. McNamara, K.J. (1995). Sexual dimorphism: the role of heterochrony. In *Evolutionary Change and Heterochrony* (K.J. McNamara, ed.). Wiley, Chichester, pp. 65–89.

1443. McNamara, K.J. (2012). Heterochrony: the evolution of development. *Evo. Edu. Outreach* **5**, 203–218.

1444. McNamara, K.J. and McKinney, M.L. (2005). Heterochrony, disparity, and macroevolution. *Paleobiology* **31**, 17–26.

1445. McNulty, K.P. (2010). Apes and tricksters: the evolution and diversification of humans' closest relatives. *Evo. Edu. Outreach* **3**, 322–332. [*See also* Stevens, N.J., *et al.* (2013). Palaeontological evidence for an Oligocene divergence between Old World monkeys and apes. *Nature* **497**, 611–614.]

1446. McNulty, K.P. (2012). Evolutionary development in *Australopithecus africanus*. *Evol. Biol.* **39**, 488–498.

1447. McPherron, A.C., Lawler, A.M., and Lee, S.-J. (1999). Regulation of anterior/posterior patterning of the axial skeleton by growth/differentiation factor 11. *Nature Genet.* **22**, 260–264.

1448. McShea, D.W. and Hordijk, W. (2013). Complexity by subtraction. *Evol. Biol.* **40**, in press. doi 10.1007/s11692-013-9227-6.

1449. Meachen-Samuels, J.A. and van Valkenburgh, B. (2010). Radiographs reveal exceptional forelimb strength in the sabertooth cat, *Smilodon fatalis*. *PLoS ONE* **5** #7, e11412.

1450. Mead, J.G. (1975). Anatomy of the external nasal passages and facial complex in the Delphinidae (Mammalia: Cetacea). *Smithsonian Contrib. Zool.* **207**, 72 pp.

1451. Medeiros, D.M. and Crump, J.G. (2012). New perspectives on pharyngeal dorsoventral patterning in development and evolution of the vertebrate jaw. *Dev. Biol.* **371**, 121–135.

1452. Méhes, E., Mones, E., Németh, V., and Vicsek, T. (2012). Collective motion of cells mediates segregation and pattern formation in co-cultures. *PLoS ONE* **7** #2, e31711.

1453. Mehta, R.S. and Wainwright, P.C. (2007). Raptorial jaws in the throat help moray eels swallow large prey. *Nature* **449**, 79–82.

1454. Mehta, R.S., Ward, A.B., Alfaro, M.E., and Wainwright, P.C. (2010). Elongation of the body in eels. *Integr. Comp. Biol.* **50**, 1091–1105.

1455. Meier, P., Finch, A., and Evan, G. (2000). Apoptosis in development. *Nature* **407**, 796–801.

1456. Meik, J.M. and Pires-daSilva, A. (2009). Evolutionary morphology of the rattlesnake style. *BMC Evol. Biol.* **9**, Article 35 (9 pp.).

1457. Meinhardt, H. (1982). *Models of Biological Pattern Formation*. Academic Press, New York, NY.

1458. Meinhardt, H. (1995). Dynamics of stripe formation. *Nature* **376**, 722–723.

1459. Meinhardt, H. (2004). Different strategies for midline formation in bilaterians. *Nature Rev. Neurosci.* **5**, 502–510.

1460. Meinhardt, H. (2004). Models for the generation of the embryonic body axes: ontogenetic and evolutionary aspects. *Curr. Opin. Gen. Dev.* **14**, 446–454.

1461. Meinhardt, H. and Gierer, A. (1974). Applications of a theory of biological pattern formation based on lateral inhibition. *J. Cell Sci.* **15**, 321–346.

1462. Meinhardt, H. and Gierer, A. (2000). Pattern formation by local self-activation and lateral inhibition. *BioEssays* **22**, 753–760.

1463. Mendelson, T.C. and Shaw, K.L. (2012). The (mis)concept of species recognition. *Trends Ecol. Evol.* **27**, 421–427.

1464. Menegaz, R.A. and Kirk, E.C. (2009). Septa and processes: convergent evolution of the orbit in haplorhine primates and strigiform birds. *J. Hum. Evol.* **57**, 672–687.

1465. Merabet, S. and Hudry, B. (2013). *Hox* transcriptional specificity despite a single class of cofactors: are flexible interaction modes the key? *BioEssays* **35**, 88–92.

1466. Mercader, N., Leonardo, E., Azpiazu, N., Serrano, A., Morata, G., Martínez, C., and Torres, M. (1999). Conserved regulation of proximodistal limb axis development by Meis1/Hth. *Nature* **402**, 425–429.

1467. Mercola, M. and Levin, M. (2001). Left-right asymmetry determination in vertebrates. *Annu. Rev. Cell Dev. Biol.* **17**, 779–805.

1468. Meredith, R.W., Gatesy, J., Emerling, C.A., York, V.M., and Springer, M.S. (2013). Rod monochromacy and the coevolution of cetacean retinal opsins. *PLoS Genet.* **9** #4, e1003432.

1469. Merino, R., Rodriguez-Leon, J., Macias, D., Gañan, Y., Economides, A.N., and Hurle, J.M. (1999). The BMP antagonist Gremlin regulates outgrowth, chondrogenesis and programmed cell death in the developing limb. *Development* **126**, 5515–5522.

1470. Merritt, D.J. (2007). The organule concept of insect sense organs: sensory transduction and organule evolution. *Adv. Insect Physiol.* **33**, 192–241.

1471. Meulemans, D. and Bronner-Fraser, M. (2005). Central role of gene cooption in neural crest evolution. *J. Exp. Zool. (Mol. Dev. Evol.)* **304B**, 298–303.

1472. Michaelson, J. (1987). Cell selection in development. *Biol. Rev.* **62**, 115–139.

1473. Mikhailov, A.T. (2005). Putting evo-devo into focus. *Int. J. Dev. Biol.* **49**, 9–16.

1474. Mikó, I., Friedrich, F., Yoder, M.J., Hines, H.M., Deitz, L.L., Bertone, M.A., Seltmann, K.C., Wallace, M.S., and Deans, A.R. (2012). On dorsal prothoracic appendages in treehoppers (Hemiptera: Membracidae) and the nature of morphological evidence. *PLoS ONE* **7** #1, e30137.

1475. Milewski, A.V. and Dierenfeld, E.S. (2013). Structural and functional comparison of the proboscis between tapirs and other extant and extinct vertebrates. *Integr. Zool.* **8**, 84–94.

1476. Milinkovitch, M.C., Manukyan, L., Debry, A., Di-Poï, N., Martin, S., Singh, D., Lambert, D., and Zwicker, M. (2013). Crocodile head scales are not developmental units but emerge from physical cracking. *Science* **339**, 78–81.

1477. Miller, G. (2007). Fruit fly fight club. *Science* **315**, 180–182.

1478. Miller, G. (2009). On the origin of the nervous system. *Science* **325**, 24–26.

1479. Miller, G.S., Jr. (1931). Human hair and primate patterning. *Smithsonian Misc. Coll. (Publ. No. 3130)* **85** #10, 1–13 (plus 5 plates).

1480. Mills, M.G. and Patterson, L.B. (2009). Not just black and white: pigment pattern development and evolution in vertebrates. *Semin. Cell Dev. Biol.* **20**, 72–81.

1481. Milner, M.J. and Haynie, J.L. (1979). Fusion of *Drosophila* eye-antennal imaginal discs during differentiation in vitro. *Wilhelm Roux's Arch.* **185**, 363–370.

1482. Min, M.S., Yang, S.Y., Bonett, R.M., Vieites, D.R., Brandon, R.A., and Wake, D.B. (2005). Discovery of the first Asian plethodontid salamander. *Nature* **435**, 87–90.

1483. Minelli, A. (2000). Limbs and tail as evolutionarily diverging duplicates of the main body axis. *Evol. Dev.* **2**, 157–165.

1484. Minelli, A. (2002). Homology, limbs, and genitalia. *Evol. Dev.* **4**, 127–132.

1485. Minelli, A. (2003). *The Development of Animal Form: Ontogeny, Morphology, and Evolution.* Cambridge University Press, New York, NY.

1486. Minelli, A. (2003). The origin and evolution of appendages. *Int. J. Dev. Biol.* **47**, 573 581.

1487. Minelli, A. (2005). A morphologist's perspective on terminal growth and segmentation. *Evol. Dev.* **7**, 568–573.

1488. Minelli, A. (2009). *Forms of Becoming: The Evolutionary Biology of Development*. Princeton University Press, Princeton, NJ.

1489. Minelli, A. (2011). A principle of developmental inertia. In *Epigenetics: Linking Genotype and Phenotype in Development and Evolution* (B. Hallgrímsson and B.K. Hall, eds.). University of California Press, Berkeley, CA, pp. 116–133.

1490. Minelli, A. and Bortoletto, S. (1988). Myriapod metamerism and arthropod segmentation. *Biol. J. Linnean Soc.* **33**, 323–343.

1491. Minelli, A., Brena, C., Deflorian, G., Maruzzo, D., and Fusco, G. (2006). From embryo to adult: beyond the conventional periodization of arthropod development. *Dev. Genes Evol.* **216**, 373–383.

1492. Minelli, A. and Fusco, G. (2005). Conserved versus innovative features in animal body organization. *J. Exp. Zool. (Mol. Dev. Evol.)* **304B**, 520–525.

1493. Minguillon, C., Del Buono, J., and Logan, M.P. (2005). *Tbx5* and *Tbx4* are not sufficient to determine limb-specific morphologies but have common roles in initiating limb outgrowth. *Dev. Cell* **8**, 75–84.

1494. Minsuk, S.B. and Raff, R.A. (2002). Pattern formation in a pentameral animal: induction of early adult rudiment development in sea urchins. *Dev. Biol.* **247**, 335–350.

1495. Minsuk, S.B., Turner, F.R., Andrews, M.E., and Raff, R.A. (2009). Axial patterning of the pentaradial adult echinoderm body plan. *Dev. Genes Evol.* **219**, 89–101.

1496. Mitchell, G. and Skinner, J.D. (2003). On the origin, evolution and phylogeny of giraffes *Giraffa camelopardalis*. *Trans. Roy. Soc. S. Afr.* **58** #1, 51–73.

1497. Mitchell, G., van Sittert, S.J., and Skinner, J.D. (2009). Sexual selection is not the origin of long necks in giraffes. *J. Zool.* **278**, 281–286.

1498. Mitchell, J.S., Heckert, A.B., and Sues, H.-D. (2010). Grooves to tubes: evolution of the venom delivery system in a Late Triassic "reptile". *Naturwissenschaften* **97**, 1117–1121.

1499. Mitgutsch, C., Richardson, M.K., Jiménez, R., Martin, J.E., Kondrashov, P., de Bakker, M.A.G., and Sánchez-Villagra, M.R. (2011). Circumventing the polydactyly "constraint": the mole's 'thumb'. *Biol. Lett.* **8**, 74–77.

1500. Mito, T., Shinmyo, Y., Kurita, K., Nakamura, T., Ohuchi, H., and Noji, S. (2011). Ancestral functions of Delta/Notch signaling in the formation of body and leg segments in the cricket *Gryllus bimaculatus*. *Development* **138**, 3823–3833.

1501. Mitrophanov, A.Y. and Groisman, E.A. (2008). Positive feedback in cellular control systems. *BioEssays* **30**, 542–555.

1502. Mitsiadis, T.A., Caton, J., and Cobourne, M. (2006). Waking-up Sleeping Beauty: recovery of the ancestral bird odontogenetic program. *J. Exp. Zool. (Mol. Dev. Evol.)* **306B**, 227–233.

1503. Mitsiadis, T.A., Caton, J., De Bari, C., and Bluteau, G. (2008). The large functional spectrum of the heparin-binding cytokines MK and HB-GAM in continuously growing organs: the rodent incisor as a model. *Dev. Biol.* **320**, 256–266.

1504. Mitsiadis, T.A., Chéraud, Y., Sharpe, P., and Fontaine-Pérus, J. (2003). Development of teeth in chick embryos after mouse neural crest transplantations. *PNAS* **100** #1, 6541–6545.

1505. Mitsiadis, T.A. and Smith, M.M. (2006). How do genes make teeth to order through development? *J. Exp. Zool. (Mol. Dev. Evol.)* **306B**, 177–182.

1506. Mitteroecker, P., Gunz, P., Neubauer, S., and Müller, G. (2012). How to explore morphological integration in human evolution and development? *Evol. Biol.* **39**, 536–553.

1507. Miura, T. and Shiota, K. (2000). TGFb2 acts as an "activator" molecule in reaction-diffusion model and is involved in cell sorting phenomenon in mouse limb micromass culture. *Dev. Dynamics* **217**, 241–249.

1508. Miyazawa, S., Okamoto, M., and Kondo, S. (2010). Blending of animal colour patterns by hybridization. *Nature Commun.* **1**, Article 66 (6 pp.).

1509. Mizutani, C.M. and Bier, E. (2008). EvoD/Vo: the origins of BMP signalling in the neuroectoderm. *Nature Rev. Genet.* **9**, 663–677.

1510. Mobley, K.B., Small, C.M., and Jones, A.G. (2011). The genetics and genomics of Syngnathidae: pipefishes, seahorses and seadragons. *J. Fish* **78**, 1624–1646.

1511. Moczek, A.P. (2006). Integrating micro- and macroevolution of development through the study of horned beetles. *Heredity* **97**, 168–178.

1512. Moczek, A.P. (2007). Developmental capacitance, genetic accommodation, and adaptive evolution. *Evol. Dev.* **9**, 299–305.

1513. Moczek, A.P. (2008). On the origins of novelty in development and evolution. *BioEssays* **30**, 432–447.

1514. Moczek, A.P. (2011). The origins of novelty. *Nature* **473**, 34–35.

1515. Moczek, A.P. (2012). The nature of nurture and the future of evodevo: toward a theory of developmental evolution. *Integr. Comp. Biol.* **52**, 108–119.

1516. Moczek, A.P., Andrews, J., Kijimoto, T., Yerushalmi, Y., and Rose, D.J. (2007). Emerging model systems in evo-devo: horned beetles and the origins of diversity. *Evol. Dev.* **9**, 323–328.

1517. Moczek, A.P. and Rose, D.J. (2009). Differential recruitment of limb patterning genes during development and diversification of beetle horns. *PNAS* **106** #22, 8992–8997.

1518. Moffett, M.W. (2006). Mantids: armed and dangerous. *Natl. Geogr.* **209** #1, 102–113.

1519. Mohit, P., Bajpai, R., and Shashidhara, L.S. (2003). Regulation of Wingless and Vestigial expression in wing and haltere discs of *Drosophila*. *Development* **130**, 1537–1547.

1520. Mohit, P., Makhijani, K., Madhavi, M.B., Bharathi, V., Lal, A., Sirdesai, G., Reddy, V.R., Ramesh, P., Kannan, R., Dhawan, J., and Shashidhara, L.S. (2006). Modulation of AP and DV signaling pathways by the homeotic gene *Ultrabithorax* during haltere development in *Drosophila*. *Dev. Biol.* **291**, 356–367.

1521. Molina, M.D., Neto, A., Maeso, I., Gómez-Skarmeta, J.L., Saló, E., and Cebrià, F. (2011). Noggin and noggin-like genes control dorsoventral axis regeneration in planarians. *Curr. Biol.* **21**, 300–305.

1522. Monastersky, R. (2001). Pterosaurs: lords of the ancient skies. *Natl. Geogr.* **199** #5, 86–105.

1523. Mongera, A. and Nüsslein-Volhard, C. (2013). Scales of fish arise from mesoderm. *Curr. Biol.* **23**, R338–R339.

1524. Monod, J. (1974). On chance and necessity. In *Studies in the Philosophy of Biology: Reduction and Related Problems* (F.J. Ayala and T. Dobzhansky, eds.). University of California Press, Berkeley, CA, pp. 357–375.

1525. Montagne, J., Groppe, J., Guillemin, K., Krasnow, M.A., Gehring, W.J., and Affolter, M. (1996). The *Drosophila* Serum Response Factor gene is required for the formation of intervein tissue of the wing and is allelic to *blistered*. *Development* **122**, 2589–2597.

1526. Montavon, T., Le Garrec, J.-F., Kerszberg, M., and Duboule, D. (2008). Modeling *Hox* gene regulation in digits: reverse collinearity and the molecular origin of thumbness. *Genes Dev.* **22**, 346–359.

1527. Montavon, T., Soshnikova, N., Mascrez, B., Joye, E., Thevenet, L., Splinter, E., De Laat, W., Spitz, F., and Duboule, D. (2011). A regulatory archipelago controls *Hox* genes transcription in digits. *Cell* **147**, 1132–1145.

1528. Monteiro, A. (2008). Alternative models for the evolution of eyespots and of serial homology on lepidopteran wings. *BioEssays* **30**, 358–366.

1529. Monteiro, A. (2011). Gene regulatory networks reused to build novel traits. *BioEssays* **34**, 181–186.

1530. Monteiro, A., Brakefield, P.M., and French, V. (1997). Butterfly eyespots: the genetics and development of the color rings. *Evolution* **51**, 1207–1216.

1531. Monteiro, A., Brakefield, P.M., and French, V. (1997). The genetics and development of an eyespot pattern in the butterfly *Bicyclus anynana*: response to selection for eyespot shape. *Genetics* **146**, 287–294.

1532. Monteiro, A., Brakefield, P.M., and French, V. (1997). The relationship between eyespot shape and wing shape in the butterfly *Bicyclus anynana*: A genetic and morphometrical approach. *J. Evol. Biol.* **10**, 787–802.

1533. Monteiro, A., French, V., Smit, G., Brakefield, P.M., and Metz, J.A.J. (2001). Butterfly eyespot patterns: evidence for specification by a morphogen diffusion gradient. *Acta Biotheor.* **49**, 77–88.

1534. Monteiro, A., Glaser, G., Stockslager, S., Glansdorp, N., and Ramos, D. (2006). Comparative insights into questions of lepidopteran wing pattern homology. *BMC Dev. Biol.* **6**, Article 52 (13 pp.).

1535. Monteiro, A. and Podlaha, O. (2009). Wings, horns, and butterfly eyespots: How do complex traits evolve? *PLoS Biol.* **7** #2, e1000037.

1536. Monteiro, A., Prijs, J., Bax, M., Hakkaart, T., and Brakefield, P.M. (2003). Mutants highlight the modular control of butterfly eyespot patterns. *Evol. Dev.* **5**, 180–187.

1537. Monteiro, A. and Prudic, K.L. (2010). Multiple approaches to study color pattern evolution in butterflies. *Trends Evol. Biol.* **2**, Article e2 (7 pp.).

1538. Monteiro, A.F., Brakefield, P.M., and French, V. (1994). The evolutionary genetics and developmental basis of wing pattern variation in the butterfly *Bicyclus anynana*. *Evolution* **48**, 1147–1157.

1539. Mooallem, J. (2013). *Wild Ones: A Sometimes Dismaying, Weirdly Reassuring Story About Looking at People Looking at Animals in America*. Penguin, New York, NY.

1540. Moore, J.A. (1993). *Science as a Way of Knowing. The Foundations of Modern Biology*. Harvard University Press, Cambridge, MA.

1541. Mooseker, M.S. and Cheney, R.E. (1995). Unconventional myosins. *Annu. Rev. Cell Dev. Biol.* **11**, 633–675.

1542. Morata, G. (1975). Analysis of gene expression during development in the homeotic mutant *Contrabithorax* of *Drosophila melanogaster*. *J. Embryol. Exp. Morph.* **34**, 19–31.

1543. Morata, G., Macías, A., Urquía, N., and González-Reyes, A. (1990). Homoeotic genes. *Semin. Cell Biol.* **1**, 219–227.

1544. Morimura, S., Maves, L., and Hoffmann, F.M. (1996). *decapentaplegic* overexpression affects *Drosophila* wing and leg imaginal disc development and *wingless* expression. *Dev. Biol.* **177**, 136–151.

1545. Morin-Kensicki, E.M., Melancon, E., and Eisen, J.S. (2002). Segmental relationship between somites and vertebral column in zebrafish. *Development* **129**, 3851–3860.

1546. Morris, V.B. (2012). Early development of coelomic structures in an echinoderm larva and a similarity with coelomic structures in a chordate embryo. *Dev. Genes Evol.* **222**, 313–323.

1547. Moss, E.G. (2007). Heterochronic genes and the nature of developmental time. *Curr. Biol.* **17**, R425–R434.

1548. Motani, R. (2000). Rulers of the Jurassic seas. *Sci. Am.* **283** #6, 52–59.

1549. Motani, R. (2009). The evolution of marine reptiles. *Evo. Edu. Outreach* **2**, 224–235.

1550. Motani, R., Rothschild, B.M., and Wahl, W., Jr. (1999). Large eyeballs in diving ichthyosaurs. *Nature* **402**, 747. [*See also* Schmitz, L., *et al.* (2013). Allometry indicates giant eyes of giant squid are not exceptional. *BMC Evol. Biol.* **13**, Article 45 (9 pp.).]

1551. Mount, J.G., Muzylak, M., Allen, S., Althnaian, T., McGonnell, I.M., and Price, J.S. (2006). Evidence that the canonical Wnt signalling pathway regulates deer antler regeneration. *Dev. Dynamics* **235**, 1390–1399.

1552. Moussian, B. and Uv, A.E. (2005). An ancient control of epithelial barrier formation and wound healing. *BioEssays* **27**, 987–990.

1553. Moustakas, A. and Heldin, C.-H. (2009). The regulation of TGFβ signal transduction. *Development* **136**, 3699–3714.

1554. Muchhala, N. (2006). Nectar bat stows huge tongue in its rib cage. *Nature* **444**, 701–702.

1555. Muchhala, N. and Thomson, J.D. (2009). Going to great lengths: selection for long corolla tubes in an extremely specialized bat–flower mutualism. *Proc. R. Soc. Lond. B* **276**, 2147–2152.

1556. Mueller, T. (2010). Valley of the whales. *Natl. Geogr.* **218** #2, 118–137.

1557. Müller, G.B. (1990). Developmental mechanisms at the origin of morphological novelty: a side-effect hypothesis. In *Evolutionary Innovations* (M.H. Nitecki, ed.). University of Chicago Press, Chicago, IL, pp. 99–130.

1558. Müller, G.B. (2007). Evo-devo: extending the evolutionary synthesis. *Nature Rev. Genet.* **8**, 943–949.

1559. Müller, G.B. (2008). Evo-devo as a discipline. In *Evolving Pathways: Key Themes in Evolutionary Developmental Biology* (A. Minelli and G. Fusco, eds.). Cambridge University Press, New York, NY, pp. 5–30.

1560. Müller, G.B. and Newman, S.A., eds. (2003). *Origination of Organismal Form: Beyond the Gene in Developmental and Evolutionary Biology*. MIT Press, Cambridge, MA.

1561. Müller, G.B. and Newman, S.A. (2005). The innovation triad: an EvoDevo agenda. *J. Exp. Zool. (Mol. Dev. Evol.)* **304B**, 487–503.

1562. Müller, G.B. and Wagner, G.P. (1991). Novelty in evolution: restructuring the concept. *Annu. Rev. Ecol. Syst.* **22**, 229–256.

1563. Muller, H.J. (1939). Reversibility in evolution considered from the standpoint of genetics. *Biol. Rev.* **14**, 261–280.

1564. Müller, J., Hipsley, C.A., Head, J.J., Kardjilov, N., Hilger, A., Wuttke, M., and Reisz, R.R. (2011). Eocene lizard from Germany reveals amphisbaenian origins. *Nature* **473**, 364–367.

1565. Müller, J., Scheyer, T.M., Head, J.J., Barrett, P.M., Werneburg, I., Ericson, P.G.P., Pol, D., and Sánchez-Villagra, M.R. (2010). Homeotic effects, somitogenesis and the evolution of vertebral numbers in recent and fossil amniotes. *PNAS* **107** #5, 2118–2123.

1566. Müller, P., Rogers, K.W., Yu, S.R., Brand, M., and Schier, A.F. (2013). Morphogen transport. *Development* **140**, 1621–1638.

1567. Müller, P. and Schier, A.F. (2011). Extracellular movement of signaling molecules. *Dev. Cell* **21**, 145–158.

1568. Muneoka, K., Han, M., and Gardiner, D.M. (2008). Regrowing human limbs. *Sci. Am.* **298** #4, 56–63.

1569. Muñoz Descalzo, S. and Martinez Arias, A. (2012). The structure of Wntch signalling and the resolution of transition states in development. *Semin. Cell Dev. Biol.* **23**, 443–449.

1570. Murata, Y., Tamura, M., Aita, Y., Fujimura, K., Murakami, Y., Okabe, M., Okada, N., and Tanaka, M. (2010). Allometric growth of the trunk leads to the rostral shift of the pelvic fin in teleost fishes. *Dev. Biol.* **347**, 236–245.

1571. Murawala, P., Tanaka, E.M., and Currie, J.D. (2012). Regeneration: the ultimate example of wound healing. *Semin. Cell Dev. Biol.* **23**, 954–962.

1572. Murray, J.D. (1981). On pattern formation mechanisms for lepidopteran wing patterns and mammalian coat markings. *Phil. Trans. Roy. Soc. Lond. B* **295**, 473–496.

1573. Murray, J.D. (1981). A pre-pattern formation mechanism for animal coat markings. *J. Theor. Biol.* **88**, 161–199.

1574. Murray, J.D. (1988). How the leopard gets its spots. *Sci. Am.* **258** #3, 80–87.

1575. Murray, J.D. (1989). *Mathematical Biology*. Springer-Verlag, Berlin.

1576. Murray, J.D. (1990). Turing's theory of morphogenesis: its influence on modelling biological pattern and form. *Bull. Math. Biol.* **52**, 119–152.

1577. Murray, J.D. (2012). Vignettes from the field of mathematical biology: the application of mathematics to biology and medicine. *Interface Focus* **2**, 397–406.

1578. Murray, J.D., Deeming, D.C., and Ferguson, M.W.J. (1990). Size-dependent pigmentation-pattern formation in embryos of *Alligator mississippiensis*: time of initiation of pattern generation mechanism. *Proc. Roy. Soc. Lond. B* **239**, 279–293.

1579. Murray, J.D. and Maini, P.K. (1989). Pattern formation mechanisms–a comparison of reaction-diffusion and mechanochemical models. In *Cell to Cell Signalling: From Experiments to Theoretical Models* (A. Goldbeter, ed.). Academic Press, New York, NY, pp. 159–170.

1580. Murray, J.D. and Myerscough, M.R. (1991). Pigmentation pattern formation on snakes. *J. Theor. Biol.* **149**, 339–360.

1581. Murren, C.J. (2012). The integrated phenotype. *Integr. Comp. Biol.* **52**, 64–76.

1582. Nacu, E. and Tanaka, E.M. (2011). Limb regeneration: a new development? *Annu. Rev. Cell Dev. Biol.* **27**, 409–440.

1583. Nadrowski, B., Albert, J.T., and Göpfert, M.C. (2008). Transducer-based force generation explains active process in *Drosophila* hearing. *Curr. Biol.* **18**, 1365–1372.

1584. Nagaraj, R. and Adler, P.N. (2012). Dusky-like functions as a Rab11 effector for the deposition of cuticle during *Drosophila* bristle development. *Development* **139**, 906–916.

1585. Nagashima, H., Kuraku, S., Uchida, K., Kawashima-Ohya, Y., Narita, Y., and Kuratani, S. (2012). Body plan of turtles: an anatomical, developmental and evolutionary perspective. *Anat. Sci. Int.* **87**, 1–13.

1586. Nagorcka, B.N. (1989). Wavelike isomorphic prepatterns in development. *J. Theor. Biol.* **137**, 127–162.

1587. Nagorcka, B.N. and Mooney, J.R. (1992). From stripes to spots: prepatterns which can be produced in the skin by a reaction-diffusion system. *Math. Med. Biol.* **9**, 249–267.

1588. Nagy, L.M. and Grbic, M. (1999). Cell lineages in larval development and evolution of holometabolous insects. In *The Origin and Evolution of Larval Forms* (B.K. Hall and M.H. Wake, eds.). Academic Press, New York, NY, pp. 275–300.

1589. Nahmad, M., Glass, L., and Abouheif, E. (2008). The dynamics of developmental system drift in the gene network underlying wing polyphenism in ants: a mathematical model. *Evol. Dev.* **10**, 360–374.

1590. Nahmad, M. and Lander, A.D. (2011). Spatiotemporal mechanisms of morphgen gradient interpretation. *Curr. Opin. Gen. Dev.* **21**, 726–731.

1591. Nakamasu, A., Takahashi, G., Kanbe, A., and Kondo, S. (2009). Interactions between zebrafish pigment cells responsible for the generation of Turing patterns. *PNAS* **106** #21, 8429–8434.

1592. Nakamura, T. and Hamada, H. (2012). Left-right patterning: conserved and divergent mechanisms. *Development* **139**, 3257–3262.

1593. Naples, V.L. (1999). Morphology, evolution and function of feeding in the giant anteater (*Myrmecophaga tridactyla*). *J. Zool. Lond.* **249**, 19–41.

1594. Nardi, F., Spinsanti, G., Boore, J.L., Carapelli, A., Dallai, R., and Frati, F. (2003). Hexapod origins: monophyletic or paraphyletic? *Science* **299**, 1887–1889.

1595. Nardi, J.B. (1994). Rearrangement of epithelial cell types in an insect wing monolayer is accompanied by differential expression of a cell surface protein. *Dev. Dynamics* **199**, 315–325.

1596. Nardi, J.B. and Magee-Adams, S.M. (1986). Formation of scale spacing patterns in a moth wing. I. Epithelial feet may mediate cell rearrangement. *Dev. Biol.* **116**, 265–277.

1597. Narita, Y. and Kuratani, S. (2005). Evolution of the vertebral formulae in mammals: a perspective on developmental constraints. *J. Exp. Zool. (Mol. Dev. Evol.)* **304B**, 91–106.

1598. Natori, K., Tajiri, R., Furukawa, S., and Kojima, T. (2012). Progressive tarsal patterning in the *Drosophila* by temporally dynamic regulation of transcription factor genes. *Dev. Biol.* **361**, 450–462.

1599. Needham, J. (1933). On the dissociability of the fundamental processes in ontogenesis. *Biol. Rev.* **8**, 180–223.

1600. Negre, B. and Simpson, P. (2009). Evolution of the achaete-scute complex in insects: convergent duplication of proneural genes. *Trends Genet.* **25**, 147–152.

1601. Nelson, C. (2004). Selector genes and the genetic control of developmental modules. In *Modularity in Development and Evolution* (G. Schlosser and G.P. Wagner, eds.). University of Chicago Press, Chicago, IL, pp. 17–33.

1602. Nelson, C.E. (2012). Why don't undergraduates really "get" evolution? In *Evolution Challenges: Integrating Research and Practice in Teaching and Learning about Evolution* (K.S. Rosengren, E.M. Evans, S. Brem, and G. Sinatra, eds.). Oxford University Press, Oxford, pp. 311–347.

1603. Nelson, C.E., Morgan, B.A., Burke, A.C., Laufer, E., DiMambro, E., Murtaugh, L.C., Gonzales, E., Tessarollo, L., Parada, L.F., and Tabin, C. (1996). Analysis of *Hox* gene expression in the chick limb bud. *Development* **122**, 1449–1466.

1604. Nelson, J. and Gemmell, R. (2004). Implications of marsupial births for an understanding of behavioural development. *Int. J. Compar. Psychol.* **17**, 53–70.

1605. Nelson, J.E. and Gemmell, R.T. (2003). Birth in the northern quoll, *Dasyurus hallucatus* (Marsupialia: Dasyuridae). *Aust. J. Zool.* **51**, 187–198.

1606. Nelson, W.J. and Nusse, R. (2004). Convergence of Wnt, b-catenin, and cadherin pathways. *Science* **303**, 1483–1487.

1607. Neubauer, S. and Hublin, J.-J. (2012). The evolution of human brain development. *Evol. Biol.* **39**, 568–586.

1608. Neumann, C.J. and Cohen, S.M. (1997). Long-range action of Wingless organizes the dorsal–ventral axis of the *Drosophila* wing. *Development* **124**, 871–880.

1609. Neuweiler, G. (2000). *The Biology of Bats*. Oxford University Press, New York, NY.

1610. Newman, C. (1997). Cats: nature's masterwork. *Natl. Geogr.* **191** #6, 54–85.

1611. Newman, C., Buesching, C.D., and Wolff, J.O. (2005). The function of facial masks in "midguild" carnivores. *Oikos* **108**, 623–633.

1612. Newman, S.A. and Comper, W.D. (1990). "Generic" physical mechanisms of morphogeneiss and pattern formation. *Development* **110**, 1–18.

1613. Newman, S.A., Forgacs, G., and Müller, G.B. (2006). Before programs: The physical origination of multicellular forms. *Int. J. Dev. Biol.* **50**, 289–299.

1614. Newman, S.A. and Frisch, H.L. (1979). Dynamics of skeletal pattern formation in developing chick limb. *Science* **205**, 662–668.

1615. Newman, S.A. and Müller, G.B. (2000). Epigenetic mechanisms of character origination. *J. Exp. Zool. (Mol. Dev. Evol.)* **288**, 304–317.

1616. Newton, A. (1896). *A Dictionary of Birds*. Adam & Charles Black, London.

1617. Ng, C.S. and Kopp, A. (2008). Sex combs are important for male mating success in *Drosophila melanogaster*. *Behav. Genet.* **38**, 195–201.

1618. Ng, M., Diaz-Benjumea, F.J., Vincent, J.P., Wu, J., and Cohen, S.M. (1996). Specification of the wing by localized expression of *wingless* protein. *Nature* **381**, 316–318.

1619. Nicklen, P. (2007). Arctic ivory: hunting the narwhal. *Natl. Geogr.* **212** #2, 110–129.

1620. Nicolson, S.W. and Human, H. (2008). Bees get a head start on honey production. *Biol. Lett.* **4**, 299–301.

1621. Nie, J., Mahato, S., Mustill, W., Tipping, C., Bhattacharya, S.S., and Zelhof, A.C. (2012). Cross species analysis of Prominin reveals a conserved cellular role in invertebrate and vertebrate photoreceptor cells. *Dev. Biol.* **371**, 312–320.

1622. Niehrs, C. (2010). On growth and form: a Cartesian coordinate system of Wnt and BMP signaling specifies bilaterian body axes. *Development* **137**, 845–857.

1623. Niehuis, O., Hartig, G., Grath, S., Pohl, H., Lehmann, J., Tafer, H., Donath, A., Krauss, V., Eisenhardt, C., Hertel, J., Petersen, M., Mayer, C., Meusemann, K., Peters, R.S., Stadler, P.F., Beutel, R.G., Bornberg-Bauer, E., McKenna, D.D., and Misof, B. (2012). Genomic and morphological evidence converge to resolve the enigma of Strepsiptera. *Curr. Biol.* **22**, 1309–1313.

1624. Nielsen, C. (2001). *Animal Evolution: Interrelationships of the Living Phyla*, 2nd edn. Oxford University Press, New York, NY.

1625. Nielsen, C. (2008). Six major steps in animal evolution: are we derived sponge larvae? *Evol. Dev.* **10**, 241–257.

1626. Niimura, Y. and Nei, M. (2007). Extensive gains and losses of olfactory receptor genes in mammalian evolution. *PLoS ONE* **8**, e708.

1627. Nijhout, H.F. (1978). Wing pattern formation in Lepidoptera: a model. *J. Exp. Zool.* **206**, 119–136.

1628. Nijhout, H.F. (1980). Ontogeny of the color pattern on the wings of *Precis coenia* (Lepidoptera: Nymphalidae). *Dev. Biol.* **80**, 275–288.

1629. Nijhout, H.F. (1980). Pattern formation on lepidopteran wings: determination of an eyespot. *Dev. Biol.* **80**, 267–274.

1630. Nijhout, H.F. (1981). The color patterns of butterflies and moths. *Sci. Am.* **245** #5, 140–151.

1631. Nijhout, H.F. (1985). Independent development of homologous pattern elements in the wing patterns of butterflies. *Dev. Biol.* **108**, 146–151.

1632. Nijhout, H.F. (1986). Pattern and pattern diversity on lepidopteran wings. *BioScience* **36**, 527–533.

1633. Nijhout, H.F. (1991). *The Development and Evolution of Butterfly Wing Patterns*. Smithsonian Press, Washington, DC.

1634. Nijhout, H.F. (1996). Focus on butterfly eyespot development. *Nature* **384**, 209–210.

1635. Nijhout, H.F. (2001). Elements of butterfly wing patterns. *J. Exp. Zool. (Mol. Dev. Evol.)* **291**, 213–225.

1636. Nijhout, H.F. (2010). Molecular and physiological basis of colour pattern formation. *Adv. Insect Physiol.* **38**, 219–265.

1637. Nijhout, H.F. and Emlen, D.J. (1998). Competition among body parts in the development and evolution of insect morphology. *PNAS* **95** #7, 3685–3689.

1638. Nijhout, H.F. and German, R.Z. (2012). Developmental causes of allometry: new models and implications for phenotypic plasticity and evolution. *Integr. Comp. Biol.* **52**, 43–52.

1639. Nilson, G. and Andren, C. (1988). A new subspecies of the subalpine meadow viper, *Vipera ursinii* Bonaparte (Reptilia, Viperidae), from Greece. *Zool. Scripta* **17**, 311–314.

1640. Nilsson, D.-E. (2004). Eye evolution: a question of genetic promiscuity. *Curr. Opin. Neurobiol.* **14**, 407–414.

1641. Nilsson, D.-E. and Arendt, D. (2009). Eye evolution: the blurry beginning. *Curr. Biol.* **18**, R1096–R1098.

1642. Nilsson, D.-E., Warrant, E.J., Johnsen, S., Hanlon, R., and Shashar, N. (2012). A unique advantage for giant eyes in giant squid. *Curr. Biol.* **22**, 683–688.

1643. Niskanen, M. and Mappes, J. (2005). Significance of the dorsal zigzag pattern of *Vipera latastei gaditana* against avian predators. *J. Anim. Ecol.* **74**, 1091–1101.

1644. Nitecki, M.H. (1990). The plurality of evolutionary innovations. In *Evolutionary Innovations* (M.H. Nitecki, ed.). University of Chicago Press, Chicago, IL, pp. 3–18.

1645. Nitzan, E., Krispin, S., Pfaltzgraff, E.R., Klar, A., Labosky, P.A., and Kalcheim, C. (2013). A dynamic code of dorsal neural tube genes regulates the segregation between neurogenic and melanogenic neural crest cells. *Development* **140**, 2269–2279.

1646. Niven, J.E. and Chittka, L. (2010). Reuse of identified neurons in multiple neural circuits. *Behav. Brain Sci.* **33**, 285.

1647. Niwa, N., Akimoto-Kato, A., Niimi, T., Tojo, K., Machida, R., and Hayashi, S. (2010). Evolutionary origin of the insect wing via integration of two developmental modules. *Evol. Dev.* **12**, 168–176.

1648. Niwa, N., Inoue, Y., Nozawa, A., Saito, M., Misumi, Y., Ohuchi, H., Yoshioka, H., and Noji, S. (2000). Correlation of diversity of leg morphology in *Gryllus bimaculatus* (cricket) with divergence in *dpp* expression pattern during leg development. *Development* **127**, 4373–4381.

1649. Nolte, M.J., Hockman, D., Cretekos, C.J., Behringer, R.R., and Rasweiler, J.J., IV (2009). Embryonic staging system for the black mastiff bat, *Molossus rufus* (Molossidae), correlated with structure-function relationships in the adult. *Anat. Rec.* **292**, 155–168.

1650. Noordermeer, D. and Duboule, D. (2013). Chromatin architectures and *Hox* gene collinearity. *Curr. Top. Dev. Biol.* **104**, 113–148.

1651. Noordermeer, D., Leleu, M., Splinter, E., Rougemont, J., De Laat, W., and Duboule, D. (2011). The dynamic architecture of *Hox* gene clusters. *Science* **334**, 222–225.

1652. Norell, M., Chiappe, L., and Clark, J. (1993). New limb on the avian family tree. *Nat. Hist.* **102** #9, 38–42.

1653. Northcutt, R.G. (2012). Evolution of centralized nervous systems: two schools of evolutionary thought. *PNAS* **109** (Suppl. 1), 10626–10633.

1654. Nottebohm, E., Usui, A., Therianos, S., Kimura, K.-i., Dambly-Chaudière, C., and Ghysen, A. (1994). The gene *poxn* controls different steps of the formation of chemosensory organs in *Drosophila*. *Neuron* **12**, 25–34.

1655. Nowak, R.M. (1999). *Walker's Mammals of the World*, 6th edn. Johns Hopkins University Press, Baltimore, MD.

1656. Nowicki, J.L. and Burke, A.C. (2000). *Hox* genes and morphological identity: axial versus lateral patterning in the vertebrate mesoderm. *Development* **127**, 4265–4275.

1657. Nussbaumer, U., Halder, G., Groppe, J., Affolter, M., and Montagne, J. (2000). Expression of the *blistered/DSRF* gene is controlled by different morphogens during *Drosophila* trachea and wing development. *Mechs. Dev.* **96**, 27–36.

1658. Nusse, R. (2003). Wnts and Hedgehogs: lipid-modified proteins and similarities in signaling mechanisms at the cell surface. *Development* **130**, 5297–5305.

1659. Nüsslein-Volhard, C. (1996). Gradients that organize embryo development. *Sci. Am.* **275** #2, 54–61.

1660. Nweeia, M.T., Eichmiller, F.C., Hauschka, P.V., Tyler, E., Mead, J.G., Potter, C.W., Angnatsiak, D.P., Richard, P.R., Orr, J.R., and Black, S.R. (2012). Vestigial tooth anatomy and tusk nomenclature for *Monodon monoceros*. *Anat. Rec.* **295**, 1006–1016.

1661. Nweeia, M.T., Eichmiller, F.C., Nutarak, C., Eidelman, N., Giuseppetti, A.A., Quinn, J., Mead, J.G., K'issuk, K., Hauschka, P.V., Tyler, E.M., Potter, C., Orr, J.R., Avike, R., Nielsen, P., and Angnatsiak, D. (2009). Considerations of anatomy, morphology, evolution, and function for narwhal dentition. In *Smithsonian at the Poles: Contributions to International Polar Year Science* (I. Krupnik, M.A. Lang, and S.E. Miller, eds.). Smithsonian Institution Scholarly Press, Washington, DC, pp. 223–240.

1662. Nyakatura, J.A., Petrovitch, A., and Fischer, M.S. (2010). Limb kinematics during locomotion in the two-toed sloth (*Choloepus didactylus*, Xenarthra) and its implications for the evolution of the sloth locomotor apparatus. *Zoology* **113**, 221–234.

1663. Nyholt, D.R., Gillespie, N.A., Heath, A.C., and Martin, N.G. (2003). Genetic basis of male pattern baldness. *J. Invest. Dermatol.* **121**, 1561–1564.

1664. O'Connor, J.K., Zhang, Y., Chiappe, L.M., Meng, Q., Quanguo, L., and Di, L. (2013). A new enantiornithine from the Yixian Formation with the first recognized avian enamel specialization. *J. Vert. Paleo.* **33**, 1–12.

1665. O'Dor, R., Stewart, J., Gilly, W., Payne, J., Borges, T.C., and Thys, T. (2012). Squid rocket science: how squid launch into air. *Deep-Sea Res. II*, in press. doi 10.1016/j.dsr2.2012.07.002. [*See also* O'Dor, R.K. (2013). How squid swim and fly. *Can. J. Zool.* **91**, 413–419.]

1666. O'Leary, M.A., Bloch, J.I., Flynn, J.J., Gaudin, T.J., Giallombardo, A., Giannini, N.P., Goldgerb, S.L., Kraatz, B.P., Luo, Z.-X., Meng, J., Ni, X., Novacek, M.J., Perini, F.A., Randall, Z.S., Rougier, G.W., Sargis, E.J., Silcox, M.T., Simmons, N.B., Spaulding, M., Velazco, P.M., Weksler, M., Wible, J.R., and Cirranello, A.L. (2013). The placental mammal ancestor and the post-K-Pg radiation of placentals. *Science* **339**, 662–667.

1667. O'Toole, B. (2002). Phylogeny of the species of the superfamily Echeneoidea (Perciformes: Carangoidei: Echeneidae, Rachycentridae, and Coryphaenidae), with an interpretation of echeneid hitchhiking behaviour. *Can. J. Zool.* **80**, 596–623.

1668. Oates, A.C., Morelli, L.G., and Ares, S. (2012). Patterning embryos with oscillations: structure, function, and dynamics of the vertebrate segmentation clock. *Development* **139**, 625–639.

1669. Ober, K.A. and Jockusch, E.L. (2006). The roles of *wingless* and *decapentaplegic* in axis and appendage development in the red flour beetle, *Tribolium castaneum*. *Dev. Biol.* **294**, 391–405.

1670. Oetting, W.S. and King, R.A. (1999). Molecular basis of albinism: mutations and polymorphisms of pigmentation genes associated with albinism. *Hum. Mut.* **13**, 99–115.

1671. Offen, N., Blum, N., Meyer, A., and Begemann, G. (2008). Fgfr1 signalling in the development of a sexually selected trait in vertebrates, the sword of the swordtail fish. *BMC Dev. Biol.* **8**, Article 98 (18 pp.).

1672. Offen, N., Meyer, A., and Begemann, G. (2009). Identification of novel genes involved in the development of the sword and gonopodium in swordtail fish. *Dev. Dynamics* **238**, 1674–1687.

1673. Ogura, A., Ikeo, K., and Gojobori, T. (2005). Estimation of ancestral gene set of bilaterian animals and its implication to dynamic change of gene content in bilaterian evolution. *Gene* **345**, 65–71.

1674. Ohde, T., Yaginuma, T., and Niimi, T. (2013). Insect morphological diversification through the modification of wing serial homologs. *Science* **340**, 495–498.

1675. Oka, K., Yoshiyama, N., Tojo, K., Machida, R., and Hatakeyama, M. (2010). Characterization of abdominal appendages in the sawfly, *Athalia rosae* (Hymenoptera), by morphological and gene expression analyses. *Dev. Genes Evol.* **220**, 53–59.

1676. Okamoto, K.W. and Grether, G.F. (2013). The evolution of species recognition in competitive and mating contexts: the relative efficacy of alternative mechanisms of character displacement. *Ecol. Letters* **16**, 670–678.

1677. Oldham, S., Stocker, H., Laffargue, M., Wittwer, F., Wymann, M., and Hafen, E. (2002). The *Drosophila* insulin/IGF receptor controls growth and size by modulating PtdIns*P*3 levels. *Development* **129**, 4103–4109.

1678. Oliver, G., Wright, C.V.E., Hardwicke, J., and De Robertis, E.M. (1988). Differential antero-posterior expression of two proteins encoded by a homeobox gene in *Xenopus* and mouse embryos. *EMBO J.* **7**, 3199–3209.

1679. Oliver, G., Wright, C.V.E., Hardwicke, J., and De Robertis, E.M. (1988). A gradient of homeodomain protein in developing forelimbs of *Xenopus* and mouse embryos. *Cell* **55**, 1017–1024.

1680. Oliver, J.C. and Monteiro, A. (2011). On the origins of sexual dimorphism in butterflies. *Proc. R. Soc. Lond. B* **278**, 1981–1988.

1681. Oliver, J.C., Tong, X.-L., Gall, L.F., Piel, W.H., and Monteiro, A. (2012). A single origin for Nymphalid butterfly eyespots followed by widespread loss of associated gene expression. *PLoS Genet.* **8** #8, e1002893.

1682. Olofsson, M., Jakobsson, S., and Wiklund, C. (2013). Bird attacks on a butterfly with marginal eyespots and the role of prey concealment against the background. *Biol. J. Linnean Soc.* **109**, 290–297.

1683. Olofsson, M., Vallin, A., Jakobsson, S., and Wiklund, C. (2010). Marginal eyespots on butterfly wings deflect bird attacks under low light intensities with UV wavelengths. *PLoS ONE* **5** #5, e10798.

1684. Olson, M.E. (2012). The developmental renaissance in adaptationism. *Trends Ecol. Evol.* **27**, 278–287.

1685. Olson, S.L. and Feduccia, A. (1980). Relationships and evolution of flamingos (Aves: Phoenicopteridae). *Smithson. Contr. Zool.* No. **316**, 1–73.

1686. Olson-Manning, C.F., Wagner, M.R., and Mitchell-Olds, T. (2012). Adaptive evolution: evaluating empirical support for theoretical predictions. *Nature Rev. Genet.* **13**, 867–877.

1687. Olsson, L. (2011). Morphogenesis of pigment patterns. In *Epigenetics: Linking Genotype and Phenotype in Development and Evolution* (B. Hallgrímsson and B.K. Hall, eds.). University of California Press, Berkeley, CA, pp. 164–180.

1688. Olsson, M., Meadows, J.R.S., Truvé, K., Pielberg, G.R., Puppo, F., Mauceli, E., Quilez, J., Tonomura, N., Zanna, G., Docampo, M.J., Bassols, A., Avery, A.C., Karlsson, E.K., Thomas, A., Kastner, D.L., Bongcam-Rudloff, E., Webster, M.T., Sanchez, A., Hedhammar, A., Remmers, E.F., Andersson, L., Ferrer, L., Tintle, L., and Lindblad-Toh, K. (2011). A novel unstable duplication upstream of *HAS2* predisposes to a breed-defining skin phenotype and a periodic fever syndrome in Chinese Shar-Pei dogs. *PLoS Genet.* **7** #3, e1001332.

1689. Oosterveen, T., Kurdija, S., Alekseenko, Z., Uhde, C.W., Bergsland, M., Sandberg, M., Andersson, E.R., Dias, J.M., Muhr, J., and Ericson, J. (2012). Mechanistic differences in the transcriptional interpretation of local and long-range Shh morphogen signaling. *Dev. Cell* **23**, 1006–1019.

1690. Oppenheimer, P. (1989). The artificial menagerie. In *Artificial Life* (C.G. Langton, ed.). Addison-Wesley, New York, NY, pp. 251–274.

1691. Orenic, T.V., Held, L.I., Jr., Paddock, S.W., and Carroll, S.B. (1993). The spatial organization of epidermal structures: *hairy* establishes the geometrical pattern of *Drosophila* leg bristles by delimiting the domains of *achaete* expression. *Development* **118**, 9–20.

1692. Ortolani, A. (1999). Spots, stripes, tail tips and dark eyes: Predicting the function of carnivore colour patterns using the comparative method. *Biol. J. Linnean Soc.* **67**, 433–476.

1693. Oster, G. (1988). Lateral inhibition models of developmental processes. *Math. Biosci.* **90**, 265–286.

1694. Oster, G. (2005). George Oster (interview). *Curr. Biol.* **15**, R5–R7.

1695. Oster, G.F., Murray, J.D., and Harris, A.K. (1983). Mechanical aspects of mesenchymal morphogenesis. *J. Embryol. Exp. Morph.* **78**, 83–125.

1696. Oster, G.F., Murray, J.D., and Maini, P.K. (1985). A model for chondrogenic condensations in the developing limb: the role of extracellular matrix and cell tractions. *J. Embryol. Exp. Morph.* **89**, 93–112.

1697. Oster, G.F., Shubin, N., Murray, J.D., and Alberch, P. (1988). Evolution and morphogenetic rules: the shape of the vertebrate limb in ontogeny and phylogeny. *Evolution* **42**, 862–884.

1698. Osterauer, R., Marschner, L., Betz, O., Gerberding, M., Sawasdee, B., Cloetens, P., Haus, N., Sures, B., Triebskorn, R., and Köhler, H.-R. (2010). Turning snails into slugs: induced body plan changes and formation of an internal shell. *Evol. Dev.* **12**, 474–483.

1699. Othmer, H.G., Painter, K., Umulis, D., and Xue, C. (2009). The intersection of theory and application in elucidating pattern formation in developmental biology. *Math. Model. Nat. Phenom.* **4** #4, 3–82.

1700. Outomuro, D., Adams, D.C., and Johansson, F. (2013). The evolution of wing shape in ornamented-winged damselflies (Calopterygidae, Odonata). *Evol. Biol.* **40**, 300–309.

1701. Ouweneel, W.J. (1976). Developmental genetics of homoeosis. *Adv. Genet.* **18**, 179–248.

1702. Pabo, C.O. and Sauer, R.T. (1992). Transcription factors: structural families and principles of DNA recognition. *Annu. Rev. Biochem.* **61**, 1053–1095.

1703. Packard, A. (1972). Cephalopods and fish: the limits of convergence. *Biol. Rev.* **47**, 241–307.

1704. Packer, C. (2010). Lions. *Curr. Biol.* **20**, R590–R591.

1705. Page, D.T. (2002). Inductive patterning of the embryonic brain in *Drosophila*. *Development* **129**, 2121–2128.

1706. Pajni-Underwood, S., Wilson, C.P., Elder, C., Mishina, Y., and Lewandoski, M. (2007). BMP signals control limb bud interdigital programmed cell death by regulating FGF signaling. *Development* **134**, 2359–2368.

1707. Palmer, A.R. (2005). Antisymmetry. In *Variation: A Central Concept in Biology* (B. Hallgrímsson and B.K. Hall, eds.). Elsevier Academic Press, New York, NY, pp. 359–397.

1708. Palmer, A.R. (2011). Developmental plasticity and the origin of novel forms: unveiling cryptic genetic variation via "use" and "disuse". *J. Exp. Zool. (Mol. Dev. Evol.)* **318B**, 466–479.

1709. Palmer, C. and Dyke, G.J. (2010). Biomechanics of the unique pterosaur pteroid. *Proc. R. Soc. Lond. B* **277**, 1121–1127.

1710. Palopoli, M.F. and Patel, N.H. (1998). Evolution of the interaction between *Hox* genes and a downstream target. *Curr. Biol.* **8**, 587–590.

1711. Panchen, A.L. (2001). Étienne Geoffroy St.-Hilaire: father of "evo-devo"? *Evol. Dev.* **3**, 41–46.

1712. Pang, K., Ryan, J.F., Baxevanis, A.D., and Martindale, M.Q. (2011). Evolution of the TGF-β signaling pathway and its potential role in the ctenophore, *Mnemiopsis leidyi*. *PLoS ONE* **6** #9, e24152.

1713. Panganiban, G. (2000). *Distal-less* function during *Drosophila* appendage and sense organ development. *Dev. Dynamics* **218**, 554–562.

1714. Panganiban, G., Nagy, L., and Carroll, S.B. (1994). The role of the *Distal-less* gene in the development and evolution of insect limbs. *Curr. Biol.* **4**, 671–675.

1715. Panganiban, G. and Rubenstein, J.L.R. (2002). Developmental functions of the *Distal-less*/*Dlx* homeobox genes. *Development* **129**, 4371–4386.

1716. Panhuis, T.M., Butlin, R., Zuk, M., and Tregenza, T. (2001). Sexual selection and speciation. *Trends Ecol. Evol.* **16**, 364–371.

1717. Pantalacci, S., Sémon, M., Martin, A., Chevret, P., and Laudet, V. (2009). Heterochronic shifts explain variations in a sequentially developing repeated pattern: palatal ridges of muroid rodents. *Evol. Dev.* **11**, 422–433.

1718. Pantin, C.F.A. (1951). Organic design. *Advancement of Science (London)* **8**, 138–150.

1719. Papa, R., Martin, A., and Reed, R.D. (2009). Genomic hotspots of adaptation in butterfly wing pattern evolution. *Curr. Opin. Gen. Dev.* **18**, 559–564.

1720. Papatsenko, D. (2009). Stripe formation in the early fly embryo: principles, models, and networks. *BioEssays* **31**, 1172–1180.

1721. Parichy, D.M. (2003). Pigment patterns: fish in stripes and spots. *Curr. Biol.* **13**, R947–R950.

1722. Paris, M., Escriva, H., Schubert, M., Brunet, F., Brtko, J., Ciesielski, F., Roecklin, D., Vivat-Hannah, V., Jamin, E.L., Cravedi, J.-P., Scanlan, T.S., Renaud, J.-P., Holland, N.D., and Laudet, V. (2008). Amphioxus postembryonic development reveals the homology of chordate metamorphosis. *Curr. Biol.* **18**, 825–830.

1723. Parker, G.H. (1928). Vestigial organs. In *Creation by Evolution: A Consensus of Present-day Knowledge as Set Forth by Leading Authorities in Non-Technical Language That All May Understand* (F. Mason, ed.). Macmillan, New York, pp. 34–48.

1724. Parsons, K.J. and Albertson, R.C. (2009). Roles for BMP4 and CaM1 in shaping the jaw: evo-devo and beyond. *Annu. Rev. Genet.* **43**, 369–388.

1725. Parsons, R., Aldous-Mycock, C., and Perrin, M.R. (2007). A genetic index for stripe-pattern reduction in the zebra: the quagga project. *S. Afr. J. Wildlife Res.* **37** #2, 105–116.

1726. Partridge, J.C. (2012). Sensory ecology: giant eyes for giant predators? *Curr. Biol.* **22**, R268–R270.

1727. Passalacqua, K.D., Hrycaj, S., Mahfooz, N., and Popadic, A. (2010). Evolving expression patterns of the homeotic gene *Scr* in insects. *Int. J. Dev. Biol.* **54**, 897–904.

1728. Paterson, J.R., García-Bellido, D.C., Lee, M.S.Y., Brock, G.A., Jago, J.B., and Edgecombe, G.D. (2011). Acute vision in the giant Cambrian predator *Anomalocaris* and the origin of compound eyes. *Nature* **480**, 237–240.

1729. Patwari, P., Emilsson, V., Schadt, E.E., Chutkow, W.A., Lee, S., Marsili, A., Zhang, Y., Dobrin, R., Cohen, D.E., Larsen, P.R., Zavacki, A.M., Fong, L.G., Young, S.G., and Lee, R.T. (2011). The arrestin domain-containing 3 protein regulates body mass and energy expenditure. *Cell Metab.* **14**, 671–683.

1730. Paulsen, S.M. and Nijhout, H.F. (1993). Phenotypic correlation structure among elements of the color pattern in *Precis coenia* (Lepidoptera: Nymphalidae). *Evolution* **47**, 593–618.

1731. Pavan, W.J. and Raible, D.W. (2012). Specification of neural crest into sensory neuron and melanocyte lineages. *Dev. Biol.* **366**, 55–63.

1732. Pavlicev, M. and Wagner, G.P. (2012). Coming to grips with evolvability. *Evo. Edu. Outreach* **5**, 231–244.

1733. Pavlopoulos, A. and Akam, M. (2011). Hox gene *Ultrabithorax* regulates distinct sets of target genes at successive stages of *Drosophila* haltere morphogenesis. *PNAS* **108** #7, 2855–2860.

1734. Pavlopoulos, A. and Averof, M. (2002). Developmental evolution: Hox proteins ring the changes. *Curr. Biol.* **12**, R291–R293.

1735. Payankaulam, S., Li, L.M., and Arnosti, D.N. (2010). Transcriptional repression: conserved and evolved features. *Curr. Biol.* **20**, R764–R771.

1736. Pearl, R. (1913). On the correlation between the number of mammae of the dam and size of litter in mammals. I. Interracial correlation. *Proc. Soc. Exp. Biol. Med.* **11**, 27–30.

1737. Pearl, R. (1913). On the correlation between the number of mammae of the dam and size of litter in mammals. II. Intraracial correlation in swine. *Proc. Soc. Exp. Biol. Med.* **11**, 31–32.

1738. Pearson, J.C., Lemons, D., and McGinnis, W. (2005). Modulating *Hox* gene functions during animal body patterning. *Nature Rev. Genet.* **6**, 893–904.

1739. Pearson, J.E. (1993). Complex patterns in a simple system. *Science* **261**, 189–192.

1740. Pecoits, E., Konhauser, K.O., Aubet, N.R., Heaman, L.M., Veroslavsky, G., Stern, R.A., and Gingras, M.K. (2012). Bilaterian burrows and grazing behavior at >585 million years ago. *Science* **336**, 1693–1696.

1741. Peery, M.Z. and Pauli, J.N. (2012). The mating system of a "lazy" mammal, Hoffmann's two-toed sloth. *Anim. Behav.* **84**, 555–562.

1742. Peichl, L., Behrmann, G., and Kröger, R.H.H. (2001). For whales and seals the ocean is not blue: a visual pigment loss in marine mammals. *Eur. J. Neurosci.* **13**, 1520–1528.

1743. Pelaz, S., Urquía, N., and Morata, G. (1993). Normal and ectopic domains of the homeotic gene *Sex combs reduced* of *Drosophila*. *Development* **117**, 917–923.

1744. Peng, Y., Han, C., and Axelrod, J.D. (2012). Planar polarized protrusions break the symmetry of EGFR signaling during *Drosophila* bract cell fate induction. *Dev. Cell* **23**, 507–518.

1745. Pennacchio, L.A., Bickmore, W., Dean, A., Nobrega, M.A., and Bejerano, G. (2013). Enhancers: five essential questions. *Nature Rev. Genet.* **14**, 288–295.

1746. Perry, S.F., Similowski, T., Klein, W., and Codd, J.R. (2010). The evolutionary origin of the mammalian diaphragm. *Respir. Physiol. Neurobiol.* **171**, 1–16.

1747. Peterkova, R., Lesot, H., and Peterka, M. (2006). Phylogenetic memory of developing mammalian dentition. *J. Exp. Zool. (Mol. Dev. Evol.)* **306B**, 234–250.

1748. Peterson, A. (1962). *Larvae of Insects: An Introduction to Nearctic Species. Part I. Lepidoptera and Plant Infesting Hymenoptera*, 4th edn. Edwards Bros., Columbus, OH.

1749. Peterson, A.A. and Peterson, A.T. (1992). Aztec exploitation of cloud forests: tributes of liquidambar resin and quetzal feathers. *Global Ecol. Biogeogr. Lett.* **2**, 165–173.

1750. Petzoldt, A.G., Coutelis, J.-B., Géminard, C., Spéder, P., Suzanne, M., Cerezo, D., and Noselli, S. (2012). DE-Cadherin regulates unconventional Myosin ID and Myosin IC in *Drosophila* left-right asymmetry establishment. *Development* **139**, 1874–1884.

1751. Pfennig, D.W. and Pfennig, K.S. (2012). *Evolution's Wedge: Competition and the Origins of Diversity*. University of California Press, Berkeley, CA.

1752. Philippe, H., Brinkmann, H., Copley, R.R., Moroz, L.L., Nakano, H., Poustka, A.J., Wallberg, A., Peterson, K.J., and Telford, M.J. (2011). Acoelomorph flatworms are deuterostomes related to *Xenoturbella*. *Nature* **470**, 255–258.

1753. Philippe, H., Derelle, R., Lopez, P., Pick, K., Borchiellini, C., Boury-Esnault, N., Vacelet, J., Renard, E., Houliston, E., Quéinnec, E., Da Silva, C., Wincker, P., Le Guyader, H., Leys, S., Jackson, D.J., Schreiber, F., Erpenbeck, D., Morgenstern, B., Wörheide, G., and Manuel, M. (2009). Phylogenomics revives traditional views on deep animal relationships. *Curr. Biol.* **19**, 706–712.

1754. Phinchongsakuldit, J., MacArthur, S., and Brookfield, J.F.Y. (2004). Evolution of developmental genes: Molecular microevolution of enhancer sequences at the *Ubx* locus in *Drosophila* and its impact on developmental phenotypes. *Mol. Biol. Evol.* **21**, 348–363.

1755. Piasecka, B., Lichocki, P., Moretti, S., Bergmann, S., and Robinson-Rechavi, M. (2013). The hourglass and the early conservation models: co-existing patterns of developmental constraints in vertebrates. *PLoS Genet.* **9** #4, e1003476.

1756. Pick, L. and Heffer, A. (2012). *Hox* gene evolution: multiple mechanisms contributing to evolutionary novelties. *Ann. N. Y. Acad. Sci.* **1256**, 15–32.

1757. Picken, L.E.R. (1949). Shape and molecular orientation in lepidopteran scales. *Phil. Trans. R. Soc. Lond. B* **234** #608, 1–28.

1758. Pigliucci, M. (2008). Is evolvability evolvable? *Nature Rev. Gen.* **9**, 75–82.

1759. Pinney, R. (1981). *The Snake Book*. Doubleday, Garden City, NY.

1760. Pires-daSilva, A. and Sommer, R.J. (2003). The evolution of signalling pathways in animal development. *Nature Rev. Genet.* **4**, 39–49.

1761. Pitsouli, C. and Perrimon, N. (2008). Our fly cousins' gut. *Nature* **454**, 592–593.

1762. Platt, J.R. (1964). Strong inference. *Science* **146**, 347–353.

1763. Plikus, M. and Chuong, C.-M. (2004). Making waves with hairs. *J. Invest. Dermatol.* **122**, vii–ix.

1764. Plikus, M.V., Baker, R.E., Chen, C.-C., Fare, C., de la Cruz, D., Andl, T., Maini, P.K., Millar, S.E., Widelitz, R., and Chuong, C.-M. (2011). Self-organizing and stochastic behaviors during the regeneration of hair stem cells. *Science* **332**, 586–589.

1765. Plikus, M.V. and Chuong, C.-M. (2008). Complex hair cycle domain patterns and regenerative hair waves in living rodents. *J. Invest. Dermatol.* **128**, 1071–1080.

1766. Polak, M., Starmer, W.T., and Wolf, L.L. (2004). Sexual selection for size and symmetry in a diversifying secondary sexual character in *Drosophila bipectinata* Duda (Diptera: Drosophilidae). *Evolution* **58**, 597–607.

1767. Policansky, D. (1982). The asymmetry of flounders. *Sci. Am.* **246** #5, 116–122.

1768. Polly, P.D. (2007). Development with a bite. *Nature* **449**, 413–415.

1769. Polly, P.D., Head, J.J., and Cohn, M.J. (2001). Testing modularity and dissociation: the evolution of regional proportions in snakes. In *Beyond Heterochrony: The Evolution of Development* (M.L. Zelditch, ed.). Wiley-Liss, New York, NY, pp. 305–335.

1770. Pool, R. (1991). Did Turing discover how the leopard got its spots? *Science* **251**, 627.

1771. Popadic, A., Abzhanov, A., Rusch, D., and Kaufman, T.C. (1998). Understanding the genetic basis of morphological evolution: the role of homeotic genes in the diversification of the arthropod bauplan. *Int. J. Dev. Biol.* **42**, 453–461.

1772. Popadic, A., Panganiban, G., Rusch, D., Shear, W.A., and Kaufman, T.C. (1998). Molecular evidence for the gnathobasic derivation of arthropod mandibles and for the appendicular origin of the labrum and other structures. *Dev. Genes Evol.* **208**, 142–150.

1773. Porcher, A. and Dostatni, N. (2010). The Bicoid morphogen system. *Curr. Biol.* **20**, R249–R254.

1774. Porter, M.L. and Crandall, K.A. (2003). Lost along the way: the significance of evolution in reverse. *Trends Ecol. Evol.* **18**, 541–547.

1775. Portmann, A. (1967). *Animal Forms and Patterns: A Study of the Appearance of Animals.* Schocken Books, New York, NY.

1776. Posnien, N., Bashasab, F., and Bucher, G. (2009). The insect upper lip (labrum) is a nonsegmental appendage-like structure. *Evol. Dev.* **11**, 480–488.

1777. Poss, K.D. (2010). Advances in understanding tissue regenerative capacity and mechanisms in animals. *Nature Rev. Genet.* **11**, 710–722.

1778. Pottin, K., Hinaux, H., and Rétaux, S. (2011). Restoring eye size in *Astyanax mexicanus* blind cavefish embryos through modulation of the *Shh* and *Fgf8* forebrain organising centers. *Development* **138**, 2467–2476.

1779. Pough, F.H., Janis, C.M., and Heiser, J.B. (2009). *Vertebrate Life.* Benjamin Cummings, New York, NY.

1780. Pourquié, O. (2011). Vertebrate segmentation: from cyclic gene networks to scoliosis. *Cell* **145**, 650–663. [*See also* Bénazéraf, B. and Pourquié, O. (2013). Formation and segmentation of the vertebrate body axis. *Annu. Rev. Cell Dev. Biol.* **29**, 1–26.]

1781. Powell, B.C. and Rogers, G.E. (1990). Cyclic hair-loss and regrowth in transgenic mice overexpressing an intermediate filament gene. *EMBO J.* **9**, 1485–1493.

1782. Preston, J.C. and Hileman, L.C. (2012). Parallel evolution of TCP and B-class genes in Commelinaceae flower bilateral symmetry. *EvoDevo* **3**, Article 6 (14 pp.).

1783. Prince, V.E. and Pickett, F.B. (2002). Splitting pairs: the diverging fates of duplicated genes. *Nature Rev. Genet.* **3**, 827–837.

1784. Pringle, J.W.S. (1948). The gyroscopic mechanism of the halteres of diptera. *Phil. Trans. Roy. Soc. Lond. B* **233**, 347–384.

1785. Pringle, J.W.S. (1957). *Insect Flight.* Cambridge Monographs in Experimental Biology, Vol. **9**. Cambridge University Press, New York, NY.

1786. Pringle, J.W.S. (1975). Insect flight. *Oxford Biol. Readers* **52**, 16 pp.

1787. Proske, U. and Gregory, E. (2003). Electrolocation in the platypus–some speculations. *Comp. Biochem. Physiol. A: Mol. & Integr. Physiol.* **136**, 821–825.

1788. Protas, M.E. and Patel, N.H. (2008). Evolution of color patterns. *Annu. Rev. Cell Dev. Biol.* **24**, 425–446. [*See also* Ruxton, G.D. (2013). Hide and seek and other sensory games. *Curr. Biol.* **23**, R465–R467.]

1789. Prothero, D.R. (1987). The rise and fall of the American rhino. *Nat. Hist.* **96** #8, 26–33.

1790. Prothero, D.R. (2009). Evolutionary transitions in the fossil record of terrestrial hoofed mammals. *Evo. Edu. Outreach* **2**, 289–302.

1791. Prothero, D.R. and Schoch, R.M. (2002). *Horns, Tusks, and Flippers: The Evolution of Hoofed Mammals*. Johns Hopkins University Press, Baltimore, MD.

1792. Prpic, N.-M. (2008). Arthropod appendages: a prime example for the evolution of morphological diversity and innovation. In *Evolving Pathways: Key Themes in Evolutionary Developmental Biology* (A. Minelli and G. Fusco, eds.). Cambridge University Press, New York, NY, pp. 381–398.

1793. Prud'homme, B. and Gompel, N. (2010). Genomic hourglass. *Nature* **468**, 768–769.

1794. Prud'homme, B., Gompel, N., and Carroll, S.B. (2007). Emerging principles of regulatory evolution. *PNAS* **104** (Suppl. 1) 8605–8612.

1795. Prud'homme, B., Gompel, N., Rokas, A., Kassner, V.A., Williams, T.M., Yeh, S.-D., True, J.R., and Carroll, S.B. (2006). Repeated morphological evolution through *cis*-regulatory changes in a pleiotropic gene. *Nature* **440**, 1050–1053.

1796. Prud'homme, B., Minervino, C., Hocine, M., Cande, J.D., Aouane, A., Dufour, H.D., Kassner, V.A., and Gompel, N. (2011). Body plan innovation in treehoppers through the evolution of an extra wing-like appendage. *Nature* **473**, 83–86.

1797. Prudic, K.L., Jeon, C., Cao, H., and Monteiro, A. (2011). Developmental plasiticity in sexual roles of butterfly species drives mutual sexual ornamentation. *Science* **331**, 73–75.

1798. Pruvosta, M., Bellonec, R., Beneckeb, N., Sandoval-Castellanosd, E., Cieslaka, M., Kuznetsovae, T., Morales-Muñizf, A., O'Connorg, T., Reissmannh, M., Hofreiteri, M., and Ludwiga, A. (2011). Genotypes of predomestic horses match phenotypes painted in Paleolithic works of cave art. *PNAS* **108** #46, 18626–18630.

1799. Prykhozhij, S.V. and Neumann, C.J. (2008). Distinct roles of Shh and Fgf signaling in regulating cell proliferation during zebrafish pectoral fin development. *BMC Dev. Biol.* **8**, Article 91 (11 pp.).

1800. Pueyo, J.I. and Couso, J.P. (2005). Parallels between the proximal-distal development of vertebrate and arthropod appendages: homology without an ancestor? *Curr. Opin. Gen. Dev.* **15**, 439–446.

1801. Pueyo, J.I. and Couso, J.P. (2011). Tarsal-less peptides control Notch signalling through the Shavenbaby transcription factor. *Dev. Biol.* **355**, 183–193.

1802. Purves, W.K., Orians, G.H., and Heller, H.C. (1992). *Life: The Science of Biology*. Sinauer, Sunderland, MA.

1803. Pyron, R.A., Burbrink, F.T., and Wiens, J.J. (2013). A phylogeny and revised classification of Squamata, including 4161 species of lizards and snakes. *BMC Evol. Biol.* **13**, Article 93 (53 pp.).

1804. Quan, X.-J. and Hassan, B.A. (2005). From skin to nerve: flies, vertebrates and the first helix. *Cell. Mol. Life Sci.* **62**, 2036–2049.

1805. Quigley, I.K., Manuel, J.L., Roberts, R.A., Nuckels, R.J., Herrington, E.R., MacDonald, E.L., and Parichy, D.M. (2004). Evolutionary diversification of pigment pattern in *Danio* fishes: differential *fms* dependence and stripe loss in *D. albolineatus*. *Development* **132**, 89–104.

1806. Quigley, I.K., Turner, J.M., Nuckels, R.J., Manuel, J.L., Budi, E.H., MacDonald, E.L., and Parichy, D.M. (2004). Pigment pattern evolution by differential deployment of neural crest and post-embryonic melanophore lineages in *Danio* fishes. *Development* **131**, 6053–6069.

1807. Rachlow, J.L. and Berger, J. (1997). Conservation implications of patterns of horn regeneration in dehorned white rhinos. *Conserv. Biol.* **11**, 84–91.

1808. Radinsky, L. and Emerson, S. (1982). The late, great sabertooths. *Nat. Hist.* **91** #4, 50–57.

1809. Raff, E.C. and Raff, R.A. (2000). Dissociability, modularity, evolvability. *Evol. Dev.* **2**, 235–237.

1810. Raff, R.A. (1996). *The Shape of Life: Genes, Development, and the Evolution of Animal Form*. University of Chicago Press, Chicago, IL.

1811. Raff, R.A. and Kaufman, T.C. (1983). *Embryos, Genes, and Evolution: The Developmental-Genetic Basis of Evolutionary Change*. Macmillan, New York, NY.

1812. Raff, R.A. and Sly, B.J. (2000). Modularity and dissociation in the evolution of gene expression territories in development. *Evol. Dev.* **2**, 102–113.

1813. Raible, F. and Arendt, D. (2004). Metazoan evolution: some animals are more equal than others. *Curr. Biol.* **14**, R106–R108.

1814. Ramachandran, V.S., Tyler, C.W., Gregory, R.L., Rogers-Ramachandran, D., Duensing, S., Pillsbury, C., and Ramachandran, C. (1996). Rapid adaptive camouflage in tropical flounders. *Nature* **379**, 815–818.

1815. Ramel, M.-C. and Hill, C.S. (2013). The ventral to dorsal BMP activity gradient in the early zebrafish embryo is determined by graded expression of BMP ligands. *Dev. Biol.* **378**, 170–182.

1816. Ramos, D.M. and Monteiro, A. (2007). Transgenic approaches to study wing color pattern development in Lepidoptera. *Mol. BioSyst.* **3**, 530–535.

1817. Ramsden, C.A., Bankier, A., Brown, T.J., Cowen, P.S.J., Frost, G.I., McCallum, D.D., Studdert, V.P., and Fraser, J.R.E. (2000). A new disorder of hyaluronan metabolism associated with generalized folding and thickening of the skin. *J. Ped.* **136**, 62–68.

1818. Rancourt, D.E., Tsuzuki, T., and Capecchi, M.R. (1995). Genetic interaction between *hoxb-5* and *hoxb-6* is revealed by nonallelic noncomplementation. *Genes Dev.* **9**, 108–122.

1819. Randall, V.A. (2007). Hormonal regulation of hair follicles exhibits a biological paradox. *Semin. Cell Dev. Biol.* **18**, 274–285.

1820. Randsholt, N.B. and Santamaria, P. (2008). How *Drosophila* change their combs: the Hox gene *Sex combs reduced* and sex comb variation among *Sophophora* species. *Evol. Dev.* **10**, 121–133.

1821. Rasmussen, A.R., Murphy, J.C., Ompi, M., Gibbons, J.W., and Uetz, P. (2011). Marine reptiles. *PLoS ONE* **6** #11, e27373.

1822. Rastegar, S., Hess, I., Dickmeis, T., Nicod, J.C., Ertzer, R., Hadzhiev, Y., Thies, W.-G., Scherer, G., and Strähle, U. (2008). The words of the regulatory code are arranged in a variable manner in highly conserved enhancers. *Dev. Biol.* **318**, 366–377.

1823. Raubenheimer, E.J. (2000). Early development of the tush and the tusk of the African elephant (*Loxodonta africana*). *Arch. Oral Biol.* **45**, 983–986.

1824. Raup, D.M. (1966). Geometric analysis of shell coiling: general problems. *J. Paleontol.* **40**, 1178–1190.

1825. Raup, D.M. and Michelson, A. (1965). Theoretical morphology of the coiled shell. *Science* **147**, 1294–1295.

1826. Rawls, J.F., Mellgren, E.M., and Johnson, S.L. (2001). How the zebrafish gets its stripes. *Dev. Biol.* **240**, 301–314.

1827. Raynaud, A. (1985). Development of limbs and embryonic limb reduction. In *Biology of the Reptilia*, Vol. 15: Development, Part B (C. Gans and F. Billett, eds.). Wiley, New York, NY, pp. 60–148.

1828. Razzell, W. and Martin, P. (2012). Embryonic clutch control. *Science* **335**, 1181–1182.

1829. Reaka, M.L. (1981). The hole shrimp story. *Nat. Hist.* **90** #7, 36–43.

1830. Rebeiz, M., Stone, T., and Posakony, J.W. (2005). An ancient transcriptional regulatory linkage. *Dev. Biol.* **281**, 299–308.

1831. Rebollo, R., Romanish, M.T., and Mager, D.L. (2012). Transposable elements: an abundant and natural source of regulatory sequences for host genes. *Annu. Rev. Genet.* **46**, 21–42.

1832. Reed, R.D. (2004). Evidence for Notch-mediated lateral inhibition in organizing butterfly wing scales. *Dev. Genes Evol.* **214**, 43–46.

1833. Reed, R.D. and Nagy, L.M. (2005). Evolutionary redeployment of a biosynthetic module: expression of eye pigment genes *vermilion*, *cinnabar*, and *white* in butterfly wing development. *Evol. Dev.* **7**, 301–311.

1834. Reed, R.D., Papa, R., Martin, A., Hines, H.M., Counterman, B.A., Pardo-Diaz, C., Jiggins, C.D., Chamberlain, N.L., Kronforst, M.R., Chen, R., Halder, G., Nijhout, H.F., and McMillan, W.O. (2011). *optix* drives the repeated convergent evolution of butterfly wing pattern mimicry. *Science* **333**, 1137–1141.

1835. Reed, R.D. and Serfas, M.S. (2004). Butterfly wing pattern evolution is associated with changes in a Notch/Distal-less temporal pattern formation process. *Curr. Biol.* **14**, 1159–1166.

1836. Reeder, D.M., Helgen, K.M., Vodzak, M.E., Lunde, D.P., and Ejotre, I. (2013). A new genus for a rare African vespertilionid bat: insights from South Sudan. *ZooKeys* **285**, 89–115.

1837. Reeves, R.R. and Mitchell, E. (1981). The whale behind the tusk. *Nat. Hist.* **90** #8, 50–57.

1838. Rehorn, K.-P., Thelen, H., Michelson, A.M., and Reuter, R. (1996). A molecular aspect of hematopoiesis and endoderm development common to vertebrates and *Drosophila*. *Development* **122**, 4023–4031.

1839. Reik, E.F. (1976). Four-winged diptera from the upper Permian of Australia. *Proc. Linnean Soc. New South Wales* **101** #4, 250–255.

1840. Reinius, B., Saetre, P., Leonard, J.A., Blekhman, R., Merino-Martinez, R., Gilad, Y., and Jazin, E. (2008). An evolutionarily conserved sexual signature in the primate brain. *PLoS Genet.* **4** #6, e1000100.

1841. Reiss, J.O. (2003). Time. In *Keywords and Concepts in Evolutionary Developmental Biology* (B.K. Hall and W.M. Olson, eds.). Harvard University Press, Cambridge, MA, pp. 358–368.

1842. Reisz, R.R. (2006). Origin of dental occlusion in tetrapods: signal for terrestrial vertebrate evolution? *J. Exp. Zool. (Mol. Dev. Evol.)* **306B**, 261–277.

1843. Reisz, R.R. and Head, J.J. (2008). Turtle origins out to sea. *Nature* **456**, 450–451.

1844. Renaud, S., Pantalacci, S., and Auffray, J.-C. (2011). Differential evolvability along lines of least resistance of upper and lower molars in island house mice. *PLoS ONE* **6** #5, e18951.

1845. Renoult, J.P., Schaefer, H.M., Sallé, B., and Charpentier, M.J.E. (2011). The evolution of the multicolored face of mandrills: insights from the perceptual space of colour vision. *PLoS ONE* **6** #12, e29117.

1846. Rhinn, M. and Dollé, P. (2012). Retinoic acid signalling during development. *Development* **139**, 843–858.

1847. Rice, S.H. (2002). The role of heterochrony in primate brain evolution. In *Human Evolution Through Developmental Change* (N. Minugh-Purvis and K.J. McNamara, eds.). Johns Hopkins University Press, Baltimore, MD, pp. 154–170.

1848. Richards, A.G. (1951). *The Integument of Arthropods*. University of Minnesota Press, Minneapolis, MN.

1849. Richardson, M.K. and Chipman, A.D. (2003). Developmental constraints in a comparative framework: a test case using variations in phalanx number during amniote evolution. *J. Exp. Zool. (Mol. Dev. Evol.)* **296B**, 8–22.

1850. Richardson, M.K., Gobes, S.M.H., van Leeuwen, A.C., Polman, J.A.E., Pieau, C., and Sánchez-Villagra, M.R. (2009). Heterochrony in limb evolution: developmental mechanisms and natural selection. *J. Exp. Zool. (Mol. Dev. Evol.)* **312B**, 639–664.

1851. Richardson, M.K., Hornbruch, A., and Wolpert, L. (1990). Mechanisms of pigment pattern formation in the quail embryo. *Development* **109**, 81–89.

1852. Richardson, M.K., Jeffery, J.E., and Tabin, C.J. (2004). Proximodistal patterning of the limb: insights from evolutionary morphology. *Evol. Dev.* **6**, 1–5.

1853. Richardson, M.K. and Keuck, G. (2002). Haeckel's ABC of evolution and development. *Biol. Rev.* **77**, 495–528.

1854. Richardson, M.K. and Oelschläger, H.H.A. (2002). Time, pattern, and heterochrony: a study of hyperphalangy in the dolphin embryo flipper. *Evol. Dev.* **4**, 435–444.

1855. Richmond, D.L. and Oates, A.C. (2012). The segmentation clock: inherited trait or universal design principle? *Curr. Opin. Gen. Dev.* **22**, 600–606.

1856. Ridley, M. (2004). *Evolution*, 3rd edn. Blackwell, Malden, MA.

1857. Riedl, R. (1978). *Order in Living Organisms: A Systems Analysis of Evolution*. Wiley, New York, NY.

1858. Rieppel, O. (2001). Turtles as hopeful monsters. *BioEssays* **23**, 987–991.

1859. Rieppel, O. (2009). How did the turtle get its shell? *Science* **325**, 154–155.

1860. Rincón-Limas, D.E., Lu, C.-H., Canal, I., Calleja, M., Rodríguez-Esteban, C., Izpisúa-Belmonte, J.C., and Botas, J. (1999). Conservation of the expression and function of *apterous* orthologs in *Drosophila* and mammals. *PNAS* **96** #5, 2165–2170.

1861. Ripple, J. (1999). *Manatees and Dugongs of the World*. Voyageur Press, Stillwater, MN.

1862. Riskin, D.K. and Hermanson, J.W. (2005). Independent evolution of running in vampire bats. *Nature* **434**, 292.

1863. Riskin, D.K., Parsons, S., Schutt, W.A., Jr., Carter, G.G., and Hermanson, J.W. (2006). Terrestrial locomotion of the New Zealand short-tailed bat *Mystacina tuberculata* and the common vampire bat *Desmodus rotundus*. *J. Exp. Biol.* **209**, 1725–1736.

1864. Rivera, A.S. and Weisblat, D.A. (2009). And Lophotrochozoa makes three: Notch/Hes signaling in annelid segmentation. *Dev. Genes Evol.* **219**, 37–43.

1865. Rizo, J. and Gierasch, L.M. (1992). Constrained peptides: models of bioactive peptides and protein substructures. *Annu. Rev. Biochem.* **61**, 387–418.

1866. Robbins, D.L. (2012). KISS and the veterans. *Boomer Magazine* **6** #2, 92–97.

1867. Robert, B. and Lallemand, Y. (2006). Anteroposterior patterning in the limb and digit specification: contributions of mouse genetics. *Dev. Dynamics* **235**, 2337–2352.

1868. Robertson, K.A. and Monteiro, A. (2005). Female *Bicyclus anynana* butterflies choose males on the basis of their dorsal UV-reflective eyespot pupils. *Proc. R. Soc. Lond. B* **272**, 1541–1546.

1869. Robertson, R.M., Pearson, K.G., and Reichert, H. (1982). Flight interneurons in the locust and the origin of insect wings. *Science* **217**, 177–179.

1870. Robinett, C.C., Vaughan, A.G., Knapp, J.-M., and Baker, B.S. (2010). Sex and the single cell. II. There is a time and place for sex. *PLoS Biol.* **8** #5, e1000365.

1871. Robinson, B.G., Khurana, S., Kuperman, A., and Atkinson, N.S. (2012). Neural adaptation leads to cognitive ethanol dependence. *Curr. Biol.* **22**, 2338–2341.

1872. Roch, F. and Akam, M. (2000). *Ultrabithorax* and the control of cell morphology in *Drosophila* halteres. *Development* **127**, 97–107.

1873. Roch, F., Baonza, A., Martín-Blanco, E., and García-Bellido, A. (1998). Genetic interactions and cell behaviour in *blistered* mutants during proliferation and differentiation of the *Drosophila* wing. *Development* **125**, 1823–1832.

1874. Roellig, D., Morelli, L.G., Ares, S., Jülicher, F., and Oates, A.C. (2011). Snapshot: The segmentation clock. *Cell* **145**, 800.

1875. Roemmich, D. (2007). Twice bitten. *Nature* **449**, 33–34.

1876. Roff, D.A. (1994). The evolution of flightlessness: is history important? *Evol. Ecol.* **8**, 639–657.

1877. Rogers, B.T., Peterson, M.D., and Kaufman, T.C. (1997). Evolution of the insect body plan as revealed by the *Sex combs reduced* expression pattern. *Development* **124**, 149–157.

1878. Rogers, B.T., Peterson, M.D., and Kaufman, T.C. (2002). The development and evolution of insect mouthparts as revealed by the expression patterns of gnathocephalic genes. *Evol. Dev.* **4**, 96–110.

1879. Rogers, K.A.C. and D'Emic, M.D. (2012). Triumph of the titans. *Sci. Am.* **306** #5, 48–55.

1880. Rogers, K.W. and Schier, A.F. (2011). Morphogen gradients: From generation to interpretation. *Annu. Rev. Cell Dev. Biol.* **27**, 377–407.

1881. Rokas, A. (2008). The origins of multicellularity and the early history of the genetic toolkit for animal development. *Annu. Rev. Genet.* **42**, 235–251.

1882. Rolf, H.J., Kierdorf, U., Kierdorf, H., Schulz, J., Seymour, N., Schliephake, H., Napp, J., Niebert, S., Wölfel, H., and Wiese, K.G. (2008). Localization and characterization of STRO-1+ cells in the deer pedicle and regenerating antler. *PLoS ONE* **3** #4, e2064.

1883. Rolian, C. (2008). Developmental basis of limb length in rodents: evidence for multiple divisions of labor in mechanisms of endochondral bone growth. *Evol. Dev.* **10**, 15–28.

1884. Rolian, C. and Willmore, K.E. (2009). Morphological integration at 50: patterns and processes of integration in biological anthropology. *Evol. Biol.* **36**, 1–4.

1885. Rome, L.C. (1997). Testing a muscle's design. *Am. Sci.* **85**, 356–363.

1886. Romer, A.S. (1933). *Vertebrate Paleontology*. University of Chicago Press, Chicago, IL.

1887. Romer, A.S. (1959). *The Vertebrate Story*. University of Chicago Press, Chicago, IL. [*See also* Knüsel, J., *et al.* (2013). A salamander's flexible spinal network for locomotion, modeled at two levels of abstraction. *Integr. Comp. Biol.* **53**, 269–282.]

1888. Romer, A.S. (1970). *The Vertebrate Body*, 4th edn. W. B. Saunders, Philadelphia, PA.

1889. Ronshaugen, M., McGinnis, N., and McGinnis, W. (2002). Hox protein mutation and macroevolution of the insect body plan. *Nature* **415**, 914–917.

1890. Roos, G., Van Wassenbergh, S., Aerts, P., Herrel, A., and Adriaens, D. (2011). Effects of snout dimensions on the hydrodynamics of suction feeding in juvenile and adult seahorses. *J. Theor. Biol.* **269**, 307–317.

1891. Roos, G., Van Wassenbergh, S., Herrel, A., Adriaens, D., and Aerts, P. (2010). Snout allometry in seahorses: insights on optimisation of pivot feeding performance during ontogeny. *J. Exp. Biol.* **213**, 2184–2193.

1892. Roos, G., Van Wassenbergh, S., Herrel, A., and Aerts, P. (2009). Kinematics of suction feeding in the seahorse *Hippocampus reidi*. *J. Exp. Biol.* **212**, 3490–3498.

1893. Rosenberg, M.I. and Desplan, C. (2010). Hiding in plain sight. *Science* **329**, 284–285.

1894. Rosenberger, A.L. and Preuschoft, H. (2012). Evolutionary morphology, cranial biomechanics and the origins of tarsiers and anthropoids. *Palaeobiol. Palaeoenv.* **92**, 507–525.

1895. Rosenqvist, G. and Berglund, A. (2011). Sexual signals and mating patterns in Syngnathidae. *J. Fish Biol.* **78**, 1647–1661.

1896. Ross, E.S. (1984). Mantids: the praying predators. *Natl. Geogr.* **165** #2, 268–279.

1897. Roth, C., Rastogi, S., Arvestad, L., Dittmar, K., Light, S., Ekman, D., and Liberles, D.A. (2007). Evolution after gene duplication: models, mechanisms, sequences, systems, and organisms. *J. Exp. Zool. (Mol. Dev. Evol.)* **308B**, 58–73.

1898. Roth, S. and Lynch, J. (2012). Axis formation: microtubules push in the right direction. *Curr. Biol.* **22**, R537–R539.

1899. Roth, V.L. (1984). On homology. *Biol. J. Linnean Soc.* **22**, 13–29.

1900. Röttinger, E. and Lowe, C.J. (2012). Evolutionary crossroads in developmental biology: hemichordates. *Development* **139**, 2463–2475.

1901. Roush, W. (1995). Wing scales may help beat the heat. *Science* **269**, 1816.

1902. Rousso, T., Lynch, J., Yogev, S., Roth, S., Schejter, E.D., and Shilo, B.-Z. (2010). Generation of distinct signaling modes via diversification of the EGFR ligand-processing cassette. *Development* **137**, 3427–3427.

1903. Roy, S. and VijayRaghavan, K. (2012). Developmental biology: taking flight. *Curr. Biol.* **22**, R63–R65.

1904. Rozowski, M. (2002). Establishing character correspondence for sensory organ traits in flies: sensory organ development provides insights for reconstructing character evolution. *Mol. Phylogenet. Evol.* **24**, 400–411.

1905. Rozowski, M. and Akam, M. (2002). Hox gene control of segment-specific bristle patterns in *Drosophila*. *Genes Dev.* **16**, 1150–1162.

1906. Rubenstein, M., Sai, Y., Chuong, C.-M., and Shen, W.-M. (2009). Regenerative patterning in *Swarm Robots:* mutual benefits of research in robotics and stem cell biology. *Int. J. Dev. Biol.* **53**, 869–881.

1907. Rudel, D. and Sommer, R.J. (2003). The evolution of developmental mechanisms. *Dev. Biol.* **264**, 15–37.

1908. Rundshagen, U., Zühlke, C., Opitz, S., Schwinger, E., and Käsmann-Kellner, B. (2004). Mutations in the *MATP* gene in five German patients affected by oculocutaneous albinism type 4. *Hum. Mut.* **23**, 106–110.

1909. Ruppert, E.E. (2005). Key characters uniting hemichordates and chordates: homologies or homoplasies? *Can. J. Zool.* **83**, 8–23.

1910. Rushlow, C.A. and Shvartsman, S.Y. (2012). Temporal dynamics, spatial range, and transcriptional interpretation of the Dorsal morphogen gradient. *Curr. Opin. Gen. Dev.* **22**, 542–546.

1911. Russell, F.E. (1984). Snake venoms. *Symp. Zool. Soc. Lond.* **52**, 469–480.

1912. Russo, G.A. and Shapiro, L.J. (2011). Morphological correlates of tail length in the catarrhine sacrum. *J. Hum. Evol.* **61**, 223–232.

1913. Russwurm, A.D.A. (1978). *Aberrations of British Butterflies*. Classey, Faringdon.

1914. Rusten, T.E., Cantera, R., Kafatos, F.C., and Barrio, R. (2002). The role of TGFb signaling in the formation of the dorsal nervous system is conserved between *Drosophila* and chordates. *Development* **129**, 3575–3584.

1915. Rusting, R.L. (2001). Hair: Why it grows, why it stops. *Sci. Am.* **284** #6, 70–79.

1916. Ruvinsky, I. and Gibson-Brown, J.J. (2000). Genetic and developmental bases of serial homology in vertebrate limb evolution. *Development* **127**, 5233–5244.

1917. Ruxton, G.D. (2002). The possible fitness benefits of striped coat coloration for zebra. *Mammal Rev.* **32** #4, 237–244.

1918. Saele, Ø., Smáradóttir, H., and Pittman, K. (2006). Twisted story of eye migration in flatfish. *J. Morph.* **267**, 730–738.

1919. Saenko, S.V., Brakefield, P.M., and Beldade, P. (2010). Single locus affects embryonic segment polarity and multiple aspects of an adult evolutionary novelty. *BMC Biol.* **8**, Article 111 (13 pp.).

1920. Saenko, S.V., French, V., Brakefield, P.M., and Beldade, P. (2008). Conserved developmental processes and the formation of evolutionary novelties: examples from butterfly wings. *Phil. Trans. R. Soc. Lond. B* **363**, 1549–1555.

1921. Saenko, S.V., Marialva, M.S.P., and Beldade, P. (2011). Involvement of the conserved *Hox* gene *Antennapedia* in the development and evolution of a novel trait. *EvoDevo* **2**, Article 9 (9 pp.).

1922. Saffo, M.B. (2005). Accidental elegance. *Am. Scholar* **74**, 18–27. [*See also* Gee, H. (2013). *The Accidental Species*. University of Chicago Press, Chicago, IL.]

1923. Salazar-Ciudad, I. (2006). On the origins of morphological disparity and its diverse developmental bases. *BioEssays* **28**, 1112–1122.

1924. Salazar-Ciudad, I. (2007). On the origins of morphological variation, canalization, robustness, and evolvability. *Integr. Comp. Biol.* **47**, 390–400.

1925. Salazar-Ciudad, I. (2010). Morphological evolution and embryonic developmental diversity in metazoa. *Development* **137**, 531–539.

1926. Salazar-Ciudad, I. (2012). Tooth patterning and evolution. *Curr. Opin. Gen. Dev.* **22**, 585–592.

1927. Salazar-Ciudad, I. and Jernvall, J. (2002). A gene network model accounting for development and evolution of mammalian teeth. *PNAS* **99** #12, 8116–8120.

1928. Salazar-Ciudad, I. and Jernvall, J. (2004). How different types of pattern formation mechanisms affect the evolution of form and development. *Evol. Dev.* **6**, 6–16.

1929. Salazar-Ciudad, I. and Jernvall, J. (2010). A computational model of teeth and the developmental origins of morphological variation. *Nature* **464**, 583–586.

1930. Salazar-Ciudad, I., Jernvall, J., and Newman, S.A. (2003). Mechanisms of pattern formation in development and evolution. *Development* **130**, 2027–2037.

1931. Salazar-Ciudad, I. and Marín-Riera, M. (2013). Adaptive dynamics under development-based genotype-phenotype maps. *Nature* **497**, 361–364.

1932. Sallé, J., Campbell, S.D., Gho, M., and Audibert, A. (2012). CycA is involved in the control of endoreplication dynamics in the *Drosophila* bristle lineage. *Development* **139**, 547–557.

1933. Salzberg, A. and Bellen, H.J. (1996). Invertebrate versus vertebrate neurogenesis: variations on the same theme? *Dev. Genet.* **18**, 1–10.

1934. Sambrani, N., Hudry, B., Maurel-Zaffran, C., Zouaz, A., Mishra, R., Merabet, S., and Graba, Y. (2013). Distinct molecular strategies for Hox-mediated limb suppression in *Drosophila*: from cooperativity to dispensability/antagonism in TALE partnership. *PLoS Genet.* **9** #3, e1003307.

1935. Sampson, S.D. (1995). Horns, herds, and hierarchies. *Nat. Hist.* **104** #6, 36–40.

1936. Sánchez, L. and Guerrero, I. (2001). The development of the *Drosophila* genital disc. *BioEssays* **23**, 698–707.

1937. Sander, K. (2002). Ernst Haeckel's ontogenetic recapitulation: irritation and incentive from 1866 to our time. *Ann. Anat.* **184**, 523–533.

1938. Sander, P.M., Christian, A., Clauss, M., Fechner, R., Gee, C.T., Griebeler, E.-M., Gunga, H.-C., Hummel, J., Mallison, H., Perry, S.F., Preuschoft, H., Rauhut, O.W.M., Remes, K., Tütken, T., Wings, O., and Witzel, U. (2011). Biology of the sauropod dinosaurs: the evolution of gigantism. *Biol. Rev.* **86**, 117–155.

1939. Sanderson, S.L. and Wasserug, R.J. (1990). Suspension-feeding vertebrates. *Sci. Am.* **262** #3, 96–101.

1940. Sane, S.P. and McHenry, M.J. (2009). The biomechanics of sensory organs. *Integr. Comp. Biol.* **49** #6, i8–i23.

1941. Sanger, T.J. (2012). The emergence of squamates as model systems for integrative biology. *Evol. Dev.* **14**, 231–233.

1942. Sanger, T.J. and Gibson-Brown, J.J. (2004). The developmental basis of limb reduction and body elongation in squamates. *Evolution* **58**, 2103–2106.

1943. Santana, S.E., Alfaro, J.L., and Alfaro, M.E. (2012). Adaptive evolution of facial colour patterns in Neotropical primates. *Proc. R. Soc. Lond. B* **279**, 2204–2211.

1944. Sato, T., Hasegawa, Y., and Manabe, M. (2006). A new elasmosaurid plesiosaur from the Upper Cretaceous of Fukushima, Japan. *Palaeontology* **49**, 467–484.

1945. Sato, Y., Yasuda, K., and Takahashi, Y. (2002). Morphological boundary forms by a novel inductive event mediated by Lunatic fringe and Notch during somitic segmentation. *Development* **129**, 3633–3644.

1946. Saunders, F. (2009). What's all the flap about? *Am. Sci.* **97**, 23–24.

1947. Savard, J., Marques-Souza, H., Aranda, M., and Tautz, D. (2006). A segmentation gene in *Tribolium* produces a polycistronic mRNA that codes for multiple conserved peptides. *Cell* **126**, 559–569.

1948. Savic, D. (1995). Model of pattern formation in animal coatings. *J. Theor. Biol.* **172**, 299–303.

1949. Savriama, Y. and Klingenberg, C.P. (2011). Beyond bilateral symmetry: geometric morphometric methods for any type of symmetry. *BMC Evol. Biol.* **11**, Article 280 (24 pp.).

1950. Sawyer, G.J. and Deak, V. (2007). *The Last Human: A Guide to Twenty-two Species of Extinct Humans*. Yale University Press, New Haven, CT.

1951. Sayed-Ahmed, A., Rudas, P., and Bartha, T. (2005). Partial cloning and localisation of leptin and its receptor in the one-humped camel (*Camelus dromedarius*). *Vet. J.* **170**, 264–269.

1952. Scanlon, J.D. and Lee, M.S.Y. (2000). The Pleistocene serpent *Wonambi* and the early evolution of snakes. *Nature* **403**, 416–420.

1953. Schier, A.F. and Needleman, D. (2009). Rise of the source-sink model. *Nature* **461**, 480–481.

1954. Schierwater, B. and DeSalle, R. (2007). Can we ever identify the Urmetazoan? *Integr. Comp. Biol.* **47**, 670–676.

1955. Schilling, T.F. and Knight, R.D. (2001). Origins of anteroposterior patterning and *Hox* gene regulation during chordate evolution. *Phil. Trans. R. Soc. Lond. B* **356**, 1599–1613.

1956. Schilling, T.F., Nie, Q., and Lander, A.D. (2012). Dynamics and precision in retinoic acid morphogen gradients. *Curr. Opin. Gen. Dev.* **22**, 562–569.

1957. Schilman, P.E., Kaiser, A., and Lighton, J.R.B. (2008). Breathe softly, beetle: Continuous gas exchange, water loss and the role of the subelytral space in the tenebrionid beetle, *Eleodes obscura*. *J. Insect Physiol.* **54**, 192–203.

1958. Schlosser, G. and Wagner, G.P. (2004). Introduction: The modularity concept in developmental and evolutionary biology. In *Modularity in Development and Evolution* (G. Schlosser and G.P. Wagner, eds.). University of Chicago Press, Chicago, IL, pp. 1–15.

1959. Schmidt, E.R. (1978). Chromatophore development and cell interactions in the skin of Xiphophorine fish. *Wilhelm Roux's Arch.* **184**, 115–134.

1960. Schmidt, J., Francois, V., Bier, E., and Kimelman, D. (1995). *Drosophila short gastrulation* induces an ectopic axis in *Xenopus*: evidence for conserved mechanisms of dorsal–ventral patterning. *Development* **121**, 4319–4328.

1961. Schmidt-Nielsen, K. (1972). *How Animals Work*. Cambridge University Press, New York, NY.

1962. Schmidt-Nielsen, K. (1984). *Scaling: Why Is Animal Size So Important?* Cambridge University Press, New York, NY.

1963. Schmidt-Nielsen, K. (1998). *The Camel's Nose: Memoirs of a Curious Scientist.* Island Press, Washington, DC.

1964. Schneider, I. and Shubin, N. (2012). Making limbs from fins. *Dev. Cell* **23**, 1121–1122.

1965. Schneider, R.A. (2005). Developmental mechanisms facilitating the evolution of bills and quills. *J. Anat.* **207**, 563–573.

1966. Schneider, R.A. (2006). How to tweak a beak: molecular techniques for studying the evolution of size and shape in Darwin's finches and other birds. *BioEssays* **29**, 1–6.

1967. Schneider, R.A. and Helms, J.A. (2003). The cellular and molecular origins of beak morphology. *Science* **299**, 565–568.

1968. Schoch, R.R. (2010). Riedl's burden and the body plan: selection, constraint, and deep time. *J. Exp. Zool. (Mol. Dev. Evol.)* **314B**, 1–10.

1969. Schoenebeck, J.J. and Ostrander, E.A. (2013). The genetics of canine skull shape variation. *Genetics* **193**, 317–325.

1970. Schoenemann, B., Castellani, C., Clarkson, E.N.K., Haug, J.T., Maas, A., Haug, C., and Waloszek, D. (2012). The sophisticated visual system of a tiny Cambrian crustacean: analysis of a stalked fossil compound eye. *Proc. Roy. Soc. Lond. B* **279**, 1335–1340.

1971. Schoenwolf, G.C., Bleyl, S.B., Brauer, P.R., and Francis-West, P.H. (2009). *Larsen's Human Embryology*, 4th edn. Churchill Livingstone, Philadelphia, PA.

1972. Schoenwolf, G.C. and Smith, J.L. (1990). Mechanisms of neurulation: traditional viewpoint and recent advances. *Development* **109**, 243–270.

1973. Scholtz, G. (2005). Homology and ontogeny: pattern and process in comparative developmental biology. *Theory Biosci.* **124**, 121–143.

1974. Scholtz, G. and Edgecombe, G.D. (2006). The evolution of arthropod heads: reconciling morphological, developmental and palaentological evidence. *Dev. Genes Evol.* **216**, 395–415.

1975. Schönbauer, C., Distler, J., Jährling, N., Radolf, M., Dodt, H.-U., Frasch, M., and Schnorrer, F. (2011). Spalt mediates an evolutionarily conserved switch to fibrillar muscle fate in insects. *Nature* **479**, 406–409.

1976. Schoppmeier, M., Fischer, S., Schmitt-Engel, C., Löhr, U., and Klingler, M. (2009). An ancient anterior patterning system promotes *caudal* repression and head formation in Ecdysozoa. *Curr. Biol.* **19**, 1811–1815.

1977. Schreiber, A.M. (2006). Asymmetric craniofacial remodeling and lateralized behavior in larval flatfish. *J. Exp. Biol.* **209**, 610–621.

1978. Schreiber, A.M. (2013). Flatfish: an asymmetric perspective on metamorphosis. *Curr. Top. Dev. Biol.* **103**, 167–194.

1979. Schröter, C. and Oates, A.C. (2010). Segment number and axial identity in a segmentation clock period mutant. *Curr. Biol.* **20**, 1254–1258.

1980. Schubert, M., Yu, J.-K., Holland, N.D., Escriva, H., Laudet, V., and Holland, L.Z. (2005). Retinoic acid signaling acts via *Hox1* to establish the posterior limit of the pharynx in the chordate amphioxus. *Development* **132**, 61–73.

1981. Schulp, A.S., Mulder, E.W.A., and Schwenk, K. (2005). Did mosasaurs have forked tongues? *Netherlands J. Geosci.* **84**, 359–371.

1982. Schutt, B. (2008). The curious, bloody lives of vampire bats. *Nat. Hist.* **117** #9, 22–27.

1983. Schwanwitsch, B.N. (1924). On the ground-plan of wing-pattern in Nymphalids and certain other families of the rhopalocerous Lepidoptera. *Proc. Zool. Soc. Lond.* **34**, 509–528.

1984. Schwanwitsch, B.N. (1926). On the modes of evolution of the wing-pattern in Nymphalids and certain other families of the Rhopalocerous Lepidoptera. *Proc. Zool. Soc. Lond.* **33**, 493–508 (plus 3 plates).

1985. Schwanwitsch, B.N. (1929). Two schemes of the wing-pattern of butterflies. *Zoomorphology* **14**, 36–58.

1986. Schwanwitsch, B.N. (1930). Studies upon the wing-pattern of *Prepona* and *Agrias*, two genera of South-American Nyphalid butterflies. *Acta Zool.* **11**, 289–410.

1987. Schwartz, J.H. (2007). Recognizing William Bateson's contributions. *Science* **315**, 1077.

1988. Schweitzer, R., Zelzer, E., and Volk, T. (2010). Connecting muscles to tendons: tendons and musculoskeletal development in flies and vertebrates. *Development* **137**, 2807–2817.

1989. Schwenk, K. (1994). A utilitarian approach to evolutionary constraint. *Zoology* **98**, 251–262.

1990. Schwenk, K. (1994). Why snakes have forked tongues. *Science* **263**, 1573–1577.

1991. Schwenk, K. (1995). Of tongues and noses: chemoreception in lizards and snakes. *Trends Ecol. Evol.* **10**, 7–12.

1992. Schwenk, K. (1995). The serpent's tongue. *Nat. Hist.* **104** #4, 48–55.

1993. Schwenk, K., ed. *Feeding: Form, Function and Evolution in Tetrapod Vertebrates*. Academic Press, New York, NY.

1994. Schwenk, K. (2001). Functional units and their selection. In *The Character Concept in Evolutionary Biology* (G.P. Wagner, ed.). Academic Press, San Diego, CA, pp. 165–198.

1995. Schwenk, K. and Wagner, G.P. (2004). The relativism of constraints on phenotypic evolution. In *Phenotypic Integration. Studying the Ecology and Evolution of Complex Phenotypes*. (M. Pigliucci and K. Preston, eds.). Oxford University Press, New York, NY, pp. 390–408.

1996. Scotland, R.W. (2010). Deep homology: a view from systematics. *BioEssays* **32**, 438–449.

1997. Sears, K.E. (2011). Novel insights into the regulation of limb development from 'natural' mammalian mutants. *BioEssays* **33**, 327–331.

1998. Sears, K.E., Behringer, R.R., Rasweiler, J.J., IV, and Niswander, L.A. (2006). Development of bat flight: morphologic and molecular evolution of bat wing digits. *PNAS* **103** #17, 6581–6586.

1999. Sears, K.E., Bormet, A.K., Rockwell, A., Powers, L.E., Cooper, L.N., and Wheeler, M.B. (2011). Developmental basis of mammalian digit reduction: a case study in pigs. *Evol. Dev.* **13**, 533–541.

2000. Seher, T.D., Ng, C.S., Signor, S.A., Podlaha, O., Barmina, O., and Kopp, A. (2012). Genetic basis of a violation of Dollo's Law: re-evolution of rotating sex combs in *Drosophila bipectinata*. *Genetics* **192**, 1465–1475.

2001. Sehnal, F., Svácha, P., and Zrzavy, J. (1996). Evolution of insect metamorphosis. In *Metamorphosis: Postembryonic Reprogramming of Gene Expression in Amphibian and Insect Cells* (L.I. Gilbert, J.R. Tata, and B.G. Atkinson, eds.). Academic Press, New York, NY, pp. 3–58.

2002. Seifert, A.W., Kiama, S.G., Seifert, M.G., Goheen, J.R., Palmer, T.M., and Maden, M. (2012). Skin shedding and tissue regeneration in African spiny mice (*Acomys*). *Nature* **489**, 561–565.

2003. Seifert, A.W., Monaghan, J.R., Smith, M.D., Pasch, B., Stier, A.C., Michonneau, F., and Maden, M. (2012). The influence of fundamental traits on mechanisms controlling appendage regeneration. *Biol. Rev.* **87**, 330–345.

2004. Seifert, A.W., Monaghan, J.R., Voss, S.R., and Maden, M. (2012). Skin regeneration in adult axolotls: a blueprint for scar-free healing in vertebrates. *PLoS ONE* **7** #4, e32875.

2005. Sekimura, T., Maini, P.K., Nardi, J.B., Zhu, M., and Murray, J.D. (1998). Pattern formation in lepidopteran wings. *Comments Theor. Biol.* **5**, 69–87.

2006. Semenov, M.V., Habas, R., MacDonald, B.T., and He, X. (2007). Snapshot: Noncanonical Wnt signaling pathways. *Cell* **131**, 1378.

2007. Senn, D.G. and Northcutt, R.G. (1973). The forebrain and midbrain of some squamates and their bearing on the origin of snakes. *J. Morph.* **140**, 135–151.

2008. Senter, P. (2007). Necks for sex: sexual selection as an explanation for sauropod dinosaur neck elongation. *J. Zool.* **271**, 45–53.

2009. Serena, M. (2000). Duck-billed platypus: Australia's urban oddity. *Natl. Geogr.* **197** #4, 118–129.

2010. Sereno, P.C. (1999). The evolution of dinosaurs. *Science* **284**, 2137–2147.

2011. Shankland, M. and Seaver, E.C. (2000). Evolution of the bilaterian body plan: What have we learned from annelids? *PNAS* **97** #9, 4434–4437.

2012. Shapiro, M.D., Bell, M.A., and Kingsley, D.M. (2006). Parallel genetic origins of pelvic reduction in vertebrates. *PNAS* **103** #37, 13753–13758.

2013. Shapiro, M.D. and Domyan, E.T. (2013). Domestic pigeons. *Curr. Biol.* **23**, R302–R303.

2014. Shapiro, M.D., Kronenberg, Z., Li, C., Domyan, E.T., Pan, H., Campbell, M., Tan, H., Huff, C.D., Hu, H., Vickrey, A.I., Nielsen, S.C.A., Stringham, S.A., Hu, H., Willerslev, E., Thomas, M., Gilbert, P., Yandell, M., Zhang, G., and Wang, J. (2013). Genomic diversity and evolution of the head crest in the rock pigeon. *Science* **339**, 1063–1067.

2015. Shapiro, M.D., Marks, M.E., Peichel, C.L., Blackman, B.K., Nereng, K.S., Jónsson, B., Schluter, D., and Kingsley, D.M. (2004). Genetic and developmental basis of evolutionary pelvic reduction in threespine sticklebacks. *Nature* **428**, 717–723.

2016. Shapiro, M.D., Shubin, N.H., and Downs, J.P. (2007). Limb diversity and digit reduction in reptilian evolution. In *Fins into Limbs: Evolution, Development, and Transformation* (B.K. Hall, ed.). University of Chicago Press, Chicago, IL, pp. 225–244.

2017. Sharon, E., Marder, M., and Swinney, H.L. (2004). Leaves, flowers and garbage bags: making waves. *Am. Sci.* **92**, 254–261.

2018. Shashidhara, L.S., Agrawal, N., Bajpai, R., Bharathi, V., and Sinha, P. (1999). Negative regulation of dorsoventral signaling by the homeotic gene *Ultrabithorax* during haltere development in *Drosophila*. *Dev. Biol.* **212**, 491–502.

2019. Shaywitz, D.A. and Melton, D.A. (2005). The molecular biography of the cell. *Cell* **120**, 729–731.

2020. Shbailat, S.J. and Abouheif, E. (2013). The wing-patterning network in the wingless castes of myrmicine and formicine ant species is a mix of evolutionarily labile and non-labile genes. *J. Exp. Zool. (Mol. Dev. Evol.)* **320B**, 74–83.

2021. Shbailat, S.J., Khila, A., and Abouheif, E. (2010). Correlations between spatiotemporal changes in gene expression and apoptosis underlie wing polyphenism in the ant *Pheidole morrisi*. *Evol. Dev.* **12**, 580–591.

2022. Shear, W.A. (1999). Millipeds. *Am. Sci.* **87**, 232–239.

2023. Shear, W.A. (2012). An insect to fill the gap. *Nature* **488**, 34–35.

2024. Sherman, P.W., Braude, S., and Jarvis, J.U.M. (1999). Litter sizes and mammary numbers of naked mole-rats: Breaking the one-half rule. *J. Mammalogy* **80**, 720–733.

2025. Sheth, R., Grégoire, D., Dumouchel, A., Scotti, M., Pham, J.M.T., Nemec, S., Bastida, M.F., Ros, M.A., and Kmita, M. (2013). Decoupling the function of *Hox* and *Shh* in developing limb reveals multiple inputs of *Hox* genes on limb growth. *Development* **140**, 2130–2138.

2026. Sheth, R., Marcon, L., Bastida, M.F., Junco, M., Quintana, L., Dahn, R., Kmita, M., Sharpe, J., and Ros, M.A. (2012). *Hox* genes regulate digit patterning by controlling the wavelength of a Turing-type mechanism. *Science* **338**, 1476–1480.

2027. Shevtsova, E., Hansson, C., Janzen, D.H., and Kjaerandsen, J. (2011). Stable structural color patterns displayed on transparent insect wings. *PNAS* **108** #2, 668–673.

2028. Shidlovskiy, F.K., Kirillova, I.V., and Wood, J. (2011). Horns of the woolly rhinoceros *Coelodonta antiquitatis* (Blumenbach, 1799) in the Ice Age Museum collection (Moscow, Russia). *Quarternary Internat.* **255**, 125–129.

2029. Shilo, B.-Z., Haskel-Ittah, M., Ben-Zvi, D., Schejter, E.D., and Barkai, N. (2013). Creating gradients by morphogen shuttling. *Trends Genet.* **29**, 339–347.

2030. Shimeld, S.M., Gaunt, S.J., Coletta, P.L., Geada, A.M.C., and Sharpe, P.T. (1993). Spatial localisation of transcripts of the *Hox-C6* gene. *J. Anat.* **183**, 515–523.

2031. Shimozono, S., Iimura, T., Kitaguchi, T., Higashijima, S., and Miyawaki, A. (2013). Visualization of an endogenous retinoic acid gradient across embryonic development. *Nature* **496**, 363–366.

2032. Shine, R. and Wall, M. (2008). Interactions between locomotion, feeding, and bodily elongation during the evolution of snakes. *Biol. J. Linnean Soc.* **95**, 293–304.

2033. Shingleton, A.W. and Frankino, W.A. (2013). New perspectives on the evolution of exaggerated traits. *BioEssays* **35**, 100–107.

2034. Shingleton, A.W., Frankino, W.A., Flatt, T., Nijhout, H.F., and Emlen, D.J. (2007). Size and shape: the developmental regulation of static allometry in insects. *BioEssays* **29**, 536–548.

2035. Shirai, L.T., Saenko, S.V., Keller, R.A., Jeronimo, M.A., Brakefield, P.M., Descimon, H., Wahlberg, N., and Beldade, P. (2012). Evolutionary history of the recruitment of conserved developmental genes in association to the formation and diversification of a novel trait. *BMC Evol. Biol.* **12**, Article 21 (11 pp.).

2036. Shohat-Ophir, G., Kaun, K.R., Azanchi, R., and Heberlein, U. (2012). Sexual deprivation increases ethanol intake in *Drosophila*. *Science* **335**, 1351–1355.

2037. Shoji, H. and Iwasa, Y. (2005). Labyrinthine versus stright-striped patterns generated by two-dimensional Turing systems. *J. Theor. Biol.* **237**, 104–116.

2038. Shoji, H., Iwasa, Y., and Kondo, S. (2003). Stripes, spots, or reversed spots in two-dimensional Turing systems. *J. Theor. Biol.* **224**, 339–350.

2039. Shoji, H., Mochizuki, A., Iwasa, Y., Hirata, M., Watanabe, T., Hioki, S., and Kondo, S. (2003). Origin of directionality in the fish stripe pattern. *Dev. Dynamics* **226**, 627–633.

2040. Sholtis, S.J. and Noonan, J.P. (2010). Gene regulation and the origins of human biological uniqueness. *Trends Genet.* **26**, 110–118.

2041. Shorrocks, B. and Croft, D.P. (2009). Necks and networks: a preliminary study of population structure in the reticulated giraffe (*Giraffa camelopardalis reticulata* de Winston). *Afr. J. Ecol.* **47**, 374–381.

2042. Shou, S., Carlson, H.L., Perez, W.D., and Stadler, H.S. (2013). *HOXA13* regulates *Aldh1a2* expression in the autopod to facilitate interdigital programmed cell death. *Dev. Dynamics* **242**, 687–698.

2043. Shroff, S., Joshi, M., and Orenic, T.V. (2007). Differential Delta expression underlies the diversity of sensory organ patterns among the legs of the *Drosophila* adult. *Mechs. Dev.* **124**, 43–58.

2044. Shubin, N., Tabin, C., and Carroll, S. (1997). Fossils, genes and the evolution of animal limbs. *Nature* **388**, 639–648.

2045. Shubin, N., Tabin, C., and Carroll, S. (2009). Deep homology and the origins of evolutionary novelty. *Nature* **457**, 818–823.

2046. Shubin, N.H. and Alberch, P. (1986). A morphogenetic approach to the origin and basic organization of the tetrapod limb. *Evol. Biol.* **20**, 319–387.

2047. Shubin, N.H. and Dahn, R.D. (2004). Lost and found. *Nature* **428**, 703–704.

2048. Sikes, N.E. (1999). Plio-Pleistocene floral context and habitat preferences of sympatric hominid species in East Africa. In *African Biogeography, Climate Change, and Human Evolution* (T.G. Bromage and F. Schrenk, eds.). Oxford University Press, New York, NY, pp. 301–315.

2049. Silverman, H.B. and Dunbar, M.J. (1980). Aggressive tusk use by the narwhal (*Monodon monocerous* L.). *Nature* **284**, 57–58.

2050. Simmonds, A.J. and Bell, J.B. (1998). A genetic and molecular analysis of an *invected-Dominant* mutation in *Drosophila melanogaster*. *Genome* **41**, 381–390.

2051. Simmonds, A.J., Brook, W.J., Cohen, S.M., and Bell, J.B. (1995). Distinguishable functions for *engrailed* and *invected* in anterior–posterior patterning in the *Drosophila* wing. *Nature* **376**, 424–427.

2052. Simmons, N.B. (2008). Taking wing. *Sci. Am.* **299** #6, 96–103.

2053. Simmons, N.B., Seymour, K.L., Habersetzer, J., and Gunnell, G.F. (2008). Primitve Early Eocene bat from Wyoming and the evolution of flight and echolocation. *Nature* **451**, 818–821.

2054. Simmons, R.E. and Scheepers, L. (1996). Winning by a neck: sexual selection in the evolution of giraffe. *Am. Nat.* **148**, 771–786.

2055. Simões-Costa, M.S., Azambuja, A.P., and Xavier-Neto, J. (2008). The search for non-chordate retinoic acid signaling: lessons from chordates. *J. Exp. Zool. (Mol. Dev. Evol.)* **310B**, 54–72.

2056. Simon, C.A. (1982). Masters of the tongue flick. *Nat. Hist.* **91** #9, 58–67.

2057. Simonnet, F. and Moczek, A.P. (2011). Conservation and diversification of gene function during mouthpart development in *Onthophagus* beetles. *Evol. Dev.* **13**, 280–289.

2058. Simpson, S.J., Sword, G.A., and Lo, N. (2011). Polyphenism in insects. *Curr. Biol.* **21**, R738–R749.

2059. Sinclair, R. (1998). Male pattern androgenetic alopecia. *BMJ* **317**, 865–869.

2060. Sinclair, R. (2007). Female pattern hair loss, dandruff and greying of hair in twins. *J. Investig. Dermatol.* **127**, 2680.

2061. Singh, A., Kango-Singh, M., Parthasarathy, R., and Gopinathan, K.P. (2007). Larval legs of mulberry silkworm *Bombyx mori* are prototypes for the adult legs. *Genesis* **45**, 169–176.

2062. Singh, N.D., Larracuente, A.M., Sackton, T.B., and Clark, A.G. (2009). Comparative genomics on the *Drosophila* phylogenetic tree. *Annu. Rev. Ecol. Evol. Syst.* **40**, 459–480.

2063. Singh, N.P. and Mishra, R.K. (2008). A double-edged sword to force posterior dominance of *Hox* genes. *BioEssays* **30**, 1058–1061.

2064. Sinigaglia, C., Busengdal, H., Leclère, L., Technau, U., and Rentzsch, F. (2013). The bilaterian head patterning gene *six3/6* controls aboral domain development in a cnidarian. *PLoS Biol.* **11** #2, e1001488.

2065. Sipper, M. (1995). Studying artificial life using a simple, general cellular model. *Artif. Life* **2**, 1–35.

2066. Sire, J.-Y., Delgado, S.C., and Girondot, M. (2008). Hen's teeth with enamel cap: from dream to impossibility. *BMC Evol. Biol.* **8**, Article 246 (11 pp.).

2067. Sites, J.W., Jr., Reeder, T.W., and Wiens, J.J. (2011). Phylogenetic insights on evolutionary novelties in lizards and snakes: sex, birth, bodies, niches, and venom. *Annu. Rev. Ecol. Evol. Syst.* **42**, 227–244.

2068. Sivak, J.G. (1977). The role of the spectacle in the visual optics of the snake eye. *Vision Res.* **17**, 293–298.

2069. Sivinski, J. (1997). Ornamentation in the Diptera. *Florida Entomologist* **80**, 142–164.

2070. Skinner, A., Lee, M.S.Y., and Hutchinson, M.N. (2008). Rapid and repeated limb loss in a clade of scincid lizards. *BMC Evol. Biol.* **8**, e310.

2071. Slack, F. and Ruvkun, G. (1997). Temporal pattern formation by heterochronic genes. *Annu. Rev. Genet.* **31**, 611–634.

2072. Slack, J.M.W., Holland, P.W.H., and Graham, C.F. (1993). The zootype and the phylotypic stage. *Nature* **361**, 490–492.

2073. Smith, A.B. (2008). Deuterostomes in a twist: the origins of a radical new body plan. *Evol. Dev.* **10**, 493–503.

2074. Smith, C.F., Schwenk, K., Earley, R.L., and Schuett, G.W. (2008). Sexual size dimorphism of the tongue in a North American pitviper. *J. Zool.* **274**, 367–374.

2075. Smith, J.V., Braun, E.L., and Kimball, R.T. (2013). Ratite nonmonophyly: independent evidence from 40 novel loci. *Syst. Biol.* **62**, 35–49.

2076. Smith, K.K. (2003). Time's arrow: heterochrony and the evolution of development. *Int. J. Dev. Biol.* **47**, 613–621.

2077. Smith, R. (2012). Cheetahs on the edge. *Natl. Geogr.* **222** #5, 110–123.

2078. Smits, P.D. and Evans, A.R. (2012). Functional constraints on tooth morphology in carnivorous mammals. *BMC Evol. Biol.* **12**, Article 146 (11 pp.).

2079. Sniegowski, P.D. and Murphy, H.A. (2006). Evolvability. *Curr. Biol.* **16**, R831–R834.

2080. Snodgrass, R.E. (1935). *Principles of Insect Morphology*. McGraw-Hill, New York, NY.

2081. Snodgrass, R.E. (1954). Insect metamorphosis. *Smithsonian Misc. Coll. Pub. 4144* **122** #9, iii–124.

2082. Snodgrass, R.E. (1958). Evolution of arthropod mechanisms. *Smithsonian Misc. Coll. Pub. 4347* **138** #2, i–77.

2083. Snodgrass, R.E. (1961). The caterpillar and the butterfly. *Smithsonian Misc. Coll. Pub. 4472* **143** #6, 1–51.

2084. Sokol, N.S. (2012). Small temporal RNAs in animal development. *Curr. Opin. Gen. Dev.* **22**, 368–373.

2085. Soligo, C. and Müller, A.E. (1999). Nails and claws in primate evolution. *J. Hum. Evol.* **36**, 97–114.

2086. Solounias, N. (1999). The remarkable anatomy of the giraffe's neck. *J. Zool.* **247**, 257–268.

2087. Sommer, R.J. (2008). Homology and the hierarchy of biological systems. *BioEssays* **30**, 653–658.

2088. Sommer, R.J. (2009). The future of evo-devo: model systems and evolutionary theory. *Nature Rev. Genet.* **10**, 416–422.

2089. Sommer, S., Whittington, C.M., and Wilson, A.B. (2012). Standardised classification of pre-release development in male-brooding pipefish, seahorses, and seadragons (Family Syngnathidae). *BMC Dev. Biol.* **12**, Article 39 (6 pp.).

2090. Soshnikova, N., Dewaele, R., Janvier, P., Krumlauf, R., and Duboule, D. (2013). Duplications of *Hox* gene clusters and the emergence of vertebrates. *Dev. Biol.* **378**, 194–199.

2091. Soto, I.M., Carreira, V.P., Soto, E.M., Márquez, F., Lipko, P., and Hasson, E. (2013). Rapid divergent evolution of male genitalia among populations of *Drosophila buzzatii*. *Evol. Biol.* **40**, 395–407.

2092. Spéder, P., Ádám, G., and Noselli, S. (2006). Type ID unconventional myosin controls left-right asymmetry in *Drosophila*. *Nature* **440**, 803–807.

2093. Spéder, P. and Noselli, S. (2007). Left-right asymmetry: class I myosins show the direction. *Curr. Opin. Cell Biol.* **19**, 82–87.

2094. Spitz, F. and Furlong, E.E.M. (2012). Transcription factors: from enhancer binding to developmental control. *Nature Rev. Gen.* **13**, 613–626.

2095. Spoon, J.M. (2001). Situs inversus totalis. *Neonatal Netw.* **20**, 59–63.

2096. Springer, K., Brown, M., and Stulberg, D.L. (2003). Common hair loss disorders. *Am. Fam. Phys.* **68**, 93–102.

2097. Springer, M.S., Kirsch, J.A.W., and Case, J.A. (1997). The chronicle of marsupial evolution. In *Molecular Evolution and Adaptive Radiation* (T.J. Givnish and K.J. Sytsma, eds.). Cambridge University Press, New York, NY, pp. 129–157.

2098. Sprinzak, D., Lakhanpal, A., LeBon, L., Santat, L.A., Fontes, M.E., Anderson, G.A., Garcia-Ojalvo, J., and Elowitz, M.B. (2010). *Cis*-interactions between Notch and Delta generate mutually exclusive signalling states. *Nature* **465**, 86–90.

2099. Srygley, R.B. and Thomas, A.L.R. (2002). Unconventional lift-generating mechanisms in free-flying butterflies. *Nature* **420**, 660–664.

2100. Stafford, P. (2000). *Snakes*. Smithsonian Institution Press, Washington, DC.

2101. Stahl, R., Walcher, T., Romero, C.D.J., Pilz, G.A., Cappello, S., Irmler, M., Sanz-Aquela, J.M., Beckers, J., Blum, R., Borrell, V., and Götz, M. (2013). Trnp1 regulates expansion and folding of the mammalian cerebral cortex by control of radial glial fate. *Cell* **153**, 535–549.

2102. Standen, E.M. (2008). Pelvic fin locomotion function in fishes: three-dimensional kinematics in rainbow trout (*Oncorhynchus mykiss*). *J. Exp. Biol.* **211**, 2931–2942.

2103. Stanek, C. and Hantak, E. (1986). Bilateral atavistic polydactyly in a colt and its dam. *Equine Vet. J.* **18**, 76–79.

2104. Stauber, M., Prell, A., and Schmidt-Ott, U. (2002). A single *Hox3* gene with composite *bicoid* and *zerknüllt* expression characteristics in non-Cyclorrhaphan flies. *PNAS* **99** #1, 274–279.

2105. Stearns, S.C. (2002). Less would have been more. *Evolution* **56**, 2339–2345.

2106. Stebbins, G.L. (1983). Mosaic evolution: an integrating principle for the modern synthesis. *Experientia* **39**, 823–834.

2107. Stegmann, U.E. (2005). John Maynard Smith's notion of animal signals. *Biol. Philos.* **20**, 1011–1025.

2108. Steinberg, M.S. (2007). Differential adhesion in morphogenesis: a modern view. *Curr. Opin. Gen. Dev.* **17**, 281–286.

2109. Stern, C. (1968). *Genetic Mosaics and Other Essays*. Harvard University Press, Cambridge, MA.

2110. Stern, D. (2006). Morphing into shape. *Science* **313**, 50–51.

2111. Stern, D.L. (1998). A role of *Ultrabithorax* in morphological differences between *Drosophila* species. *Nature* **396**, 463–466.

2112. Stern, D.L. (2000). Evolutionary developmental biology and the problem of variation. *Evolution* **54**, 1079–1091.

2113. Stern, D.L. (2003). The Hox gene *Ultrabithorax* modulates the shape and size of the third leg of *Drosophila* by influencing diverse mechanisms. *Dev. Biol.* **256**, 355–366.

2114. Stern, D.L. and Emlen, D.J. (1999). The developmental basis for allometry in insects. *Development* **126**, 1091–1101.

2115. Stevens, M. (2005). The role of eyespots as anti-predator mechanisms, principally demonstrated in the Lepidoptera. *Biol. Rev.* **80**, 573–588.

2116. Stevens, M. (2007). Predator perception and the interrelation between different forms of protective coloration. *Proc. R. Soc. Lond. B* **274**, 1457–1464.

2117. Stevens, M., Hardman, C.J., and Stubbins, C.L. (2008). Conspicuousness, not eye mimicry, makes "eyespots" effective antipredator signals. *Behav. Ecol.* **19**, 525–531.

2118. Stevens, M. and Merilaita, S., eds. (2011). *Animal Camouflage: Mechanisms and Function*. Cambridge University Press, New York, NY.

2119. Stevens, M., Winney, I.S., Cantor, A., and Graham, J. (2009). Outline and surface disruption in animal camouflage. *Proc. R. Soc. Lond. B* **276**, 781–786.

2120. Stevens, M., Yule, D.H., and Ruxton, G.D. (2008). Dazzle coloration and prey movement. *Proc. R. Soc. Lond. B* **275**, 2639–2643.

2121. Stewart, A.J. and Plotkin, J.B. (2012). Why transcription factor binding sites are ten nucleotides long. *Genetics* **192**, 973–985.

2122. Stiassny, M.L.J. (2003). Atavism. In *Keywords and Concepts in Evolutionary Developmental Biology* (B.K. Hall and W.M. Olson, eds.). Harvard University Press, Cambridge, MA, pp. 10–14.

2123. Stock, D.W. (2001). The genetic basis of modularity in the development and evolution of the vertebrate dentition. *Phil. Trans. Roy. Soc. Lond. B* **356**, 1633–1653.

2124. Stock, G.B. and Bryant, S.V. (1981). Studies of digit regeneration and their implications for theories of development and evolution of vertebrate limbs. *J. Exp. Zool.* **216**, 423–433.

2125. Stockard, C.R. (1930). The presence of a factorial basis for characters lost in evolution: the atavistic reappearance of digits in mammals. *Am. J. Anat.* **45**, 345–375.

2126. Stocum, D.L. (2012). *Regenerative Biology and Medicine*, 2nd edn. Academic Press, New York, NY.

2127. Stocum, D.L. and Cameron, J.A. (2011). Looking proximally and distally: 100 years of limb regeneration and beyond. *Dev. Dynamics* **240**, 943–968.

2128. Stoehr, A.M., Walker, J.F., and Monteiro, A. (2013). Spalt expression and the development of melanic color patterns in pierid butterflies. *EvoDevo* **4**, Article 6 (11 pp.).

2129. Stoick-Cooper, C.L., Moon, R.T., and Weidinger, G. (2007). Advances in signaling in vertebrate regeneration as a prelude to regenerative medicine. *Genes Dev.* **21**, 1292–1315.

2130. Stokely, P.S. (1947). Limblessness and correlated changes in the girdles of a comparative morphological series of lizards. *Am. Midland Nat.* **38**, 725–754.

2131. Stokes, M.D. and Holland, N. (1998). The lancelet. *Am. Sci.* **86** #6, 552–560.

2132. Stokstad, E. (2001). Early tyrannosaur was small but well armed. *Science* **292**, 1278–1279.

2133. Stokstad, E. (2004). *T. rex* clan evolved head first. *Science* **306**, 211.

2134. Stokstad, E. (2007). Jaw shows platypus goes way back. *Science* **318**, 1237.

2135. Stopper, G.F. and Wagner, G.P. (2005). Of chicken wings and frog legs: a smorgasbord of evolutionary variation in mechanisms of tetrapod limb development. *Dev. Biol.* **288**, 21–39.

2136. Strausfeld, N.J. and Hirth, F. (2013). Deep homology of arthropod central complex and vertebrate basal ganglia. *Science* **340**, 157–161.

2137. Strugnell, J., Norman, M., Drummond, A.J., and Cooper, A. (2004). Neotenous origins for pelagic octopuses. *Curr. Biol.* **14**, R300–R301.

2138. Stuart-Fox, D. and Moussalli, A. (2008). Selection for social signalling drives the evolution of chameleon colour change. *PLoS Biol.* **6** #1, e25.

2139. Stuart-Fox, D. and Moussalli, A. (2009). Camouflage, communication and thermoregulation: lessons from colour changing organisms. *Phil. Trans. R. Soc. Lond. B* **364**, 463–470.

2140. Stumpke, H. (1981). *The Snouters: Form and Life of the Rhinogrades*. University of Chicago Press, Chicago, IL.

2141. Sudarsan, V., Anant, S., Guptan, P., VijayRaghavan, K., and Skaer, H. (2001). Myoblast diversification and ectodermal signaling in *Drosophila*. *Dev. Cell* **1**, 829–839.

2142. Sues, H.-D. (1991). Venom-conducting teeth in a Triassic reptile. *Nature* **351**, 141–143.

2143. Sumbre, G., Fiorito, G., Flash, T., and Hochner, B. (2005). Motor control of flexible octopus arms. *Nature* **433**, 595–596.

2144. Sumbre, G., Fiorito, G., Flash, T., and Hochner, B. (2006). Octopuses use a human-like strategy to control precise point-to-point arm movements. *Curr. Biol.* **16**, 767–772.

2145. Summers, A. (2002). Fast food joints. *Nat. Hist.* **111** #4, 84–85.

2146. Summers, A. (2008). Jaws two. *Nat. Hist.* **117** #1, 22–23.

2147. Sun, Y., Kanekar, S.L., Vetter, M.L., Gorski, S., Jan, Y.-N., Glaser, T., and Brown, N.L. (2003). Conserved and divergent functions of *Drosophila atonal*, amphibian, and mammalian *Ath5* genes. *Evol. Dev.* **5**, 532–541. [*See also* Jarman, A.P., and Groves, A.K. (2013). The role of Atonal transcription factors in the development of mechanosensitive cells. *Semin. Cell Dev. Biol.* **24**, 438–447.]

2148. Sundaram, M.V. (2005). The love-hate relationship between Ras and Notch. *Genes Dev.* **19**, 1825–1839.

2149. Superina, M. and Loughry, W.J. (2012). Life on the half-shell: consequences of a carapace in the evolution of armadillos (Xenarthra: Cingulata). *J. Mamm. Evol.* **19**, 217–224.

2150. Sustaita, D., Pouydebat, E., Manzano, A., Abdala, V., Hertel, F., and Herrel, A. (2013). Getting a grip on tetrapod grasping: form, function, and evolution. *Biol. Rev.* **88**, 380–405.

2151. Suzanne, M., Petzoldt, A.G., Spéder, P., Coutelis, J.-B., Steller, H., and Noselli, S. (2010). Coupling of apoptosis and L/R patterning controls stepwise organ looping. *Curr. Biol.* **20**, 1773–1778.

2152. Suzuki, N., Hirata, M., and Kondo, S. (2003). Traveling stripes on the skin of a mutant mouse. *PNAS* **100** #17, 9680–9685.

2153. Suzuki, T., Washio, Y., Aritaki, M., Fujinami, Y., Shimizu, D., Uji, S., and Hashimoto, H. (2009). Metamorphic *pitx2* expression in the left habenula correlated with lateralization of eye-sidedness in flounder. *Develop. Growth Differ.* **51**, 797–808.

2154. Suzuki, Y. and Palopoli, M.F. (2001). Evolution of insect abdominal appendages: are prolegs homologous or convergent traits? *Dev. Genes Evol.* **211**, 486–492.

2155. Swalla, B.J. (2006). Building divergent body plans with similar genetic pathways. *Heredity* **97**, 235–243.

2156. Swanson, C.I., Schwimmer, D.B., and Barolo, S. (2011). Rapid evolutionary rewiring of a structurally constrained eye enhancer. *Curr. Biol.* **21**, 1186–1196.

2157. Szeto, D.P., Rodriguez-Esteban, C., Ryan, A.K., O'Connell, S.M., Liu, F., Kioussi, C., Gleiberman, A.S., Izpisúa-Belmonte, J.C., and Rosenfeld, M.G. (1999). Role of the Bicoid-related homeodomain factor Pitx1 in specifying hindlimb morphogenesis and pituitary development. *Genes Dev.* **13**, 484–494.

2158. Tabata, T. and Takei, Y. (2004). Morphogens, their identification and regulation. *Development* **131**, 703–712.

2159. Tabin, C.J. (1992). Why we have (only) five fingers per hand: *Hox* genes and the evolution of paired limbs. *Development* **116**, 289–296.

2160. Tabin, C.J., Carroll, S.B., and Panganiban, G. (1999). Out on a limb: parallels in vertebrate and invertebrate limb patterning and the origin of appendages. *Am. Zool.* **39**, 650–663.

2161. Taborsky, M. and Taborsky, B. (1993). The kiwi's parental burden. *Nat. Hist.* **102** #12, 50–57.

2162. Tajiri, R., Misaki, K., Yonemura, S., and Hayashi, S. (2011). Joint morphology in the insect leg: evolutionary history inferred from *Notch* loss-of-function phenotypes in *Drosophila*. *Development* **138**, 4621–4626.

2163. Takahashi, G. and Kondo, S. (2008). Melanophores in the stripes of adult zebrafish do not have the nature to gather, but disperse when they have the space to move. *Pigment Cell Melanoma Res.* **21**, 677–686.

2164. Takahashi, M., Arita, H., Hiraiwa-Hasegawa, M., and Hasegawa, T. (2008). Peahens do not prefer peacocks with more elaborate trains. *Anim. Behav.* **75**, 1209–1219. [*See also* Yorzinski, J.L., *et al.* (2013). Through their eyes: selective attention in peahens during courtship. *J. Exp. Biol.* **216**: 3035–3046.]

2165. Takashima, S., Mkrtchyan, M., Younossi-Hartenstein, A., Merriam, J.R., and Hartenstein, V. (2008). The behaviour of *Drosophila* adult hindgut stem cells is controlled by Wnt and Hh signalling. *Nature* **454**, 651–655.

2166. Takashima, Y., Ohtsuka, T., González, A., Miyachi, H., and Kageyama, R. (2011). Intronic delay is essential for oscillatory expression in the segmentation clock. *PNAS* **108** #8, 3300–3305.

2167. Tanaka, E.M. (2003). Regeneration: If they can do it, why can't we? *Cell* **113**, 559–562.

2168. Tanaka, E.M. (2012). Skin, heal thyself. *Nature* **489**, 508–510.

2169. Tanaka, E.M. and Reddien, P.W. (2011). The cellular basis for animal regeneration. *Dev. Cell* **21**, 172–185.

2170. Tanaka, K., Barmina, O., and Kopp, A. (2009). Distinct developmental mechanisms underlie the evolutionary diversification of *Drosophila* sex combs. *PNAS* **106** #12, 4764–4769.

2171. Tanaka, K., Barmina, O., Sanders, L.E., Arbeitman, M.N., and Kopp, A. (2011). Evolution of sex-specific traits through changes in HOX-dependent *doublesex* expression. *PLoS Biol.* **9** #8, e1001131.

2172. Tanaka, K. and Truman, J.W. (2005). Development of the adult leg epidermis in *Manduca sexta*: contribution of different larval cell populations. *Dev. Genes Evol.* **215**, 78–89.

2173. Tanaka, K. and Truman, J.W. (2007). Molecular patterning mechanism underlying metamorphosis of the thoracic leg in *Manduca sexta*. *Dev. Biol.* **305**, 539–550.

2174. Tanaka, M. (2011). Revealing the mechanisms of the rostral shift of pelvic fins among teleost fishes. *Evol. Dev.* **13**, 382–390.

2175. Taniguchi, K., Maeda, R., Ando, T., Okumura, T., Nakazawa, N., Hatori, R., Nakamura, M., Hozumi, S., Fujiwara, H., and Matsuno, K. (2011). Chirality in planar cell shape contributes to left-right asymmetric epithelial morphogenesis. *Science* **333**, 339–341.

2176. Taylor, G.K. (2001). Mechanics and aerodynamics of insect flight control. *Biol. Rev.* **76**, 449–471.

2177. Taylor, H.L., Cole, C.J., Dessauer, H.C., and Parker, E.D., Jr. (2003). Congruent patterns of genetic and morphological variation in the parthenogenetic lizard *Aspidoscelis tesselata* (Squamata: Teiidae) and the origins of color pattern classes and genotypic clones in eastern New Mexico. *Am. Mus. Novitates* **3424**, 1–40.

2178. Taylor, I.W. (2008). Snapshot: The TGFβ pathway interactome. *Cell* **133**, 378.

2179. Taylor, M.P., Hone, D.W.E., Wedel, M.J., and Naish, D. (2011). The long necks of sauropods did not evolve primarily through sexual selection. *J. Zool.* **285**, 150–161.

2180. Taylor, M.P. and Wedel, M.J. (2013). Why sauropods had long necks; and why giraffes have short necks. *PeerJ* **1**, e36. doi: 10.7717/peerj.36.

2181. Tchernov, E., Rieppel, O., Zaher, H., Polcyn, M.J., and Jacobs, L.L. (2000). A fossil snake with legs. *Science* **287**, 2010–2012.

2182. Technau, U. and Steele, R.E. (2011). Evolutionary crossroads in developmental biology: Cnidaria. *Development* **138**, 1447–1458.

2183. Telford, M.J. (2007). A single origin of the central nervous system? *Cell* **129**, 237–239.

2184. Telford, M.J. (2009). Animal evolution: once upon a time. *Curr. Biol.* **19**, R339–R341.

2185. ten Berge, D., Brouwer, A., El Bahi, S., Guenet, J.-L., Robert, B., and Meijlink, F. (1998). Mouse *Alx3:* an *aristaless*-like homeobox gene expressed during embryogenesis in ectomesenchyme and lateral plate mesoderm. *Dev. Biol.* **199**, 11–25.

2186. ten Broek, C.M.A., Bakker, A.J., Varela-Lasheras, I., Bugiani, M., Van Dongen, S., and Galis, F. (2012). Evo-devo of the human vertebral column: on homeotic transformations, pathologies and prenatal selection. *Evol. Biol.* **39**, 456–471.

2187. Tennant, A. (1984). *The Snakes of Texas*. Texas Monthly Press, Austin, TX.

2188. Teotónio, H. and Rose, M.R. (2001). Reversible evolution. *Evolution* **55**, 653–660.

2189. Terra, W.R. (1990). Evolution of digestive systems of insects. *Annu. Rev. Entomol.* **35**, 181–200.

2190. Teske, P.R. and Beheregaray, L.B. (2009). Evolution of seahorses' upright posture was linked to Oligocene expansion of seagrass habitats. *Biol. Lett.* **5**, 521–523.

2191. Theissen, G. (2006). The proper place of hopeful monsters in evolutionary biology. *Theory Biosci.* **124**, 349–369.

2192. Théry, M. and Gomez, D. (2010). Insect colours and visual appearance in the eyes of their predators. *Adv. Insect Physiol.* **38**, 267–353.

2193. Theveneau, E. and Mayor, R. (2012). Neural crest delamination and migration: from epithelium-to-mesenchyme transition to collective cell migration. *Dev. Biol.* **366**, 34–54.

2194. Thewissen, J.G.M. (2007). Aquatic adaptations in the limbs of amniotes. In *Fins into Limbs: Evolution, Development, and Transformation* (B.K. Hall, ed.). University of Chicago Press, Chicago, IL, pp. 310–322.

2195. Thewissen, J.G.M., Cohn, M.J., Stevens, L.S., Bajpai, S., Heyning, J., and Horton, W.E., Jr. (2006). Developmental basis for hind-limb loss in dolphins and origin of the cetacean bodyplan. *PNAS* **103** #22, 8414–8418.

2196. Thewissen, J.G.M., Cooper, L.N., Clementz, M.T., Bajpai, S., and Tiwari, B.N. (2007). Whales originated from aquatic artiodactyls in the Eocene epoch of India. *Nature* **450**, 1190–1194.

2197. Thewissen, J.G.M., Cooper, L.N., George, J.C., and Bajpai, S. (2009). From land to water: the origin of whales, dolphins, and porpoises. *Evo. Edu. Outreach* **2**, 272–288.

2198. Thewissen, J.G.M., Williams, E.M., Roe, L.J., and Hussain, S.T. (2001). Skeletons of terrestrial cetaceans and the relationship of whales to artiodactyls. *Nature* **413**, 277–281.

2199. Thistle, R., Cameron, P., Ghorayshi, A., Dennison, L., and Scott, K. (2012). Contact chemoreceptors mediate male-male repulsion and male-female attraction during *Drosophila* courtship. *Cell* **149**, 1140–1151.

2200. Thomas, R.D.K. and Reif, W.-E. (1993). The skeleton space: a finite set of organic designs. *Evolution* **47**, 341–360.

2201. Thompson, B.J. (2013). Cell polarity: models and mechanisms from yeast, worms, and flies. *Development* **140**, 13–21.

2202. Thompson, C.W. (2013). Implications of hybridization between the Rio Grande ground squirrel (*Ictidomys parvidens*) and the thirteen-lined ground squirrel (*I. tridecemlineatus*). Ph.D. dissertation. Department of Biological Sciences, Texas Tech University, Lubbock, TX.

2203. Thompson, D.W. (1917). *On Growth and Form*. Cambridge University Press, Cambridge.

2204. Thor, S. and Thomas, J.B. (2002). Motor neuron specification in worms, flies and mice: conserved and "lost" mechanisms. *Curr. Opin. Gen. Dev.* **12**, 558–564.

2205. Thorington, R.W., Jr., Koprowski, J.L., Steele, M.A., and Whatton, J.F. (2012). *Squirrels of the World*. Johns Hopkins University Press, Baltimore, MD.

2206. Thorington, R.W., Jr., Pitassy, D., and Jansa, S.A. (2002). Phylogenies of flying squirrels (Pteromyinae). *J. Mamm. Evol.* **9**, 99–135.

2207. Thorington, R.W., Jr. and Santana, E.M. (2007). How to make a flying squirrel: *Glaucomys* anatomy in phylogenetic perspective. *J. Mammalogy* **88**, 882–896.

2208. Thurber, A.R., Jones, W.J., and Schnabel, K. (2011). Dancing for food in the deep sea: bacterial farming by a new species of yeti crab. *PLoS ONE* **6** #11, e26243.

2209. Tobias, J.A., Montgomerie, R., and Lyon, B.E. (2012). The evolution of female ornaments and weaponry: social selection, sexual selection and ecological competition. *Phil. Trans. R. Soc. Lond. B* **367**, 2274–2293.

2210. Tobin, A.J. and Dusheck, J. (2004). *Asking About Life*, 3rd edn. Brooks/Cole, Belmont, CA.

2211. Toda, H., Zhao, X., and Dickson, B.J. (2012). The *Drosophila* female aphrodisiac pheromone activates *ppk23+* sensory neurons to elicit male courtship behavior. *Cell Reports* **1**, 599–607.

2212. Toh, Y. (1985). Structure of campaniform sensilla on the haltere of *Drosophila* prepared by cryofixation. *J. Ultrastruct. Res.* **93**, 92–100.

2213. Tokita, M., Chaeychomsri, W., and Siruntawineti, J. (2012). Developmental basis of toothlessness in turtles: insight into convergent evolution of vertebrate morphology. *Evolution* **67**, 260–273.

2214. Tokita, M., Kiyoshi, T., and Armstrong, K.N. (2007). Evolution of craniofacial novelty in parrots through developmental modularity and heterochrony. *Evol. Dev.* **9**, 590–601.

2215. Tokita, M. and Schneider, R.A. (2009). Developmental origins of species-specific muscle pattern. *Dev. Biol.* **331**, 311–325.

2216. Tokunaga, C. (1961). The differentiation of a secondary sex comb under the influence of the gene engrailed in *Drosophila melanogaster*. *Genetics* **46**, 157–176.

2217. Tokunaga, C. (1962). Cell lineage and differentiation on the male foreleg of *Drosophila melanogaster*. *Dev. Biol.* **4**, 489–516.

2218. Tokunaga, C. (1978). Genetic mosaic studies of pattern formation in *Drosophila melanogaster*, with special reference to the prepattern hypothesis. In *Genetic Mosaics and Cell Differentiation* (W.J. Gehring, ed.). Results and Problems in Cell Differentiation, Vol. 9. Springer-Verlag, Berlin, pp. 157–204.

2219. Tomita, S. and Kikuchi, A. (2009). *Abd-B* suppresses lepidopteran proleg development in posterior abdomen. *Dev. Biol.* **328**, 403–409.

2220. Tomoyasu, Y., Arakane, Y., Kramer, K.J., and Denell, R.E. (2009). Repeated co-options of exoskeleton formation during wing-to-elytron evolution in beetles. *Curr. Biol.* **19**, 2057–2065.

2221. Tomoyasu, Y., Wheeler, S.R., and Denell, R.E. (2005). *Ultrabithorax* is required for membranous wing identity in the beetle *Tribolium casteneum*. *Nature* **433**, 643–647.

2222. Tong, X., Lindemann, A., and Monteiro, A. (2012). Differential involvement of Hedgehog signaling in butterfly wing and eyespot development. *PLoS ONE* **7** #12, e51087.

2223. Toro, R. (2012). On the possible shapes of the human brain. *Evol. Biol.* **39**, 600–612.

2224. Tour, E. and McGinnis, W. (2006). Gap peptides? A new way to control embryonic patterning? *Cell* **126**, 448–449.

2225. Trainor, P. (2003). The bills of qucks and duails. *Science* **299**, 523–524.

2226. Treisman, J.E. (2004). How to make an eye. *Development* **131**, 3823–3827.

2227. True, J.R. (2003). Insect melanism: the molecules matter. *Trends Ecol. Evol.* **18**, 640–647.

2228. True, J.R. and Carroll, S.B. (2002). Gene co-option in physiological and morphological evolution. *Annu. Rev. Cell Dev. Biol.* **18**, 53–80.

2229. True, J.R. and Haag, E.S. (2001). Developmental system drift and flexibility in evolutionary trajectories. *Evol. Dev.* **3**, 109–119.

2230. Trueb, L. (1973). Bones, frogs, and evolution. In *Evolutionary Biology of the Anurans: Contemporary Research on Major Problems* (J.L. Vial, ed.). University of Missouri Press, Columbia, MO, pp. 65–132.

2231. Trueman, J.W.H., Pfeil, B.E., Kelchner, S.A., and Yeates, D.K. (2004). Did stick insects really regain their wings? *Syst. Entomol.* **29**, 138–139.

2232. Truman, J.W. and Riddiford, L.M. (1999). The origins of insect metamorphosis. *Nature* **401**, 447–452.

2233. Tschopp, P., Frandeau, N., Béna, F., and Duboule, D. (2011). Reshuffling genomic landscapes to study the regulatory evolution of *Hox* gene clusters. *PNAS* **108** #26, 10632–10637.

2234. Tsubota, T., Saigo, K., and Kojima, T. (2008). *Hox* genes regulate the same character by different strategies in each segment. *Mechs. Dev.* **125**, 894–905.

2235. Tsuihiji, T., Kearney, M., and Rieppel, O. (2006). First report of a pectoral girdle muscle in snakes, with comments on the snake cervico-dorsal boundary. *Copeia* **2006**, 206–215.

2236. Tsuihiji, T., Kearney, M., and Rieppel, O. (2012). Finding the neck-trunk boundary in snakes: anteroposterior dissociation of myological characteristics in snakes and its implications for their neck and trunk body regionalization. *J. Morph.* **273**, 992–1009.

2237. Tsujimoto, M. and Hattori, A. (2005). The oxytocinase subfamily of M1 aminopeptidases. *Biochim. Biophys. Acta* **1751**, 9–18.

2238. Tucker, A. and Sharpe, P. (2004). The cutting-edge of mammalian development; how the embryo makes teeth. *Nature Rev. Genet.* **5**, 499–508.

2239. Tucker, R.P. and Erickson, C.A. (1986). The control of pigment cell pattern formation in the California newt, *Taricha torosa*. *J. Embryol. Exp. Morph.* **97**, 141–168.

2240. Tucker, R.P. and Erickson, C.A. (1986). Pigment cell pattern formation in *Taricha torosa*: the role of the extracellular matrix in controlling pigment cell migration and differentiation. *Dev. Biol.* **118**, 268–285.

2241. Tummers, M. and Thesleff, I. (2003). Root or crown: a developmental choice orchestrated by the differential regulation of the epithelial stem cell niche in the tooth of two rodent species. *Development* **130**, 1049–1057.

2242. Tunnicliffe, V. (1992). Hydrothermal-vent communities of the deep sea. *Am. Sci.* **80**, 336–349.

2243. Turchyn, N., Chesebro, J., Hrycaj, S., Couso, J.P., and Popadic, A. (2011). Evolution of *nubbin* function in hemimetabolous and holometabolous insect appendages. *Dev. Biol.* **357**, 83–95.

2244. Turing, A.M. (1952). The chemical basis of morphogenesis. *Phil. Trans. Roy. Soc. Lond., B* **237**, 37–72.

2245. Tuttle, R.H. (1990). Apes of the world. *Am. Sci.* **78**, 115–125.

2246. Twitty, V.C. (1944). Chromatophore migration as a response to mutual influences of the developing pigment cells. *J. Exp. Zool.* **95**, 259–290.

2247. Twitty, V.C. (1945). The developmental analysis of specific pigment patterns. *J. Exp. Zool.* **100**, 141–178.

2248. Twitty, V.C. and Niu, M.C. (1948). Causal analysis of chromatophore migration. *J. Exp. Zool.* **108**, 405–437.

2249. Twitty, V.C. and Niu, M.C. (1954). The motivation of cell migration, studied by isolation of embryonic pigment cells singly and in small groups in vitro. *J. Exp. Zool.* **125**, 541–573.

2250. Underwood, G. (1970). The eye. In *Biology of the Reptilia, Vol. 2: Morphology, Part B* (C. Gans and T.S. Parsons, eds.). Academic Press, New York, NY, pp. 1–97.

2251. Unwin, D.M. (1999). Pterosaurs: back to the traditional model? *Trends Ecol. Evol.* **14**, 263–268.

2252. Unwin, M. (2011). *Southern African Wildlife*, 2nd edn. Bradt Wildlife Explorer. Bradt Travel Guides, Chalfont St. Peter.

2253. Updike, J. and Gwin, P. (2007). Extreme dinosaurs. *Natl. Geogr.* **212** #6, 32–57.

2254. Urdy, S. (2012). On the evolution of morphogenetic models: mechano-chemical interactions and an integrated view of cell differentiation, growth, pattern formation and morphogenesis. *Biol. Rev.* **87**, 786–803.

2255. Vachon, G., Cohen, B., Pfeifle, C., McGuffin, M.E., Botas, J., and Cohen, S.M. (1992). Homeotic genes of the Bithorax Complex repress limb development in the abdomen of the *Drosophila* embryo through the target gene *Distal-less*. *Cell* **71**, 437–450.

2256. Valentine, J.W. (2004). *On the Origin of Phyla*. University of Chicago Press, Chicago, IL.

2257. Valkonen, J., Niskanen, M., Björklund, M., and Mappes, J. (2011). Disruption or aposematism? Significance of dorsal zigzag pattern of European vipers. *Evol. Ecol.* **25**, 1047–1063.

2258. Vallin, A., Dimitrova, M., Kodandaramaiah, U., and Merilaita, S. (2011). Deflective effect and the effect of prey detectability on anti-predator function of eyespots. *Behav. Ecol. Sociobiol.* **65**, 1629–1636.

2259. Vallin, A., Jakobsson, S., Lind, J., and Wiklund, C. (2006). Crypsis versus intimidation: anti-predation defence in three closely related butterflies. *Behav. Ecol. Sociobiol.* **59**, 455–459.

2260. Vallin, A., Jakobsson, S., and Wiklund, C. (2007). "An eye for an eye?" On the generality of the intimidating quality of eyespots in a butterfly and a hawkmoth. *Behav. Ecol. Sociobiol.* **61**, 1419–1424.

2261. van Amerongen, R. and Nusse, R. (2009). Towards an integrated view of Wnt signaling in development. *Development* **136**, 3205–3214.

2262. Van de Peer, Y., Maere, S., and Meyer, A. (2009). The evolutionary significance of ancient genome duplications. *Nature Rev. Genet.* **10**, 725–732.

2263. van Deusen, H.M. (1966). The seventh Archbold expedition. *BioScience* **16**, 449–455.

2264. van Doorn, K. (2012). Investigations on the reptilian spectacle. Ph.D. dissertation. School of Optometry and Vision Science, University of Waterloo, Canada.

2265. Van Essen, D.C. (1997). A tension-based theory of morphogenesis and compact wiring in the central nervous system. *Nature* **385**, 313–318.

2266. van Gelder, R.G. (1959). A taxonomic revision of the spotted skunks (genus *Spilogale*). *Bull. Am. Mus. Nat. Hist.* **117** #5, 229–392.

2267. Van Sittert, S.J., Skinner, J.D., and Mitchell, G. (2010). From fetus to adult: an allometric analysis of the giraffe vertebral column. *J. Exp. Zool. (Mol. Dev. Evol.)* **314B**, 469–479.

2268. Van Valkenburgh, B. and Wayne, R.K. (2010). Carnivores. *Curr. Biol.* **20**, R915–R919.

2269. Van Wassenbergh, S., Leysen, H., Adriaens, D., and Aerts, P. (2013). Mechanics of snout expansion in suction-feeding seahorses: musculoskeletal force transmission. *J. Exp. Biol.* **216**, 407–417.

2270. Van Wassenbergh, S., Roos, G., and Ferry, L. (2011). An adaptive explanation for the horse-like shape of seahorses. *Nature Commun.* **2**, Article 164 (5 pp.).

2271. Van Wassenbergh, S., Roos, G., Genbrugge, A., Leysen, H., Aerts, P., Adriaens, D., and Herrel, A. (2009). Suction is kid's play: extremely fast suction in newborn seahorses. *Biol. Lett.* **5**, 200–203.

2272. Vandenberg, L.N. and Levin, M. (2010). Far from solved: a perspective on what we know about early mechanisms of left-right asymmetry. *Dev. Dynamics* **239**, 3131–3146.

2273. Vandenberg, L.N. and Levin, M. (2013). A unified model for left-right asymmetry? Comparison and synthesis of molecular models of embryonic laterality. *Dev. Biol.* **379**, 1–15.

2274. Vandervorst, P. and Ghysen, A. (1980). Genetic control of sensory connections in *Drosophila*. *Nature* **286**, 65–67.

2275. Varela-Lasheras, I., Bakker, A.J., van der Mije, S.D., Metz, J.A.J., van Alphen, J., and Galis, F. (2011). Breaking evolutionary and pleiotropic constraints in mammals: on sloths, manatees and homeotic mutations. *EvoDevo* **2**, Article 11 (27 pp.).

2276. Varjosalo, M. and Taipale, J. (2008). Hedgehog: functions and mechanisms. *Genes Dev.* **22**, 2454–2472.

2277. Varricchio, D.J. (2001). Gut contents from a Cretaceous Tyrannosaurid: implications for theropod dinosaur digestive tracts. *J. Paleont.* **75**, 401–406.

2278. Vaughn, T.A., Ryan, J.M., and Czaplewski, N.J. (2000). *Mammalogy*, 4th edn. Saunders, New York, NY.

2279. Vecchione, M., Young, R.E., Guerra, A., Lindsay, D.J., Clague, D.A., Bernhard, J.M., Sager, W.W., Gonzalez, A.F., Rocha, F.J., and Segonzac, M. (2001). Worldwide observations of remarkable deep-sea squids. *Science* **294**, 2505–2506.

2280. Veeman, M.T., Newman-Smith, E., El-Nachef, D., and Smith, W.C. (2010). The ascidian mouth opening is derived from the anterior neuropore: reassessing the mouth/neural tube relationship in chordate evolution. *Dev. Biol.* **344**, 138–149.

2281. Venditti, C. and Pagel, M. (2008). Speciation and bursts of evolution. *Evo. Edu. Outreach* **1**, 274–280.

2282. Veraksa, A., Del Campo, M., and McGinnis, W. (2000). Developmental patterning genes and their conserved functions: from model organisms to humans. *Mol. Genet. Metab.* **69**, 85–100.

2283. Verhulst, J. (1996). Atavisms in *Homo sapiens*: a Bolkian heterodoxy revisited. *Acta Biotheor.* **44**, 59–73.

2284. Vickaryous, M. and Olson, W.M. (2007). Sesamoids and ossicles in the appendicular skeleton. In *Fins into Limbs: Evolution, Development, and Transformation* (B.K. Hall, ed.). University of Chicago Press, Chicago, IL, pp. 323–341.

2285. Vickaryous, M.K. (2009). The integumentary skeleton of tetrapods: origin, evolution, and development. *J. Anat.* **214**, 441–464.

2286. Vidal, N. and Hedges, S.B. (2004). Molecular evidence for a terrestrial origin of snakes. *Proc. R. Soc. Lond. B (Suppl.)* **271**, S226–S229.

2287. Viguerie, N., Montastier, E., Maoret, J.-J., Roussel, B., Combes, M., Valle, C., Villa-Vialaneix, N., Iacovoni, J.S., Martinez, J.A., Holst, C., Astrup, A., Vidal, H., Clément, K., Hager, J., Saris, W.H.M., and Langin, D. (2012). Determinants of human adipose tissue gene expression: impact of diet, sex, metabolic status, and *cis* genetic regulation. *PLoS Genet.* **8** #9, e1002959.

2288. Villee, C.A. (1942). The phenomenon of homeosis. *Am. Nat.* **76**, 494–506.

2289. Villella, A. and Hall, J.C. (2008). Neurogenetics of courtship and mating in *Drosophila*. *Adv. Genet.* **62**, 67–184.

2290. Villmoare, B. (2013). Morphological integration, evolutionary constraints, and extinction: a computer simulation-based study. *Evol. Biol.* **40**, 76–83.

2291. Vinagre, T., Moncaut, N., Carapuço, M., Nóvoa, A., Bom, J., and Mallo, M. (2010). Evidence for a myotomal Hox/Myf cascade governing nonautonomous control of rib specification within global vertebral domains. *Dev. Cell* **18**, 655–661.

2292. Vincent, J.-P. and Dubois, L. (2002). Morphogen transport along epithelia, an integrated trafficking problem. *Dev. Cell* **3**, 615–623.

2293. Vitt, L.J. and Pianka, E.R. (2006). The scaly ones. *Nat. Hist.* **115** #6, 28–35.

2294. Vogel, G. (2012). Turing pattern fingered for digit formation. *Science* **338**, 1406.

2295. von Neumann, J. (1966). *Theory of Self-Reproducing Automata*. University of Illinois Press, Urbana, IL.

2296. Vonk, F.J., Admiraal, J.F., Jackson, K., Reshef, R., de Bakker, M.A.G., Vanderschoot, K., van den Berge, I., van Atten, M., Burgerhout, E., Beck, A., Mirtschin, P.J., Kochva, E., Witte, F., Fry, B.G., Woods, A.E., and Richardson, M.K. (2008). Evolutionary origin and development of snake fangs. *Nature* **454**, 630–633.

2297. Vonk, F.J., Jackson, K., Doley, R., Madaras, F., Mirtschin, P.J., and Vidal, N. (2011). Snake venom: from fieldwork to the clinic. *BioEssays* **33**, 269–279.

2298. Vonk, F.J. and Richardson, M.K. (2008). Serpent clocks tick faster. *Nature* **454**, 282–283.

2299. Vreede, B.M.I., Lynch, J.A., Roth, S., and Sucena, E. (2013). Co-option of a coordinate system defined by the EGFr and Dpp pathways in the evolution of a morphological novelty. *EvoDevo* **4**, Article 7 (12 pp.).

2300. Waage, J.K. (1981). How the zebra got its stripes: biting flies as selective agents in the evolution of zebra colouration. *J. Entomol. Soc. S. Afr.* **44**, 351–358.

2301. Waddington, C.H. (1969). The theory of evolution today. In *Beyond Reductionism: New Perspectives in the Life Sciences* (A. Koestler and J.R. Smythies, eds.). Macmillan, New York, NY, pp. 357–395.

2302. Wagner, A. (2008). Gene duplications, robustness and evolutionary innovations. *BioEssays* **30**, 367–373.

2303. Wagner, A. (2011). The molecular origins of evolutionary innovations. *Trends Genet.* **27**, 397–410.

2304. Wagner, D.L. and Liebherr, J.K. (1992). Flightlessness in insects. *Trends Ecol. Evol.* **7**, 216–220.

2305. Wagner, G.P. (1989). The biological homology concept. *Annu. Rev. Ecol. Syst.* **20**, 51–69.

2306. Wagner, G.P. (1989). The origin of morphological characters and the biological basis of homology. *Evolution* **43**, 1157–1171.

2307. Wagner, G.P. (1993). How can a character be developmentally constrained despite variation in developmental pathways? *J. Evol. Biol.* **6**, 449–455.
2308. Wagner, G.P. (1996). Homologues, natural kinds and the evolution of modularity. *Am. Zool.* **36**, 36–43.
2309. Wagner, G.P. (2007). The developmental genetics of homology. *Nature Rev. Genet.* **8**, 473–479.
2310. Wagner, G.P. and Altenberg, L. (1996). Complex adaptations and the evolution of evolvability. *Evolution* **50**, 967–976.
2311. Wagner, G.P., Chiu, C.-H., and Laubichler, M. (2000). Developmental evolution as a mechanistic science: the inference from developmental mechanisms to evolutionary processes. *Am. Zool.* **40**, 819–831.
2312. Wagner, G.P. and Lynch, V.J. (2008). The gene regulatory logic of transcription factor evolution. *Trends Ecol. Evol.* **23**, 377–385.
2313. Wagner, G.P. and Lynch, V.J. (2010). Evolutionary novelties. *Curr. Biol.* **20**, R48–R52.
2314. Wagner, G.P., Pavlicev, M., and Cheverud, J.M. (2007). The road to modularity. *Nature Rev. Genet.* **8**, 921–931.
2315. Wagner, G.P. and Zhang, J. (2011). The pleiotropic structure of the genotype-phenotype map: the evolvability of complex organisms. *Nature Rev. Gen.* **12**, 204–213.
2316. Wagner, R.A., Tabibiazar, R., Liao, A., and Quertermous, T. (2005). Genome-wide expression dynamics during mouse embryonic development reveal similarities to *Drosophila* development. *Dev. Biol.* **288**, 595–611.
2317. Wake, D.B. and Larson, A. (1987). Multidimensional analysis of an evolving lineage. *Science* **238**, 42–48.
2318. Wake, D.B., Wake, M.H., and Specht, C.D. (2011). Homoplasy: from detecting pattern to determining process and mechanism of evolution. *Science* **331**, 1032–1035.
2319. Walker, E.P., Warnick, F., Hamlet, S.E., Lange, K.I., Davis, M.A., Uible, H.E., Wright, P.F., and Paradiso, J.L. (1975). *Mammals of the World.* Johns Hopkins University Press, Baltimore, MD.
2320. Wallace, B. (1985). Reflections on the still-"hopeful monster". *Quart. Rev. Biol.* **60**, 31–42.
2321. Walls, G.L. (1942). *The Vertebrate Eye.* Cranbrook Press, Bloomfield Hills, MI.
2322. Wang, L., Han, X., Mehren, J., Hiroi, M., Billeter, J.-C., Miyamoto, T., Amrein, H., Levine, J.D., and Anderson, D.J. (2011). Hierarchical chemosensory regulation of male-male social interactions in *Drosophila*. *Nat. Neurosci.* **14**, 757–762.
2323. Wang, S. and Samakovlis, C. (2012). Grainy head and its target genes in epithelial morphogenesis and wound healing. *Curr. Top. Dev. Biol.* **98**, 35–63.
2324. Wang, X. (2012). Passing the smell test. *Nat. Hist.* **120** #5, 22–29.
2325. Wang, X. and Zhou, Z. (2004). Pterosaur embryo from the Early Cretaceous. *Nature* **429**, 621.
2326. Wang, X.-P., Suomalainen, M., Felszeghy, S., Zelarayan, L.C., Alonso, M.T., Plikus, M.V., Maas, R.L., Chuong, C.-M., Schimmang, T., and Thesleff, I. (2007). An integrated gene regulatory network controls stem cell proliferation in teeth. *PLoS Biol.* **5** #6, e159.
2327. Wang, X.-P., Suomalainen, M., Jorgez, C.J., Matzuk, M.M., Werner, S., and Thesleff, I. (2004). Follistatin regulates enamel patterning in mouse incisors by asymmetrically inhibiting BMP signaling and ameloblast differentiation. *Dev. Cell* **7**, 719–730.
2328. Wang, Y., Gao, Y., Imsland, F., Gu, X., Feng, C., Liu, R., Song, C., Tixier-Boichard, M., Gourichon, D., Li, Q., Chen, K., Li, H., Andersson, L., Hu, X., and Li, N. (2012). The Crest

phenotype in chicken is associated with ectopic expression of *Hoxc8* in cranial skin. *PLoS ONE* **7** #4, e34012.

2329. Wang, Z., Dong, D., Ru, B., Young, R.L., Han, N., Guo, T., and Zhang, S. (2010). Digital gene expression tag profiling of bat digits provides robust candidates contributing to wing formation. *BMC Genomics* **11**, Article 619 (12 pp.).

2330. Ward, A.B. and Brainerd, E.L. (2007). Evolution of axial patterning in elongate fishes. *Biol. J. Linnean Soc.* **90**, 97–116.

2331. Ward, C.V. (1997). Functional anatomy and phyletic implications of the hominoid trunk and hindlimb. In *Function, Phylogeny, and Fossils: Miocene Hominoid Evolution and Adaptations* (D.R. Begun, C.V. Ward, and M.D. Rose, eds.). Plenum Press, New York, NY, pp. 101–130.

2332. Ward, M. (1997). Everest 1951: the footprints attributed to the Yeti – myth and reality. *Wilderness Environ. Med.* **8**, 29–32.

2333. Ward, P.D. (1991). *On Methuselah's Trail: Living Fossils and the Great Extinctions*. W. H. Freeman, New York, NY.

2334. Ward, P.S. (2007). Phylogeny, classification, and species-level taxonomy of ants (Hymenoptera: Formicidae). *Zootaxa* **1668**, 549–563.

2335. Ward, S.J. (1998). Numbers of teats and pre- and post-natal litter sizes in small diprotodont marsupials. *J. Mammalogy* **79**, 999–1008.

2336. Warner, J.F., Lyons, D.C., and McClay, D.R. (2012). Left-right asymmetry in the sea urchin embryo: BMP and the asymmetrical origins of the adult. *PLoS Biol.* **10** #10, e1001404.

2337. Warrant, E.J. (2007). Visual ecology: hiding in the dark. *Curr. Biol.* **17**, R209–R211.

2338. Warren, I. and Smith, H. (2007). Stalk-eyed flies (Diopsidae): modelling the evolution and development of an exaggerated sexual trait. *BioEssays* **29**, 300–307.

2339. Warren, R. and Carroll, S. (1995). Homeotic genes and diversification of the insect body plan. *Curr. Opin. Gen. Dev.* **5**, 459–465.

2340. Warren, R.W., Nagy, L., Selegue, J., Gates, J., and Carroll, S. (1994). Evolution of homeotic gene regulation and function in flies and butterflies. *Nature* **372**, 458–461.

2341. Warren, W.C., Hillier, L.W., Graves, J.A.M., Birney, E., Ponting, C.P., Grützner, F., Belov, K., Miller, W., Clarke, L., Chinwalla, A.T., Yang, S.-P., Heger, A., Locke, D.P., Miethke, P., Waters, P.D., Veyrunes, F., Fulton, L., Fulton, B., Graves, T., Wallis, J., Puente, X.S., López-Otín, C., Ordóñez, G.R., Eichler, E.E., Chen, L., Cheng, Z., Deakin, J.E., Alsop, A., Thompson, K., Kirby, P., Papenfuss, A.T., Wakefield, M.J., Olender, T., Lancet, D., Huttley, G.A., Smit, A.F.A., Pask, A., Temple-Smith, P., Batzer, M.A., Walker, J.A., Konkel, M.K., Harris, R.S., Whittington, C.M., Wong, E.S.W., Gemmell, N.J., Buschiazzo, E., Jentzsch, I.M.V., Merkel, A., Schmitz, J., Zemann, A., Churakov, G., Kriegs, J.O., Brosius, J., Murchison, E.P., Sachidanandam, R., Smith, C., Hannon, G.J., Tsend-Ayush, E., McMillan, D., Attenborough, R., Rens, W., Ferguson-Smith, M., Lefèvre, C.M., Sharp, J.A., Nicholas, K.R., Ray, D.A., Kube, M., Reinhardt, R., Pringle, T.H., Taylor, J., Jones, R.C., Nixon, B., Dacheux, J.-L., Niwa, H., Sekita, Y., Huang, X., Stark, A., Kheradpour, P., Kellis, M., Flicek, P., Chen, Y., Webber, C., Hardison, R., Nelson, J., Hallsworth-Pepin, K., Delehaunty, K., Markovic, C., Minx, P., Feng, Y., Kremitzki, C., Mitreva, M., Glasscock, J., Wylie, T., Wohldmann, P., Thiru, P., Nhan, M.N., Pohl, C.S., Smith, S.M., Hou, S., Nefedov, M., de Jong, P.J., Renfree, M.B., Mardis, E.R., and Wilson, R.K. (2008). Genome analysis of the platypus reveals unique signatures of evolution. *Nature* **453**, 175–183.

2342. Warrick, D., Hedrick, T., Fernández, M.J., Tobalkse, B., and Biewener, A. (2012). Hummingbird flight. *Curr. Biol.* **22**, R472–R477.

2343. Washio, Y., Aritaki, M., Fujinami, Y., Shimizu, D., Yokoi, H., and Suzuki, T. (2013). Ocular-side lateralization of adult-type chromatophore precursors: development of pigment asymmetry in metamorphosing flounder larvae. *J. Exp. Zool. (Mol. Dev. Evol.)* **320B**, 151–165.

2344. Wasik, B.R., Rose, D.J., and Moczek, A.P. (2010). Beetle horns are regulated by the *Hox* gene, *Sex combs reduced*, in a species- and sex-specific manner. *Evol. Dev.* **12**, 353–362.

2345. Watanabe, M., Iwashita, M., Ishii, M., Kurachi, Y., Kawakami, A., Kondo, S., and Okada, N. (2006). Spot pattern of *leopard Danio* is caused by mutation in the zebrafish *connexin41.8* gene. *EMBO Reports* **7**, 893–897.

2346. Watanabe, M. and Kondo, S. (2012). Changing clothes easily: *connexin41.8*regulates skin pattern variation. *Pigment Cell Melanoma Res.* **25**, 326–330.

2347. Watanabe, M., Watanabe, D., and Kondo, S. (2012). Polyamine sensitivity of gap junctions is required for skin pattern formation in zebrafish. *Sci. Reports* **2**, Article 473 (5 pp.).

2348. Weatherbee, S.D., Behringer, R.R., Rasweiler, J.J., IV, and Niswander, L.A. (2006). Interdigital webbing retention in bat wings illustrates genetic changes underlying amniote limb diversification. *PNAS* **103** #41, 15103–15107.

2349. Weatherbee, S.D. and Carroll, S.B. (1999). Selector genes and limb identity in arthropods and vertebrates. *Cell* **97**, 283–286.

2350. Weatherbee, S.D., Halder, G., Kim, J., Hudson, A., and Carroll, S. (1998). Ultrabithorax regulates genes at several levels of the wing-patterning hierarchy to shape the development of the *Drosophila* haltere. *Genes Dev.* **12**, 1474–1482.

2351. Weatherbee, S.D., Nijhout, H.F., Grunert, L.W., Halder, G., Galant, R., Selegue, J., and Carroll, S. (1999). *Ultrabithorax* function in butterfly wings and the evolution of insect wing patterns. *Curr. Biol.* **9**, 109–115.

2352. Weaver, J. (2012). Striking similarities in fly and vertebrate olfactory network formation. *PLoS Biol.* **10** #10, e1001401.

2353. Weavers, H., Prieto-Sánchez, S., Grawe, F., Garcia-López, A., Artero, R., Wilsch-Braüninger, M., Ruiz-Gómez, M., Skaer, H., and Denholm, B. (2009). The insect nephrocyte is a podocyte-like cell with a filtration slit diaphragm. *Nature* **457**, 322–326.

2354. Weber, B.H. and Depew, D.J. (2003). *Evolution and Learning: The Baldwin Effect Reconsidered.* MIT Press, Cambridge, MA.

2355. Weidauer, T., Pauluis, O., and Schumacher, J. (2010). Cloud patterns and mixing properties in shallow moist Rayleigh-Bénard convection. *New J. Physics* **12**, 105002. [*See also* Schaefer, V.J., and Day, J.A. (1981). *A Field Guide to the Atmosphere.* The Peterson Field Guide Series, Vol. 26. Houghton Mifflin, Boston, MA.]

2356. Weil, A. (2003). Teeth as tools. *Nature* **422**, 128.

2357. Weisbecker, V. (2011). Monotreme ossification sequences and the riddle of mammalian skeletal development. *Evolution* **65**, 1323–1335.

2358. Weiss, K. (2002). Good vibrations: the silent symphony of life. *Evol. Anthrop.* **11**, 176–182.

2359. Weiss, K. (2004). Doin' what comes naturally. *Evol. Anthrop.* **13**, 47–52.

2360. Weiss, K. and Sholtis, S. (2003). Dinner at Baby's: Werewolves, dinosaur jaws, hen's teeth, and horse toes. *Evol. Anthrop.* **12**, 247–251.

2361. Weiss, K.M. (2005). The phenogenetic logic of life. *Nature Rev. Genet.* **6**, 36–45.

2362. Weiss, K.M. and Buchanan, A.V. (2009). *The Mermaid's Tale: Four Billion Years of Cooperation in the Making of Living Things.* Harvard University Press, Cambridge, MA.

2363. Weiss, K.M. and Fullerton, S.M. (2000). Phenogenetic drift and the evolution of genotype-phenotype relationships. *Theor. Pop. Biol.* **57**, 187–195.
2364. Weiss, K.M., Stock, D.W., and Zhao, Z. (1998). Dynamic interactions and the evolutionary genetics of dental patterning. *Crit. Rev. Oral Biol. Med.* **9**, 369–398.
2365. Weiss, P. (1969). The living system: determinism stratified. In *Beyond Reductionism: New Perspectives in the Life Sciences* (A. Koestler and J.R. Smythies, eds.). Macmillan, New York, NY, pp. 3–55.
2366. Wellik, D.M. and Capecchi, M.R. (2003). *Hox10* and *Hox11* genes are required to globally pattern the mammalian skeleton. *Science* **301**, 363–367.
2367. Welsh, I.C. and O'Brien, T.P. (2009). Signaling integration in the rugae growth zone directs sequential SHH signaling center formation during the rostral outgrowth of the palate. *Dev. Biol.* **336**, 53–67.
2368. Werdelin, L. and Olsson, L. (1997). How the leopard got its spots: a phylogenetic view of the evolution of felid coat patterns. *Biol. J. Linnean Soc.* **62**, 383–400.
2369. Werner, T., Koshikawa, S., Williams, T.M., and Carroll, S.B. (2010). Generation of a novel wing colour pattern by the Wingless morphogen. *Nature* **464**, 1143–1148.
2370. West, P.M. and Packer, C. (2002). Sexual selection, temperature, and the lion's mane. *Science* **297**, 1339–1343.
2371. West-Eberhard, M.J. (1998). Evolution in the light of developmental and cell biology, and *vice versa*. *PNAS* **95** #15, 8417–8419.
2372. West-Eberhard, M.J. (2003). *Developmental Plasticity and Evolution.* Oxford University Press, New York, NY.
2373. West-Eberhard, M.J. (2005). Developmental plasticity and the origin of species differences. *PNAS* **102** (Suppl. 1), 6543–6549.
2374. Wharton, C.H. (1969). The cottonmouth moccasin on Sea Horse Key, Florida. *Bull. Fla. State Mus. Biol. Sci.* **14**, 227–272.
2375. White, M. (2012). Paradise found. *Natl. Geogr.* **222** #6, 70–89.
2376. White, M. (2013). When push comes to shove. *Natl. Geogr.* **223** #4.
2377. White, P.J.T., Heidemann, M., Loh, M., and Smith, J.J. (2013). Integrative cases for teaching evolution. *Evo. Edu. Outreach* **6**, Article 17 (7 pp.).
2378. White, R.A.H. and Akam, M.E. (1985). *Contrabithorax* mutations cause inappropriate expression of *Ultrabithorax* products in *Drosophila*. *Nature* **318**, 567–569.
2379. White, R.A.H. and Wilcox, M. (1985). Distribution of *Ultrabithorax* proteins in *Drosophila*. *EMBO J.* **4**, 2035–2043.
2380. White, R.A.H. and Wilcox, M. (1985). Regulation of the distribution of *Ultrabithorax* proteins in *Drosophila*. *Nature* **318**, 563–567.
2381. Whiting, M.F., Bradler, S., and Maxwell, T. (2003). Loss and recovery of wings in stick insects. *Nature* **421**, 264–267.
2382. Whiting, M.F. and Wheeler, W.C. (1994). Insect homeotic transformation. *Nature* **368**, 696.
2383. Whitlatch, T. (2010). *Animals Real and Imagined: Fantasy of What Is and What Might Be.* Design Studio Press, Culver City, CA.
2384. Whyte, W.A., Orlando, D.A., Hnisz, D., Abraham, B.J., Lin, C.Y., Kagey, M.H., Rahl, P.B., Lee, T.I., and Young, R.A. (2013). Master transcription factors and Mediator establish super-enhancers at key cell identity genes. *Cell* **153**, 307–319.
2385. Wibowo, I., Pinto-Teixeira, F., Satou, C., Higashijima, S.-i., and López-Schier, H. (2011). Compartmentalized Notch signaling sustains epithelial mirror symmetry. *Development* **138**, 1143–1152.

2386. Widelitz, R.B., Baker, R.E., Plikus, M., Lin, C.-M., Maini, P.K., Paus, R., and Chuong, C.M. (2006). Distinct mechanisms underlie pattern formation in the skin and skin appendages. *Birth Defects Res. (Part C)* **78**, 280–291.

2387. Wiegmann, B.M., Yeates, D.K., Thorne, J.L., and Kishino, H. (2003). Time flies, a new molecular time-scale for Brachyceran fly evolution without a clock. *Syst. Biol.* **52**, 745–756.

2388. Wiehs, D., Fish, F.E., and Nicastro, A.J. (2007). Mechanics of remora removal by dolphin spinning. *Marine Mamm. Sci.* **23**, 707–714.

2389. Wiens, J.J. (2001). Shape shifters: Time after time, lizards have dropped their legs in favor of a snakelike body form. *Nat. Hist.* **110** #8, 70–75.

2390. Wiens, J.J. (2001). Widespread loss of sexually selected traits: how the peacock lost its spots. *Trends Ecol. Evol.* **16**, 517–523.

2391. Wiens, J.J. (2004). Development and evolution of body form and limb reduction in squamates: a response to Sanger and Gibson-Brown. *Evolution* **58**, 2107–2108.

2392. Wiens, J.J. (2009). Estimating rates and patterns of morphological evolution from phylogenies: lessons in limb lability from Australian *Lerista* lizards. *J. Biol.* **8**, e19.

2393. Wiens, J.J. (2011). Re-evolution of lost mandibular teeth in frogs after more than 200 million years, and re-evaluating Dollo's Law. *Evolution* **65**, 1283–1296.

2394. Wiens, J.J. and Brandley, M.C. (2009). The evolution of limblessness. In *Grzimek's Animal Life Encyclopedia* (Internet edn.). Gale Cengage, Farmington Hills, MI.

2395. Wiens, J.J., Brandley, M.C., and Reeder, T.W. (2006). Why does a trait evolve multiple times within a clade? Repeated evolution of snakelike body form in squamate reptiles. *Evolution* **60**, 123–141.

2396. Wiens, J.J. and Hoverman, J.T. (2008). Digit reduction, body size, and paedomorphosis in salamanders. *Evol. Dev.* **10**, 449–463.

2397. Wiens, J.J., Hutter, C.R., Mulcahy, D.G., Noonan, B.P., Townsend, T.M., Sites, J.W., Jr., and Reeder, T.W. (2012). Resolving the phylogeny of lizards and snakes (Squamata) with extensive sampling of genes and species. *Biol. Lett.* **8**, 1043–1046.

2398. Wiens, J.J. and Slingluff, J.L. (2001). How lizards turn into snakes: a phylogenetic analysis of body-form evolution in anguid lizards. *Evolution* **55**, 2303–2318.

2399. Wigglesworth, V.B. (1940). Local and general factors in the development of "pattern" in *Rhodnius prolixus* (Hemiptera). *J. Exp. Zool.* **17**, 180–200.

2400. Wigglesworth, V.B. (1973). Evolution of insect wings and flight. *Nature* **246**, 127–129.

2401. Wigglesworth, V.B. (1976). The evolution of insect flight. In *Insect Flight.* (R.C. Rainey, ed.). Symposium of the Royal Entomological Society of London, Vol. 7. Wiley, New York, NY, pp. 255–269.

2402. Wikramanayake, A.H., Hong, M., Lee, P.N., Pang, K., Byrum, C.A., Bince, J.M., Xu, R., and Martindale, M.Q. (2003). An ancient role for nuclear β-catenin in the evolution of axial polarity and germ layer segregation. *Nature* **426**, 446–450.

2403. Wilby, O.K. and Ede, D.A. (1975). A model generating the pattern of cartilage skeletal elements in the embryonic chick limb. *J. Theor. Biol.* **52**, 199–217.

2404. Wilczynskia, B. and Furlong, E.E.M. (2010). Challenges for modeling global gene regulatory networks during development: Insights from *Drosophila*. *Dev. Biol.* **340**, 161–169.

2405. Wilder, E.L. and Perrimon, N. (1995). Dual functions of *wingless* in the *Drosophila* leg imaginal disc. *Development* **121**, 477–488.

2406. Wilkie, A.L., Jordan, S.A., and Jackson, I.J. (2002). Neural crest progenitors of the melanocyte lineage: coat colour patterns revisited. *Development* **129**, 3349–3357.

2407. Wilkins, A.S. (1989). Organizing the *Drosophila* posterior pattern: why has the fruit fly made life so complicated for itself? *BioEssays* **11**, 67–69.

2408. Wilkins, A.S. (1993). *Genetic Analysis of Animal Development*, 2nd edn. Wiley-Liss, New York, NY.

2409. Wilkins, A.S. (1997). Canalization: a molecular genetic perspective. *BioEssays* **19**, 257–262.

2410. Wilkins, A.S. (2007). Between "design" and "bricolage": Genetic networks, levels of selection, and adaptive evolution. *PNAS* **104** (Suppl. 1), 8590–8596.

2411. Wilkinson, D.M. and Ruxton, G.D. (2012). Understanding selection for long necks in different taxa. *Biol. Rev.* **87**, 616–630.

2412. Wilkinson, G.S. (1994). Female choice response to artificial selection on an exaggerated male trait in a stalk-eyed fly. *Proc. R. Soc. Lond. B* **255**, 1–6.

2413. Wilkinson, G.S. and Johns, P.M. (2005). Sexual selection and the evolution of mating systems in flies. In *The Evolutionary Biology of Flies* (D.K. Yeates and B.M. Wiegmann, eds.). Columbia University Press, New York, NY, pp. 312–339.

2414. Wilkinson, G.S., Johns, P.M., Metheny, J.D., and Baker, R.H. (2013). Sex-biased gene expression during head development in a sexually dimorphic stalk-eyed fly. *PLoS ONE* **8** #3, e59826.

2415. Wilkinson, M.T. (2007). Sailing the skies: the improbable aeronautical success of the pterosaurs. *J. Exp. Biol.* **210**, 1663–1671.

2416. Willey, A. (1911). *Convergence in Evolution*. John Murray, London.

2417. Williams, G.C. (1992). *Natural Selection: Domains, Levels, and Challenges*. Oxford Series in Ecology and Evolution, Vol. 4. Oxford University Press, New York, NY.

2418. Williston, S.W. (1914). *Water Reptiles of the Past and Present*. University of Chicago Press, Chicago, IL.

2419. Willmore, K.E. (2010). Development influences evolution. *Am. Sci.* **98**, 220–227.

2420. Willmore, K.E. (2012). The body plan concept and its centrality in evo-devo. *Evo. Edu. Outreach* **5**, 219–230.

2421. Willnow, T.E., Christ, A., and Hammes, A. (2012). Endocytic receptor-mediated control of morphogen signaling. *Development* **139**, 4311–4319.

2422. Wilson, A.B. and Orr, J.W. (2011). The evolutionary origins of Syngnathidae: pipefishes and seahorses. *J. Fish Biol.* **78**, 1603–1623.

2423. Wilson, D.E. and Mittermeier, R.A., eds. *Handbook of the Mammals of the World. Vol. 1 (Carnivores)*. Lynx Edicions, Barcelona.

2424. Wilson, D.E., Mittermeier, R.A., Ruff, S., and Martinez-Vilalta, A., eds. *Handbook of the Mammals of the World. Vol. 2 (Hoofed Mammals)*. Lynx Edicions, Barcelona.

2425. Wilting, A., Buckley-Beason, V.A., Feldhaar, H., Gadau, J., O'Brien, S.J., and Linsenmair, K.E. (2007). Clouded leopard phylogeny revisited: support for species recognition and population division between Borneo and Sumatra. *Front. Zool.* **4**, Article 15 (10 pp.).

2426. Winchell, C.J. and Jacobs, D.K. (2013). Expression of the Lhx genes *apterous* and *lim1* in an errant polychaete: implications for bilaterian appendage evolution, neural development, and muscle diversification. *EvoDevo* **4**, Article 4 (19 pp.).

2427. Winchell, C.J., Valencia, J.E., and Jacobs, D.K. (2010). Expression of *Distal-less*, *dachshund*, and *optomotor blind* in *Neanthes arenaceodentata* (Annelida, Nereididae) does not support homology of appendage-forming mechanisms across the Bilateria. *Dev. Genes Evol.* **220**, 275–295.
2428. Winfree, A.T. (1980). *The Geometry of Biological Time*. Springer-Verlag, Berlin.
2429. Winfree, A.T. (1984). The prehistory of the Belousov-Zhabotinsky oscillator. *J. Chem. Ed.* **61**, 661–663.
2430. Winfree, A.T., Winfree, E.M., and Seifert, H. (1985). Organizing centers in a cellular excitable medium. *Physica* **17D**, 109–115.
2431. Wings, O. and Sander, P.M. (2007). No gastric mill in sauropod dinosaurs: new evidence from analysis of gastrolith mass and function in ostriches. *Proc. R. Soc. Lond. B* **274**, 635–640.
2432. Winter, S. (1998). The elusive quetzal. *Natl. Geogr.* **193** #6, 34–45.
2433. Witkop, C.J., Jr., Quevedo, W.C., Jr., Fitzpatrick, T.B., and King, R.A. (1989). Albinism. In *The Metabolic Basis of Inherited Disease, Vol. 2*, 6th edn. (C.R. Scriver, A.L. Beaudet, W.S. Sly, and D. Valle, eds.). McGraw-Hill, New York, NY, pp. 2905–2947.
2434. Wittkopp, P.J. and Beldade, P. (2009). Development and evolution of insect pigmentation: genetic mechanisms and the potential consequences of pleiotropy. *Semin. Cell Dev. Biol.* **20**, 65–71.
2435. Wittkopp, P.J. and Kalay, G. (2012). *Cis*-regulatory elements: molecular mechanisms and evolutionary processes underlying divergence. *Nature Rev. Genet.* **13**, 59–69.
2436. Wolfe, J.M., Oliver, J.C., and Monteiro, A. (2010). Evolutionary reduction of the first thoracic limb in butterflies. *J. Insect Sci.* **11**, Article 66 (9 pp.).
2437. Wolfram, S. (1984). Cellular automata as models of complexity. *Nature* **311**, 419–424.
2438. Wolpert, L. (1968). The French Flag problem: a contribution to the discussion on pattern development and regulation. In *Towards a Theoretical Biology. I. Prolegomena* (C.H. Waddington, ed.). Aldine, Chicago, IL, pp. 125–133.
2439. Wolpert, L. (1969). Positional information and the spatial pattern of cellular differentiation. *J. Theor. Biol.* **25**, 1–47.
2440. Wolpert, L. (1989). Positional information revisited. *Development* **1989** Suppl., 3–12.
2441. Wolpert, L. (1990). Signals in limb development: STOP, GO, STAY and POSITION. *J. Cell Sci. Suppl.* **13**, 199–208.
2442. Wolpert, L. (1996). One hundred years of positional information. *Trends Genet.* **12**, 359–364.
2443. Wolpert, L., Tickle, C., Jessell, T., Lawrence, P., Meyerowitz, E., Robertson, E., and Smith, J. (2011). *Principles of Development*, 4th edn. Oxford University Press, New York, NY.
2444. Woltering, J.M. (2012). From lizard to snake: behind the evolution of an extreme body plan. *Curr. Genomics* **13**, 289–299.
2445. Woltering, J.M., Vonk, F.J., Müller, H., Bardine, N., Tuduce, I.L., de Bakker, M.A.G., Knöchel, W., Sirbu, I.O., Durston, A.J., and Richardson, M.K. (2009). Axial patterning in snakes and caecilians: Evidence for an alternative interpretation of the *Hox* code. *Dev. Biol.* **332**, 82–89.
2446. Wong, B.B.M. and Rosenthal, G.G. (2006). Female disdain for swords in the swordtail fish. *Am. Nat.* **167**, 136–140.
2447. Wong, K. (2002). The mammals that conquered the seas. *Sci. Am.* **286** #5, 70–79.
2448. Wong, K. (2004). Becoming behemoth. *Sci. Am.* **290** #2, 23–24.

2449. Wootton, R.J. (1976). The fossil record and insect flight. In *Insect Flight.* (R.C. Rainey, ed.). Symposium of the Royal Entomological Society of London, Vol. 7. Wiley, New York, NY, pp. 235–254.

2450. Wootton, R.J. (1992). Functional morphology of insect wings. *Annu. Rev. Entomol.* **37**, 113–140.

2451. Wootton, R.J. (1999). Invertebrate paraxial locomotory appendages: design, deformation and control. *J. Exp. Biol.* **202**, 3333–3345.

2452. Wootton, R.J. and Kukalová-Peck, J. (2000). Flight adaptations in Palaeozoic Palaeoptera (Insecta). *Biol. Rev.* **75**, 129–167.

2453. Wootton, R.J., Kukalová-Peck, J., Newman, D.J.S., and Muzon, J. (1998). Smart engineering in the mid-Carboniferous: how well could Palaeozoic dragonflies fly? *Science* **282**, 749–751.

2454. Wourms, M.K. and Wasserman, F.E. (1985). Butterfly wing markings are more advantageous during handling than during the initial strike of an avian predator. *Evolution* **39**, 845–851.

2455. Wray, G.A. and Abouheif, E. (1998). When is homology not homology? *Curr. Opin. Gen. Dev.* **8**, 675–680.

2456. Wu, C.-I. (1996). Now blows the east wind. *Nature* **380**, 105–107.

2457. Wu, D., Freund, J.B., Fraser, S.E., and Vermot, J. (2011). Mechanical basis of otolith formation during teleost inner ear development. *Dev. Cell* **20**, 271–278.

2458. Wu, M.Y. and Hill, C.S. (2009). TGF-β superfamily signaling in embryonic development and homeostasis. *Dev. Cell* **16**, 329–343.

2459. Wu, P., Hou, L., Plikus, M., Hughes, M., Scehnet, J., Suksaweang, S., Widelitz, R.B., Jiang, T.-X., and Chuong, C.-M. (2004). *Evo-Devo* of amniote integuments and appendages. *Int. J. Dev. Biol.* **48**, 249–270.

2460. Wu, P., Jiang, T.-X., Shen, J.-Y., Widelitz, R.B., and Chuong, C.-M. (2006). Morphoregulation of avian beaks: comparative mapping of growth zone activities and morphological evolution. *Dev. Dynamics* **235**, 1400–1412.

2461. Wu, P., Jiang, T.-X., Suksaweang, S., Widelitz, R.B., and Chuong, C.-M. (2004). Molecular shaping of the beak. *Science* **305**, 1465–1466.

2462. Wu, X., Jung, G., and Hammer, J.A., III (2000). Functions of unconventional myosins. *Curr. Opin. Cell Biol.* **12**, 42–51.

2463. Wueringer, B.E., Squire, L., Jr., Kajiura, S.M., Hart, N.S., and Collin, S.P. (2012). The function of the sawfish's saw. *Curr. Biol.* **22**, R150–R151.

2464. Wund, M.A. (2012). Assessing the impacts of phenotypic plasticity on evolution. *Integr. Comp. Biol.* **52**, 5–15.

2465. Wyss, A.R. (1988). Evidence from flipper structure for a single origin of pinnipeds. *Nature* **334**, 427–428.

2466. Xenia, M., Ilagan, G., and Kopan, R. (2007). Snapshot: Notch signaling pathway. *Cell* **128**, 1246.

2467. Xiang, H., Li, M.W., Guo, J.H., Jiang, J.H., and Huang, Y.P. (2011). Influence of RNAi knockdown for E-complex genes on the silkworm proleg development. *Arch. Insect Biochem. Physiol.* **76**, 1–11.

2468. Xiong, F., Tentner, A.R., Huang, P., Gelas, A., Mosaliganti, K.R., Souhait, L., Rannou, N., Swinburne, I.A., Obholzer, N.D., Cowgill, P.D., Schier, A.F., and Megason, S.G. (2013). Specified neural progenitors sort to form sharp domains after noisy Shh signaling. *Cell* **153**, 550–561.

2469. Xu, X. (2013). Modular genetic control of innate behaviors. *BioEssays* **35**, 421–424.

2470. Xu, X., Dong, G.-X., Hu, X.-S., Miao, L., Zhang, X.-L., Zhang, D.-L., Yang, H.-D., Zhang, T.-Y., Zou, Z.-T., Zhang, T.-T., Zhuang, Y., Bhak, J., Cho, Y.S., Dai, W.-T., Jiang, T.-J., Xie, C., Li, R., and Luo, S.-J. (2013). The genetic basis of white tigers. *Curr. Biol.* **23**, 1031–1035.

2471. Xu, X., Sullivan, C., Pittman, M., Choiniere, J.N., Hone, D., Upchurch, P., Tan, Q., Xiao, D., Tan, L., and Han, F. (2011). A monodactyl nonavian dinosaur and the complex evolution of the alvarezsauroid hand. *PNAS* **108** #6, 2338–2342.

2472. Yamada, G., Suzuki, K., Haraguchi, R., Miyagawa, S., Satoh, Y., Kamimura, M., Nakagata, N., Kataoka, H., Kuroiwa, A., and Chen, Y. (2006). Molecular genetic cascades for external genitalia formation: an emerging organogenesis program. *Dev. Dynamics* **235**, 1738–1752.

2473. Yamaguchi, M., Yoshimoto, E., and Kondo, S. (2007). Pattern regulation in the stripe of zebrafish suggests an underlying dynamic and autonomous mechanism. *PNAS* **104** #12, 4790–4793.

2474. Yamamoto, S., Charng, W.-L., Rana, N.A., Kakuda, S., Jaiswal, M., Bayat, V., Xiong, B., Zhang, K., Sandoval, H., David, G., Wang, H., Haltiwanger, R.S., and Bellen, H.J. (2012). A mutation in EGF repeat-8 of Notch discriminates between Serrate/Jagged and Delta family ligands. *Science* **338**, 1229–1232.

2475. Yamanoue, Y., Setiamarga, D.H.E., and Matsuura, K. (2010). Pelvic fins in teleosts: structure, function and evolution. *J. Fish Biol.* **77**, 1173–1208.

2476. Yáñez-Cuna, J.O., Kvon, E.Z., and Stark, A. (2013). Deciphering the transcriptional *cis*-regulatory code. *Trends Genet.* **29**, 11–22.

2477. Yang, A.S. (2001). Modularity, evolvability, and adaptive radiations: a comparison of the hemi- and holometabolous insects. *Evol. Dev.* **3**, 59–72.

2478. Yang, Z., Bertolucci, F., Wolf, R., and Heisenberg, M. (2013). Flies cope with uncontrollable stress by learned helplessness. *Curr. Biol.* **23**, 799–803.

2479. Yano, T. and Tamura, K. (2013). The making of differences between fins and limbs. *J. Anat.* **222**, 100–113.

2480. Yarmolinsky, D.A., Zuker, C.S., and Ryba, N.J.P. (2009). Common sense about taste: from mammals to insects. *Cell* **139**, 234–244.

2481. Yau, K.-W. and Hardie, R.C. (2009). Phototransduction motifs and variations. *Cell* **139**, 246–264.

2482. Yekta, S., Tabin, C.J., and Bartel, D.P. (2008). MicroRNAs in the Hox network: an apparent link to posterior prevalence. *Nature Rev. Genet.* **9**, 789–796.

2483. Yoshida, A. and Aoki, K. (1989). Scale arrangement pattern in a lepidopteran wing. 1. Periodic cellular pattern in the pupal wing of *Pieris rapae*. *Dev. Growth Differ.* **31**, 601–609.

2484. Young, B.A. and Kardong, K.V. (2010). The functional morphology of hooding in cobras. *J. Exp. Biol.* **213**, 1521–1528.

2485. Young, K.V., Brodie, E.D., Jr., and Brodie, E.D., III (2004). How the horned lizard got its horns. *Science* **304**, 65.

2486. Young, N.M. and Hallgrímsson, B. (2005). Serial homology and the evolution of mammalian limb covariation structure. *Evolution* **59**, 2691–2704.

2487. Young, R.L. and Wagner, G.P. (2011). Why ontogenetic homology criteria can be misleading: lessons from digit identity transformations. *J. Exp. Zool. (Mol. Dev. Evol.)* **316B**, 165–170. [*See also* Xu, X., and Makem, S. (2013). Tracing the evolution of avian wing digits. *Curr. Biol.* **23**, R538–R544. Another recent paper that is relevant here is by de Bakker, M.A.G., *et al.* (2013). Digit loss in archosaur evolution and the interplay between selection and constraints. *Nature* **500**, 445–448.]

2488. Young, T., Rowland, J.E., van de Ven, C., Bialecka, M., Novoa, A., Carapuco, M., van Nes, J., de Graaff, W., Duluc, I., Freund, J.-N., Beck, F., Mallo, M., and Deschamps, J. (2009). *Cdx* and *Hox* genes differentially regulate posterior axial growth in mammalian embryos. *Dev. Cell* **17**, 516–526.

2489. Yu, C.Q., Schwab, I.R., and Dubielzig, R.R. (2009). Feeding the vertebrate retina from the Cambrian to the Tertiary. *J. Zool.* **278**, 259–269.

2490. Yu, L., Jin, W., Zhang, X., Wang, D., Zheng, J.-s., Yang, G., Xu, S.-x., Cho, S., and Zhang, Y.-p. (2011). Evidence for positive selection on the leptin gene in cetacea and pinnipedia. *PLoS ONE* **6** #10, e26579.

2491. Yue, Z., Jiang, T.-X., Widelitz, R.B., and Chuong, C.-M. (2006). Wnt3a gradient converts radial to bilateral feather symmetry via topological arrangement of epithelia. *PNAS* **103** #4, 951–955.

2492. Zaher, H. and Rieppel, O. (1999). The phylogenetic relationships of *Pachyrhachis problematicus*, and the evolution of limblessness in snakes (Lepidosauria, Squamata). *C. R. Acad. Sci. Paris, Sciences de la terre et des planètes* **329**, 831–837.

2493. Zahradnicek, O., Horacek, I., and Tucker, A.S. (2008). Viperous fangs: Development and evolution of the venom canal. *Mechs. Dev.* **125**, 786–796.

2494. Zakany, J. and Duboule, D. (2007). The role of *Hox* genes during vertebrate limb development. *Curr. Opin. Gen. Dev.* **17**, 359–366.

2495. Zakin, L. and De Robertis, E.M. (2010). Extracellular regulation of BMP signaling. *Curr. Biol.* **20**, R89–R92.

2496. Zamora, S., Rahman, I.A., and Smith, A.B. (2012). Plated Cambrian bilaterians reveal the earliest stages of echinoderm evolution. *PLoS ONE* **7** #6, e38296. [*See also* Smith, A.B. and Zamora, S. (2013). Cambrian spiral-plated echinoderms from Gondwana reveal the earliest pentaradial body plan. *Proc. R. Soc. Lond. B* **280**. doi 20131197. N.B.: Believe it or not, some sea stars have 50 arms or more. See Janosik, A.M., *et al.* (2008) Life history of the Antarctic sea star *Labidiaster annulatus* (Asteroidea: Labidiasteridae) revealed by DNA barcoding. *Antarctic Sci.* **20**, 563–564.]

2497. Zbikowski, R. (2002). Red admiral agility. *Nature* **420**, 615–618.

2498. Zecca, M., Basler, K., and Struhl, G. (1996). Direct and long-range action of a Wingless morphogen gradient. *Cell* **87**, 833–844.

2499. Zeeman, E.C. (1974). Primary and secondary waves in developmental biology. In *Lectures on Mathematics in the Life Sciences*, Vol. 7. American Mathematical Society, Providence, RI, pp. 69–161.

2500. Zeil, J., Nalbach, G., and Nalbach, H.-O. (1986). Eyes, eye stalks and the visual world of semi-terrestrial crabs. *J. Comp. Physiol. A* **159**, 801–811.

2501. Zelditch, M.L. (2003). Space, time, and repatterning. In *Keywords and Concepts in Evolutionary Developmental Biology* (B.K. Hall and W.M. Olson, eds.). Harvard University Press, Cambridge, MA, pp. 341–348.

2502. Zelditch, M.L. and Fink, W.L. (1996). Heterochrony and heterotopy: innovation and stability in the evolution of form. *Paleobiology* **22**, 241–254.

2503. Zelenitsky, D.K., Therrien, F., Erickson, G.M., DeBuhr, C.L., Kobayashi, Y., Eberth, D.A., and Hadfield, F. (2012). Feathered non-avian dinosaurs from North America provide insight into wing origins. *Science* **338**, 510–514.

2504. Zhai, Z., Han, N., Papagiannouli, F., Hamacher-Brady, A., Brady, N., Sorge, S., Bezdan, D., and Lohmann, I. (2012). Antagonistic regulation of apoptosis and differentiation by the Cut transcription factor represents a tumor-suppressing mechanism in *Drosophila*. *PLoS Genet.* **8** #3, e1002582.

2505. Zhang, J., Tian, Y., Wang, L., and He, C. (2010). Functional evolutionary developmental biology (evo-devo) of morphological novelties in plants. *J. Syst. Evol.* **48**, 94–101.

2506. Zhang, X.-g. and Pratt, B.R. (2012). The first stalk-eyed phosphatocopine crustacean from the Lower Cambrian of China. *Curr. Biol.* **22**, 2149–2154.

2507. Zheng, X., Zhou, Z., Wang, X., Zhang, F., Zhang, X., Wang, Y., Wei, G., Wang, S., and Xu, X. (2013). Hind wings in basal birds and the evolution of leg feathers. *Science* **339**, 1309–1312.

2508. Zheng, Z., Khoo, A., Fambrough, D., Jr., Garza, L., and Booker, R. (1999). Homeotic gene expression in the wild-type and a homeotic mutant of the moth *Manduca sexta*. *Dev. Genes Evol.* **209**, 460–472.

2509. Zhong, Y.-f. and Holland, P.W.H. (2011). The dynamics of vertebrate homeobox gene evolution: gain and loss of genes in mouse and human lineages. *BMC Evol. Biol.* **11**, Article 169 (13 pp.).

2510. Zhou, Q., Yu, L., Shen, X., Li, Y., Xu, W., Yi, Y., and Zhang, Z. (2009). Homology of dipteran bristles and lepidopteran scales: requirement for the *Bombyx mori achaete-scute* homologue *ASH2*. *Genetics* **183**, 619–627.

2511. Zhou, S., Lo, W.-C., Suhalim, J.L., Digman, M.A., Gratton, E., Nie, Q., and Lander, A.D. (2012). Free extracellular diffusion creates the Dpp morphogen gradient of the *Drosophila* wing disc. *Curr. Biol.* **22**, 668–675.

2512. Zhou, Z., Clark, J., Zhang, F., and Wings, O. (2004). Gastroliths in Yanornis: an indication of the earliest radical diet-switching and gizzard plasticity in the lineage leading to living birds? *Naturwissenschaften* **91**, 571–574.

2513. Zhou, Z. and Zhang, F. (2006). A beaked basal ornithurine bird (Aves, Ornithurae) from the Lower Cretaceous of China. *Zool. Scripta* **35**, 363–373.

2514. Zhu, A.J. and Scott, M.P. (2004). Incredible journey: how do developmental signals travel through tissue? *Genes Dev.* **18**, 2985–2997.

2515. Zhu, B., Pennack, J.A., McQuilton, P., Forero, M.G., Mizuguchi, K., Sutcliffe, B., Gu, C.-J., Fenton, J.C., and Hidalgo, A. (2008). *Drosophila* neurotrophins reveal a common mechanism for nervous system formation. *PLoS Biol.* **6** #11, 2476–2495 (e284).

2516. Zhu, L., Wilken, J., Phillips, N.B., Narendra, U., Chan, G., Stratton, S.M., Kent, S.B., and Weiss, M.A. (2000). Sexual dimorphism in diverse metazoans is regulated by a novel class of intertwined zinc fingers. *Genes Dev.* **14**, 1750–1764.

2517. Zill, S.N. and Seyfarth, E.-A. (1996). Exoskeletal sensors for walking. *Sci. Am.* **275** #1, 86–90.

2518. Zimmer, C. (2002). The rise and fall of the nasal empire. *Nat. Hist.* **111** #5, 32–35.

2519. Zimmer, C. (2005). Dinosaurs: Why do we have so many questions about the most successful animals that ever lived? *Discover* **26** #4, 32–39.

2520. Zimmer, C. (2008). The evolution of extraordinary eyes: the cases of flatfishes and stalk-eyed flies. *Evo. Edu. Outreach* **1**, 487–492.

2521. Zimmer, C. (2011). The long curious extravagant evolution of feathers. *Natl. Geogr.* **219** #2, 32–57.

2522. Zimmer, C. (2012). The common hand. *Natl. Geogr.* **221** #5, 98–105.

2523. Zuber, M.E., Gestri, G., Viczian, A.S., Barsacchi, G., and Harris, W.A. (2003). Specification of the vertebrate eye by a network of eye field transcription factors. *Development* **130**, 5155–5167.

2524. Zuzarte-Luis, V. and Hurle, J.M. (2005). Programmed cell death in the embryonic vertebrate limb. *Semin. Cell Dev. Biol.* **16**, 261–269.

2525. Zwarts, L., Magwire, M.M., Carbone, M.A., Versteven, M., Herteleer, L., Anholt, R.R.H., Callaerts, P., and Mackay, T.F.C. (2011). Complex genetic architecture of *Drosophila* aggressive behavior. *PNAS* **108** #41, 17070–17075.

2526. Zylinski, S. and Johnsen, S. (2011). Mesopelagic cephalopods switch between transparency and pigmentation to optimize camouflage in the deep. *Curr. Biol.* **21**, 1937–1941.

Index

Printed in the United States
By Bookmasters